Lecture Notes in Computer Science 5533

Commenced Publication in 1973
Founding and Former Series Editors:
Gerhard Goos, Juris Hartmanis, and Jan van Leeuwen

W0235168

Orna Grumberg   Michael Kaminski
Shmuel Katz   Shuly Wintner (Eds.)

Languages: From Formal to Natural

Essays Dedicated to Nissim Francez
on the Occasion of His 65th Birthday

 Springer

Volume Editors

Orna Grumberg
Michael Kaminski
Shmuel Katz
Technion – Israel Institute of Technology
Department of Computer Science
Technion City, Haifa 32000, Israel
E-mail: {orna,kaminski,katz}@cs.technion.ac.il

Shuly Wintner
University of Haifa
Department of Computer Science
Haifa 31905, Israel
E-mail: shuly@cs.haifa.ac.il

The illustration on the cover of this book is the Tower of Babel by Pieter Brueghel the Elder. Permission for reproduction has been obtained from the Kunsthistorisches Museum in Vienna, Austria.

Library of Congress Control Number: Applied for

CR Subject Classification (1998): B.2, F.3, F.4, D.2.4

LNCS Sublibrary: SL 2 – Programming and Software Engineering

ISSN   0302-9743
ISBN-10   3-642-01747-9 Springer Berlin Heidelberg New York
ISBN-13   978-3-642-01747-6 Springer Berlin Heidelberg New York

springer.com

© Springer-Verlag Berlin Heidelberg 2009
Printed in Germany

Typesetting: Camera-ready by author, data conversion by Scientific Publishing Services, Chennai, India
Printed on acid-free paper   SPIN: 12667154   06/3180   5 4 3 2 1 0

Nissim Francez

Preface

The symposium "Languages: From Formal to Natural," celebrating the 65th birthday of Nissim Francez, was held on May 24–25, 2009 at the Technion, Haifa. The symposium consisted of two parts, a *verification* day and a *language* day, and covered all areas of Nissim's past and present research interests, areas which he has inspiringly influenced and to which he has contributed so much.

This volume comprises several papers presented at the symposium, as well as additional articles that were contributed by Nissim's friends and colleagues who were unable to attend the event. We thank the authors for their contributions. We are also grateful to the reviewers for their dedicated and timely work.

Nissim Francez was born on January 19, 1944. In 1962 he started his mathematical education at the Hebrew University. He received a BSc in Mathematics in 1965, and, after four years of military service, started his MSc studies in Computer Science at the Weizmann Institute of Science under the supervision of Amir Pnueli.

After completing the MSc program in 1971, Nissim continued his studies toward a PhD, again, at the Weizmann Institute of Science and, again, under the supervision of Amir Pnueli. Nissim was awarded a PhD in Computer Science in 1976.

During his graduate studies Nissim worked as a teaching instructor, first at Ben-Gurion University and then at Tel Aviv University. During this period he wrote his first book (along with Gideon Amir), a guide to Fortran, in Hebrew, that became a standard text in Israel for many years. After receiving his PhD, he spent one year at Queen's University in Belfast working with Tony Hoare, and then one more year at the University of Southern California.

In 1978 Nissim joined the Department of Computer Science at the Technion, where he became a professor in 1991. Since 2000 he has held the Bank Leumi Chair in Computer Science. During this time he was also a Visiting Scientist at MCC, Austin, Texas during 1989-1990 and at CWI, Amsterdam in 1992 and 1997, and held visiting positions at the Department of Computer Science, University of Utrecht (1992), the Department of Computer Science, Manchester University (1996-1997), and most recently at the School of Informatics, University of Edinburgh, in 2002.

Nissim's academic career has been exceptionally productive. To date, he has authored five books (three in English and two in Hebrew) and over 120 research papers, and advised over 30 graduate students, many of whom are now university professors in Israel and elsewhere. Nissim has given several invited talks at international conferences and symposia, has been on the program committee of numerous conferences, symposia, and workshops, reviewed submissions for 20 journals and is a member of the editorial board of two journals.

Nissim's scientific interests evolved significantly over time, but have always focused on fundamental questions, motivated by an insatiable thirst for knowledge and limitless scientific curiosity. Nissim started his career in program verification and the semantics of programming languages, especially for concurrent and distributed programming, and later expanded his interest to new language constructs for that paradigm. He has also made contributions to logic programming and the fundamentals of functional programming. His research monograph *Fairness* became a standard reference for work on assumptions about the scheduling of multiple processes that allow proving termination and other liveness properties. His book *Interacting Processes* (written with Ira Forman) presented original and influential proposals on language constructs for concurrency, and his textbook *Program Verification* summarized his approach to specifying and proving the correctness of programs.

The year 1991 signified a change of paradigm in Nissim's research career. Following a life-long fascination with natural languages, he shifted his attention from *formal* programming or specification languages to *natural* ones, but brought to computational linguistics his training and expertise from formal systems. In the nearly two decades that followed, Nissim contributed to diverse areas of computational linguistics, including formal grammars (in particular, unification grammars and more recently type-logical grammars), formal semantics of natural languages, λ-calculus, and proof theory – the areas in which he is active today. His contribution to computational linguistics is marked by rigorous, formal developments that are well-informed by actual natural language data.

Nissim's wide interests have also been reflected in his hobbies. He has collected stamps all his life, and has also developed collections of coins, telephone cards, and airline menus. He shares with his wife Tikva a love of literature–from the Bible to modern detective novels.

The occasion of Nissim's 65th birthday provided an excellent pretext to celebrate his accomplishments. We feel honored to have had a chance to know and work with Nissim. Co-organizing the symposium and co-editing this Festschrift allowed us on the one hand to acknowledge Nissim's scientific contributions and, on the other hand, to thank him for his advice, guidance, and friendship.

The response of Nissim's students, collaborators, and friends to our initiative, which was so critical for the success of this event, made the organization of the symposium and the publication of this volume much easier. We wish to thank them all. And one more time, Happy Birthday, Nissim, and many more productive years!

March 2009

Orna Grumberg
Michael Kaminski
Shmuel Katz
Shuly Wintner

Table of Contents

Languages: From Formal to Natural

Modular Verification of Recursive Programs

Krzysztof R. Apt[1,2], Frank S. de Boer[1,3], and Ernst-Rüdiger Olderog[4]

[1] Centre for Mathematics and Computer Science (CWI), Amsterdam, The Netherlands
[2] University of Amsterdam, Institute of Language, Logic and Computation, Amsterdam
[3] Leiden Institute of Advanced Computer Science, University of Leiden, The Netherlands
[4] Department of Computing Science, University of Oldenburg, Germany

Abstract. We argue that verification of recursive programs by means of the assertional method of C.A.R. Hoare can be conceptually simplified using a modular reasoning. In this approach some properties of the program are established first and subsequently used to establish other program properties. We illustrate this approach by providing a modular correctness proof of the *Quicksort* program.

1 Introduction

Program verification by means of the assertional method of Hoare (so-called Hoare's logic) is by now well-understood. One of its drawbacks is that it calls for a tedious manipulation of assertions, which is error prone. The support offered by the available by interactive proof checkers, such as PVS (Prototype Verification System), see [15], is very limited.

One way to reduce the complexity of an assertional correctness proof is by organizing it into a series of simpler proofs. For example, to prove $\{p\}\ S\ \{q_1 \land q_2\}$ we could establish $\{p\}\ S\ \{q_1\}$ and $\{p\}\ S\ \{q_2\}$ separately (and *independently*). Such an obvious approach is clearly of very limited use.

In this paper we propose a different approach that is appropriate for recursive programs. In this approach a simpler property, say $\{p_1\}\ S\ \{q_1\}$, is established first and then *used in the proof* of another property, say $\{p_2\}\ S\ \{q_2\}$. This allows us to establish $\{p_1 \land p_2\}\ S\ \{q_1 \land q_2\}$ in a modular way. It is obvious how to generalize this approach to an arbitrary sequence of program properties for which the earlier properties are used in the proofs of the latter ones. So, in contrast to the simplistic approach mentioned above, the proofs of the program properties are *not* independent but are arranged instead into an acyclic directed graph.

We illustrate this approach by providing a modular correctness proof of the *Quicksort* program due to [10]. This yields a correctness proof that is better structured and conceptually easier to understand than the original one, given in [7]. A minor point is that we use different proof rules concerning procedure calls and also provide an assertional proof of termination of the program, a property not considered in [7]. (It should be noted that termination of recursive procedures with parameters within the framework of the assertional method was considered only in the eighties, see, e.g., [2]. In these proofs some subtleties arise that necessitate a careful exposition, see [1].)

We should mention here two other references concerning formal verification of the *Quicksort* program. In [6] the proof of *Quicksort* is certified using the interactive

O. Grumberg et al. (Eds.): Francez Festschrift, LNCS 5533, pp. 1–21, 2009.

theorem prover Coq, while in [13] a correctness proof of a non-recursive version of *Quicksort* is given.

The paper is organized as follows. In the next section we introduce a small programming language that involves recursive procedures with parameters called by value and discuss its operational semantics. Then, in Section 3 we introduce a proof system for proving partial and total correctness of these programs. The presentation in these two sections is pretty standard except for the treatment of the call-by-value parameter mechanism that avoids the use of substitution.

Next, in Section 4 we discuss how the correctness proofs, both of partial and of total correctness, can be structured in a modular way. In Section 5 we illustrate this approach by proving correctness of the *Quicksort* program while in Section 6 we discuss related work and draw some conclusions. Finally, in the appendix we list the used axioms and proof rules concerned with non-recursive programs. The soundness of the considered proof systems is rigorously established in [3] using the operational semantics of [16,17].

2 A Small Programming Language

Syntax

We use *simple* variables and *array* variables. Simple variables are of a basic type (for example **integer** or **Boolean**), while array variables are of a higher type (for example **integer** × **Boolean** → **integer**). A *subscripted variable* derived from an array variable a of type $T_1 \times \ldots \times T_n \to T$ is an expression of the form $a[t_1, \ldots, t_n]$, where each expression t_i is of type T_i.

In this section we introduce a class of recursive programs as an extension of the class of **while** programs which are generated by the following grammar:

$$S ::= skip \mid u := t \mid \bar{x} := \bar{t} \mid S_1; \ S_2 \mid \textbf{if } B \textbf{ then } S_1 \textbf{ else } S_2 \textbf{ fi} \mid \textbf{while } B \textbf{ do } S_1 \textbf{ od},$$

where S stands for a typical statement or program, u for a simple or subscripted variable, t for an expression (of the same type as u), and B for a Boolean expression. Further, $\bar{x} := \bar{t}$ is a parallel assignment, with $\bar{x} = x_1, \ldots, x_n$ a non-empty list of distinct simple variables and $\bar{t} = t_1, \ldots, t_n$ a list of expressions of the corresponding types. The parallel assignment plays a crucial role in our modelling of the parameter passing. We do not discuss the types and only assume that the set of basic types includes at least the types **integer** and **Boolean**. As an abbreviation we introduce **if** B **then** S **fi** ≡ **if** B **then** S **else** *skip* **fi**.

Given an expression t, we denote by $var(t)$ the set of all simple and array variables that appear in t. Analogously, given a program S, we denote by $var(S)$ the set of all simple and array variables that appear in S, and by $change(S)$ the set of all simple and array variables that can be modified by S, i.e., the set of variables that appear on the left-hand side of an assignment in S.

We arrive at recursive programs by adding recursive procedures with call-by-value parameters. To distinguish between local and global variables, we first introduce a *block statement* by the grammar rule

$$S ::= \textbf{begin local } \bar{x} := \bar{t}; S_1 \textbf{ end}.$$

A block statement introduces a non-empty sequence $\bar{x}$ of simple local variables, all of which are explicitly initialized by means of a parallel assignment $\bar{x} := \bar{t}$, and provides an explicit scope for these simple local variables. The precise explanation of a scope is more complicated because the block statements can be nested.

Assuming $\bar{x} = x_1, \ldots, x_k$ and $\bar{t} = t_1, \ldots, t_k$, each occurrence of a local variable x_i within the statement S_1 *and not* within another block statement that is a subprogram of S_1 refers to the same variable. Each such variable x_i is initialized to the expression t_i by means of the parallel assignment $\bar{x} := \bar{t}$. Further, given a statement S' such that **begin local** $\bar{x} := \bar{t}; S_1$ **end** is a subprogram of S', all occurrences of x_i in S' outside this block statement refer to some other variable(s).

Procedure calls with parameters are introduced by the grammar rule

$$S ::= P(t_1, \ldots, t_n),$$

where P is a procedure identifier and $t_1, \ldots, t_n$, with $n \geq 0$, are expressions called *actual parameters*. The statement $P(t_1, \ldots, t_n)$ is called a *procedure call*. The resulting class of programs is then called *recursive programs*.

Procedures are defined by *declarations* of the form

$$P(u_1, \ldots, u_n) :: S,$$

where $u_1, \ldots, u_n$ are distinct simple variables, called *formal parameters* of the procedure P and S is the *body* of the procedure P.

We assume a given set of procedure declarations D such that each procedure that appears in D has a unique declaration in D. When considering recursive programs we assume that all procedures whose calls appear in the considered recursive programs are declared in D. Additionally, we assume that the procedure calls are *well-typed*, which means that the numbers of formal and actual parameters agree and that for each parameter position the types of the corresponding actual and formal parameters coincide.

Given a recursive program S, we call a variable x_i *local* if it appears within a subprogram of D or S of the form **begin local** $\bar{x} := \bar{t}; S_1$ **end** with $\bar{x} = x_1, \ldots, x_k$, and *global* otherwise.

To avoid possible name clashes between local and global variables we assume that given a set of procedure declarations D and a recursive program S, no local variable of S occurs in D. So given the procedure declaration

$$P :: \textbf{if } x = 1 \textbf{ then } b := \textbf{true else } b := \textbf{false fi}$$

the program

$$S \equiv \textbf{begin local } x := 1; P \textbf{ end}$$

is not allowed. If it were, the semantics we are about to introduce would allow us to conclude that $\{x = 0\} \, S \, \{b\}$ holds. However, the customary semantics of the programs in the presence of procedures prescribes that in this case $\{x = 0\} \, S \, \{\neg b\}$ should hold, as the meaning of a program should not depend on the choice of the names of its local variables. (This is a consequence of the so-called *static scope* of the variables that we assume here.)

This problem is trivially solved by just renaming the 'offensive' local variables to avoid name clashes, so by considering here the program **begin local** $y := 1; P$ **end** instead of S. Once we limit ourselves to recursive programs no local variable of which occurs in the considered set of procedure declarations, the semantics we introduce ensures that the names of local variables indeed do not matter. More precisely, the programs that only differ in the choice of the names of local variables and obey the above syntactic restriction have then identical meaning. In what follows, when considering a recursive program S in the context of a set of procedure declarations D we always implicitly assume that the above syntactic restriction is satisfied.

The local and global variables play an analogous role to the bound and free variables in first-order formulas or in λ-terms. In fact, the above syntactic restriction corresponds to the 'Variable Convention' of [4, page 26] according to which "all bound variables are chosen to be different from the free variables."

Note that the above definition of programs puts no restrictions on the actual parameters in procedure calls; in particular they can be formal parameters or global variables.

Semantics

For recursive programs we use a structural operational semantics in the sense of Plotkin [17]. As usual, it is defined in terms of transitions between configurations. A *configuration* C is a pair $< S, \sigma >$ consisting a statement S that is to be executed and a state σ that assigns a value to each variable (including local variables). A *transition* is written as a step $C \to C'$ between configurations. To express termination we use the empty statement E; a configuration $< E, \sigma >$ denotes termination in the state σ.

Transitions are specified by the transition axioms and rules which are defined in the context of a set D of procedure declarations. The only transition axioms that are somewhat non-standard are the ones that deal with the block statement and the procedure calls, in that they avoid the use of substitution thanks to the use of parallel assignment:

$$< \textbf{begin local } \bar{x} := \bar{t}; S \textbf{ end}, \sigma > \ \to \ < \bar{x} := \bar{t}; S; \bar{x} := \sigma(\bar{x}), \sigma >,$$

$$< P(\bar{t}); R, \sigma > \ \to \ < \textbf{begin local } \bar{u} := \bar{t}; S \textbf{ end}; R, \sigma >,$$

where $P(\bar{u}) :: S \in D$.

The first axiom ensures that the local variables are initialized as prescribed by the parallel assignment and that upon termination the global variables whose names coincide with the local variables are restored to their initial values, held at the beginning of the block statement. This is a way of implicitly modeling a *stack discipline* for (nested) blocks. So the use of the block statement in the second transition axiom ensures that prior to the execution of the procedure body the formal parameters are *simultaneously* instantiated to the actual parameters and that upon termination of a procedure call the formal parameters are restored to their initial values. Additionally, the block statement limits the scope of the formal parameters so that they are not accessible upon termination of the procedure call. So the second transition axiom describes the *call-by-value* parameter mechanism.

Based on the transition relation $\rightarrow$ we consider two variants of input/output semantics for recursive programs S refering to the set Σ of states σ, τ. The *partial correctness semantics* is a mapping $\mathcal{M}[\![S]\!] : \Sigma \rightarrow \mathcal{P}(\Sigma)$ defined by

$$\mathcal{M}[\![S]\!](\sigma) = \{\tau \mid < S, \sigma > \rightarrow^* < E, \tau >\}.$$

The *total correctness semantics* is a mapping $\mathcal{M}_{tot}[\![S]\!] : \Sigma \rightarrow \mathcal{P}(\Sigma \cup \{\bot\})$ defined by

$$\mathcal{M}_{tot}[\![S]\!](\sigma) = \mathcal{M}[\![S]\!](\sigma) \cup \{\bot \mid S \text{ can diverge from } \sigma\}.$$

Here $\bot$ is an error state signalling divergence, i.e., an infinite sequence of transitions starting in the configuration $< S, \sigma >$.

3 Proof Systems for Partial and Total Correctness

Program correctness is expressed by *correctness formulas* of the form $\{p\}\, S\, \{q\}$, where S is a program and p and q are assertions. The assertion p is the *precondition* of the correctness formula and q is the *postcondition*. A correctness formula $\{p\}\, S\, \{q\}$ is true in the sense of partial correctness if every terminating computation of S that starts in a state satisfying p terminates in a state satisfying q. And $\{p\}\, S\, \{q\}$ is true in the sense of total correctness if every computation of S that starts in a state satisfying p terminates and its final state satisfies q. Thus in the case of partial correctness, diverging computations of S are not taken into account.

Using the semantics $\mathcal{M}$ and $\mathcal{M}_{tot}$, we formalize these two interpretations of correctness formulas uniformly as set theoretic inclusions as follows (cf. [3]). For an assertion p let $[\![p]\!]$ denote the set of states satisfying p. Then we define:

(i) The correctness formula $\{p\}\, S\, \{q\}$ is true in the sense of *partial correctness*, abbreviated by $\models \{p\}\, S\, \{q\}$, if $\mathcal{M}[\![S]\!]([\![p]\!]) \subseteq [\![q]\!]$.

(ii) The correctness formula $\{p\}\, S\, \{q\}$ is true in the sense of *total correctness*, abbreviated by $\models_{tot} \{p\}\, S\, \{q\}$, if $\mathcal{M}_{tot}[\![S]\!]([\![p]\!]) \subseteq [\![q]\!]$.

Since by definition $\bot \notin [\![q]\!]$, part (ii) indeed formalizes the above intuition about total correctness.

Partial Correctness

Partial correctness of **while** programs is proven using the customary proof system *PD* consisting of the group of axioms and rules 1–7 shown in the appendix. Consider now partial correctness of recursive programs. First, we introduce the following rule that deals with the block statement.

BLOCK

$$\frac{\{p\}\, \bar{x} := \bar{t}; S\, \{q\}}{\{p\}\, \textbf{begin local}\ \bar{x} := \bar{t}; S\ \textbf{end}\, \{q\}}$$

where $var(\bar{x}) \cap free(q) = \emptyset$.

By $free(q)$ we denote here the set of all free simple and array variables that have a free occurrence in the assertion q.

The main issue is how to deal with the procedure calls. To this end, we want to adjust the proofs of 'generic' procedure calls to arbitrary ones. The definition of a generic call and the conditions for the correctness of such an adjustment process refer to the assumed set of procedure declarations D. By a generic call of a procedure P we mean a call of the form $P(\bar{x})$, where $\bar{x}$ is a sequence of fresh (w.r.t. D) variables.

First, we extend the definition of $change(S)$ to recursive programs and sets of procedure declarations as follows:

$$change(\textbf{begin local } \bar{x} := \bar{t}; S \textbf{ end}) = change(S) \setminus \{\bar{x}\},$$
$$change(P(\bar{u}) :: S) = change(S) \setminus \{\bar{u}\},$$
$$change(\{P(\bar{u}) :: S\} \cup D) = change(P(\bar{u}) :: S) \cup change(D),$$
$$change(P(\bar{t})) = \emptyset.$$

The adjustment of the generic procedure calls is then taken care of by the following proof rule that refers to the set of procedure declarations D:

INSTANTIATION
$$\frac{\{p\} \, P(\bar{x}) \, \{q\}}{\{p[\bar{x} := \bar{t}]\} \, P(\bar{t}) \, \{q[\bar{x} := \bar{t}]\}}$$
where $var(\bar{x}) \cap var(D) = var(\bar{t}) \cap change(D) = \emptyset$ and $P(\bar{u}) :: S \in D$ for some S.

In the following rule for recursive procedures with parameters we use the provability symbol $\vdash$ to refer to the proof system PD augmented with the auxiliary axiom and rules A1–A6 defined in the appendix and the above two proof rules.

RECURSION
$$\{p_1\} \, P_1(\bar{x}_1) \, \{q_1\}, \ldots, \{p_n\} \, P_n(\bar{x}_n) \, \{q_n\} \vdash \{p\} \, S \, \{q\},$$
$$\{p_1\} \, P_1(\bar{x}_1) \, \{q_1\}, \ldots, \{p_n\} \, P_n(\bar{x}_n) \, \{q_n\} \vdash$$
$$\frac{\{p_i\} \, \textbf{begin local } \bar{u}_i := \bar{x}_i; S_i \textbf{ end} \, \{q_i\}, \; i \in \{1, \ldots, n\}}{\{p\} \, S \, \{q\}}$$

where $D = P_1(\bar{u}_1) :: S_1, \ldots, P_n(\bar{u}_n) :: S_n$ and $var(\bar{x}_i) \cap var(D) = \emptyset$ for $i \in \{1, \ldots, n\}$.

The intuition behind this rule is as follows. Say that a program S is (p, q)-*correct* if $\{p\} \, S \, \{q\}$ holds in the sense of partial correctness. The second premise of the rule states that we can establish from the *assumption* of the (p_i, q_i)-correctness of the 'generic' procedure calls $P_i(\bar{x}_i)$ for $i \in \{1, \ldots, n\}$, the (p_i, q_i)-correctness of the procedure bodies S_i for $i \in \{1, \ldots, n\}$, which are adjusted as in the transition axiom that deals with the procedure calls. Then we can prove the (p_i, q_i)-correctness of the procedure calls $P_i(\bar{x}_i)$ unconditionally, and thanks to the first premise establish the (p, q)-correctness of the recursive program S.

To prove *partial* correctness of *recursive* programs with parameters we use the proof system PR that is obtained by extending the proof system PD by the block rule, the instantiation rule, the recursion rule, and the auxiliary axiom and rules A1–A6.

Note that when we deal only with one recursive procedure and use the procedure call as the considered recursive program, this rule simplifies to

$$\frac{\{p\}\ P(\bar{x})\ \{q\} \vdash \{p\}\ \textbf{begin local}\ \bar{u} := \bar{x}; S\ \textbf{end}\ \{q\}}{\{p\}\ P(\bar{x})\ \{q\}}$$

where $D = P(\bar{u}) :: S$ and $var(\bar{x}) \cap var(D) = \emptyset$.

Total Correctness

Total correctness of **while** programs is proven using the proof system TD consisting of the group of axioms and rules 1–5, 7, and 8 shown in the appendix. For total correctness of recursive programs we need a modification of the recursion rule. The provability symbol $\vdash$ refers now to the proof system TD augmented with the auxiliary rules A2–A6, the block rule and the instantiation rule. The proof rule is a minor variation of a rule originally proposed in [1] and has the following form:

RECURSION II

$$\frac{\begin{array}{l} \{p_1\}\ P_1(\bar{x}_1)\ \{q_1\}, \ldots, \{p_n\}\ P_n(\bar{x}_n)\ \{q_n\} \vdash \{p\}\ S\ \{q\}, \\ \{p_1 \wedge t < z\}\ P_1(\bar{x}_1)\ \{q_1\}, \ldots, \{p_n \wedge t < z\}\ P_n(\bar{x}_n)\ \{q_n\} \vdash \\ \quad \{p_i \wedge t = z\}\ \textbf{begin local}\ \bar{u}_i := \bar{x}_i; S_i\ \textbf{end}\ \{q_i\},\ i \in \{1, \ldots, n\} \end{array}}{\{p\}\ S\ \{q\}}$$

where $D = P_1(\bar{u}_1) :: S_1, \ldots, P_n(\bar{u}_n) :: S_n$, $var(\bar{x}_i) \cap var(D) = \emptyset$ for $i \in \{1, \ldots, n\}$, and z is an integer variable that does not occur in p_i, t, q_i and S_i for $i \in \{1, \ldots, n\}$ and is treated in the proofs as a constant, which means that in these proofs neither the $\exists$-introduction rule A4 nor the substitution rule A6 defined in the appendix is applied to z.

To prove *total* correctness of *recursive* programs with parameters we use the proof system TR that is obtained by extending the proof system TD by the block rule, the instantiation rule, the recursion rule II, and the auxiliary rules A2–A6.

As before, in the case of one recursive procedure this rule can be simplified to

$$\frac{\begin{array}{l} \{p \wedge t < z\}\ P(\bar{x})\ \{q\} \vdash \{p \wedge t = z\}\ \textbf{begin local}\ \bar{u} := \bar{x}; S\ \textbf{end}\ \{q\}, \\ p \to t \geq 0 \end{array}}{\{p\}\ P(\bar{x})\ \{q\}}$$

where $D = P(\bar{u}) :: S$, $var(\bar{x}) \cap var(D) = \emptyset$ and z is an integer variable that does not occur in p, t, q and S and is treated in the proof as a constant.

4   Modularity

Proof system TR allows us to establish total correctness of recursive programs directly. However, sometimes it is more convenient to decompose the proof of total correctness into two separate proofs, one of partial correctness and one of termination. More specifically, given a correctness formula $\{p\}\ S\ \{q\}$, we first establish its partial correctness, using proof system PR. Then, to show termination it suffices to prove the simpler correctness formula $\{p\}\ S\ \{\textbf{true}\}$ using proof system TR.

These two different proofs can be combined into one using the following general proof rule for total correctness:

DECOMPOSITION

$$\frac{\vdash_{PR} \{p\}\ S\ \{q\},}{\{p\}\ S\ \{q\}}\ \ \ \vdash_{TR} \{p\}\ S\ \{\textbf{true}\}$$

where $\vdash_{PR}$ and $\vdash_{PR}$ refer to the proofs in the proof systems PR and TR, respectively.

The decomposition rule and other auxiliary rules like A2 or A3 allow us to combine two correctness formulas derived *independently*. In some situations it is helpful to reason about procedure calls in a hierarchical way, by first deriving one correctness formula and then using it in a proof of another correctness formula. The following modification of the above simplified version of the recursion rule illustrates this principle, where we limit ourselves to a two-stage proof and one procedure:

MODULARITY

$$\frac{\{p_0\}\ P(\bar{x})\ \{q_0\} \vdash \{p_0\}\ \textbf{begin local}\ \bar{u} := \bar{x};\ S\ \textbf{end}\ \{q_0\},}{\{p\}\ P(\bar{x})\ \{q\}}$$
$$\{p_0\}\ P(\bar{x})\ \{q_0\}, \{p\}\ P(\bar{x})\ \{q\} \vdash \{p\}\ \textbf{begin local}\ \bar{u} := \bar{x};\ S\ \textbf{end}\ \{q\}$$

where $D = P(\bar{u}) :: S$ and $var(\bar{x}) \cap var(D) = \emptyset$.

So first we derive an auxiliary property, $\{p_0\}\ P(\bar{x})\ \{q_0\}$ that we subsequently use in the proof of the 'main' property, $\{p\}\ P(\bar{x})\ \{q\}$. In general, more procedures may be used and an arbitrary 'chain' of auxiliary properties may be constructed. In the next section we show that such a modular approach can lead to better structured correctness proofs.

5   Correctness Proof of the *Quicksort* Procedure

We now apply the modular proof method to verify total correctness of the *Quicksort* algorithm, originally introduced in [10]. For a given array a of type $\textbf{integer} \rightarrow \textbf{integer}$ and integers x and y this algorithm sorts the section $a[x : y]$ consisting of all elements $a[i]$ with $x \leq i \leq y$. Sorting is accomplished 'in situ', i.e., the elements of the initial (unsorted) array section are permuted to achieve the sorting property. We consider here the following version of *Quicksort* close to the one studied in [7]. It consists of a recursive procedure $Quicksort(m, n)$, where the formal parameters m, n and the local variables v, w are all of type $\textbf{integer}$:

$Quicksort(m, n)$ ::
   if $m < n$
   then $Partition(m, n)$;
      begin
      local $v, w := ri, le$;
      $Quicksort(m, v)$;
      $Quicksort(w, n)$
      end
  fi

$Quicksort$ calls a non-recursive procedure $Partition(m, n)$ which partitions the array a suitably, using global variables ri, le, pi of type **integer** standing for *pivot*, *left*, and *right* elements:

$Partition(m, n)$ ::
   $pi := a[m]$;
   $le, ri := m, n$;
   while $le \leq ri$ **do**
     while $a[le] < pi$ **do** $le := le + 1$ **od**;
     while $pi < a[ri]$ **do** $ri := ri - 1$ **od**;
     if $le \leq ri$ **then**
     $swap(a[le], a[ri])$;
     $le, ri := le + 1, ri - 1$
    fi
  od

Here for two given simple or subscripted variables u and v the program $swap(u, v)$ is used to *swap* the values of u and v. So we stipulate that the following correctness formula

$$\{x = u \wedge y = v\} \; swap(u, v) \; \{x = v \wedge y = u\}$$

holds in the sense of partial and total correctness, where x and y are fresh variables.

In the following D denotes the set of the above two procedure declarations and S_Q the body of the procedure $Quicksort(m, n)$.

Formal Problem Specification

Correctness of $Quicksort$ amounts to proving that upon termination of the procedure call $Quicksort(m, n)$ the array section $a[m : n]$ is sorted and is a permutation of the input section. To write the desired correctness formula we introduce some notation. The assertion

$$sorted(a[x : y]) \equiv \forall i, j : (x \leq i \leq j \leq y \rightarrow a[i] \leq a[j])$$

states that the integer array section $a[x : y]$ is sorted. To express the permutation property we use an auxiliary array a_0 in the section $a_0[x : y]$ of which we record the initial values of $a[x : y]$. The abbreviation

$$bij(f, x, y) \equiv f \text{ is a bijection on } \mathbb{Z} \; \wedge \; \forall i \notin [x : y] : f(i) = i$$

states that f is a bijection on $\mathbb{Z}$ which is the identity outside the interval $[x : y]$. Hence

$$perm(a, a_0, [x : y]) \equiv \exists f : (bij(f, x, y) \wedge \forall i : a[i] = a_0[f(i)])$$

specifies that the array section $a[x : y]$ is a permutation of the array section $a_0[x : y]$ and that a and a_0 are the same elsewhere.

We can now express the correctness of *Quicksort* by means of the following correctness formula:

Q1 $\{a = a_0\}$ $Quicksort(x, y)$ $\{perm(a, a_0, [x : y]) \wedge sorted(a[x : y])\}$.

To prove correctness of *Quicksort* in the sense of partial correctness we proceed in stages and follow the methodology explained in Section 4. In other words, we establish some auxiliary correctness formulas first, using among others the recursion rule. Then we use them as premises in order to derive other correctness formulas, also using the recursion rule.

Properties of *Partition*

In the proofs we shall use a number of properties of the *Partition* procedure. This procedure is non-recursive, so to verify them it suffices to prove the corresponding properties of the procedure body using the proof systems *PD* and *TD*, a task we leave to Nissim Francez.

More precisely, we assume the following properties of *Partition* in the sense of partial correctness:

P1 $\{\textbf{true}\}$ $Partition(m, n)$ $\{ri \leq n \wedge m \leq le\}$,

P2 $\{x' \leq m \wedge n \leq y' \wedge perm(a, a_0, [x' : y'])\}$
$Partition(m, n)$
$\{x' \leq m \wedge n \leq y' \wedge perm(a, a_0, [x' : y'])\}$,

P3 $\{\textbf{true}\}$
$Partition(m, n)$
$\{ le > ri \wedge$
$(\forall i \in [m : ri] : a[i] \leq pi) \wedge$
$(\forall i \in [ri + 1 : le - 1] : a[i] = pi) \wedge$
$(\forall i \in [le : n] : pi \leq a[i])\}$,

and the following property in the sense of total correctness:

P4 $\{m < n\}$
$Partition(m, n)$
$\{ri - m < n - m \wedge n - le < n - m\}$.

Property **P1** states the bounds for ri and le. We remark that $le \leq n$ and $m \leq ri$ need not hold upon termination. Property **P2** implies that the call $Partition(n, k)$ permutes the array section $a[m : n]$ and leaves other elements of a intact, but actually is a stronger statement involving an interval $[x' : y']$ that includes $[m : n]$, so that we can carry out the reasoning about the recursive calls of $Quicksort$. Finally, property **P3** captures the main effect of the call $Partition(n, k)$: the elements of the section $a[m : n]$ are rearranged into three parts, those smaller than pi (namely $a[m : ri]$), those equal to pi (namely $a[ri + 1 : le - 1]$), and those larger than pi (namely $a[le : n]$). Property **P4** is needed in the termination proof of the $Quicksort$ procedure: it states that the subsections $a[m : ri]$ and $a[le : n]$ are strictly smaller than the section $a[m : n]$.

Auxiliary proof: permutation property

In the remainder of this section we use the following abbreviation:

$$J \equiv m = x \land n = y.$$

We first extend the permutation property **P2** to the procedure $Quicksort$:

> **Q2** $\{perm(a, a_0, [x' : y']) \land x' \le x \land y \le y'\}$
> $Quicksort(x, y)$
> $\{perm(a, a_0, [x' : y'])\}$

Until further notice the provability symbol $\vdash$ refers to the proof system *PD* augmented with the the block rule, the instantiation rule and the auxiliary rules A2–A6.

The appropriate claim needed for the application of the recursion rule is:

Claim 1

> **P1, P2, Q2** $\vdash$ $\{perm(a, a_0, [x' : y']) \land x' \le x < y \le y'\}$
> **begin local** $m, n := x, y; S_Q$ **end**
> $\{perm(a, a_0, [x' : y'])\}.$

Proof. In Figure 1 a proof outline is given that uses as assumptions the correctness formulas **P1, P2,** and **Q2**. More specifically, the used correctness formula about the call of $Partition$ is derived from **P1** and **P2** by the conjunction rule. In turn, the correctness formulas about the recursive calls of $Quicksort$ are derived from **Q2** by an application of the instantiation rule and the invariance rule. This concludes the proof of Claim 1. $\square$

We can now derive **Q2** by the recursion rule. In summary, we proved

$$\textbf{P1, P2} \vdash \textbf{Q2}.$$

Auxiliary proof: sorting property

We can now verify that the call $Quicksort(x, y)$ sorts the array section $a[x : y]$, so

Q3 $\{\textbf{true}\}$ $Quicksort(x, y)$ $\{sorted(a[x : y])\}.$

The appropriate claim needed for the application of the recursion rule is:

Claim 2

> **P3, Q2, Q3** $\vdash$ $\{\textbf{true}\}$ **begin local** $m, n := x, y; S_Q$ **end** $\{sorted(a[x : y])\}.$

$\{perm(a, a_0, [x' : y']) \wedge x' \le x \wedge y \le y'\}$
begin local
  $\{perm(a, a_0, [x' : y']) \wedge x' \le x \wedge y \le y'\}$
  $m, n := x, y;$
  $\{perm(a, a_0, [x' : y']) \wedge x' \le x \wedge y \le y' \wedge J\}$
  $\{perm(a, a_0, [x' : y']) \wedge x' \le m \ \wedge n \le y'\}$
  if $m < n$ **then**
      $\{perm(a, a_0, [x' : y']) \wedge x' \le m \ \wedge n \le y'\}$
      $Partition(m, n);$
      $\{perm(a, a_0, [x' : y']) \wedge x' \le m \wedge n \le y' \wedge ri \le n \wedge m \le le\}$
      begin local
        $\{perm(a, a_0, [x' : y']) \wedge x' \le m \wedge n \le y' \wedge ri \le n \wedge m \le le\}$
        $v, w := ri, le;$
        $\{perm(a, a_0, [x' : y']) \wedge x' \le m \wedge n \le y' \wedge v \le n \wedge m \le w\}$
        $\{perm(a, a_0, [x' : y']) \wedge x' \le m \wedge v \le y' \wedge x' \le w \wedge n \le y'\}$
        $Quicksort(m, v);$
        $\{perm(a, a_0, [x' : y']) \wedge x' \le w \wedge n \le y'\}$
        $Quicksort(w, n)$
        $\{perm(a, a_0, [x' : y'])\}$
      end
      $\{perm(a, a_0, [x' : y'])\}$
  fi
  $\{perm(a, a_0, [x' : y'])\}$
end
$\{perm(a, a_0, [x' : y'])\}$

Fig. 1. Proof outline showing permutation property **Q2**

Proof. In Figure 2 a proof outline is given that uses as assumptions the correctness formulas **P3**, **Q2**, and **Q3**. In the following we justify the correctness formulas about *Partition* and the recursive calls of *Quicksort* used in this proof outline. In the postcondition of *Partition* we use the following abbreviation:

$$
\begin{aligned}
K \equiv \ &v < w \wedge \\
&(\forall i \in [m : v] : \ a[i] \le pi) \wedge \\
&(\forall i \in [v + 1 : w - 1] : \ a[i] = pi) \wedge \\
&(\forall i \in [w : n] : \ pi \le a[i]).
\end{aligned}
$$

Observe that the correctness formula

$$\{J\} \ Partition(m, n) \ \{J \wedge K[v, w := ri, le]\}$$

is derived from **P3** by the invariance rule. Next we verify the correctness formulas

$$\{J \wedge K\}Quicksort(m, v)\{sorted(a[m : v]) \wedge J \wedge K\}, \tag{1}$$

$\{\textbf{true}\}$
$\textbf{begin local}$
$\{\textbf{true}\}$
$m, n := x, y;$
$\{J\}$
$\textbf{if } m < n \textbf{ then}$
$\qquad \{J \wedge m < n\}$
$\qquad Partition(m, n);$
$\qquad \{J \wedge K[v, w := ri, le]\}$
$\qquad \textbf{begin local}$
$\qquad \{J \wedge K[v, w := ri, le]\}$
$\qquad v, w := ri, le;$
$\qquad \{J \wedge K\}$
$\qquad Quicksort(m, v);$
$\qquad \{sorted(a[m : v]) \wedge J \wedge K\}$
$\qquad Quicksort(w, n)$
$\qquad \{sorted(a[m : v] \wedge sorted(a[w : n] \wedge J \wedge K\}$
$\qquad \{sorted(a[x : v] \wedge sorted(a[w : y] \wedge K[m, n := x, y]\}$
$\qquad \{sorted(a[x : y])\}$
$\qquad \textbf{end}$
$\qquad \{sorted(a[x : y])\}$
$\textbf{fi}$
$\{sorted(a[x : y])\}$
$\textbf{end}$
$\{sorted(a[x : y])\}$

Fig. 2. Proof outline showing sorting property **Q3**

and

$$\{sorted(a[m : v]) \wedge J \wedge K\}$$
$$Quicksort(w, n) \qquad\qquad\qquad (2)$$
$$\{sorted(a[m : v] \wedge sorted(a[w : n] \wedge J \wedge K\}.$$

about the recursive calls of $Quicksort$.

Proof of (1). By applying the instantiation rule to **Q3**, we obtain

A1 $\{\textbf{true}\}$ $Quicksort(m, v)$ $\{sorted(a[m : v])\}$.

Moreover, by the invariance axiom, we have

A2 $\{J\}$ $Quicksort(m, v)$ $\{J\}$.

By applying the instantiation rule to **Q2**, we then obtain

$$\{perm(a, a_0, [x' : y']) \wedge x' \leq m \wedge v \leq y'\}$$
$$Quicksort(m, v)$$
$$\{perm(a, a_0, [x' : y'])\}.$$

Applying next the substitution rule with the substitution $[x', y' := m, v]$ yields

$$\{perm(a, a_0, [m : v]) \land m \le m \land v \le v\}$$
$$Quicksort(m, v)$$
$$\{perm(a, a_0, [m : v])\}.$$

So by a trivial application of the consequence rule, we obtain

$$\{a = a_0\} \; Quicksort(m, v) \; \{perm(a, a_0, [m : v])\}.$$

We then obtain by an application of the invariance rule

$$\{a = a_0 \land K[a := a_0]\} \; Quicksort(m, v) \; \{perm(a, a_0, [m : v]) \land K[a := a_0]\}.$$

Note now the following implications:

$$K \to \exists a_0 : (a = a_0 \land K[a := a_0]),$$
$$perm(a, a_0, [m : v]) \land K[a := a_0] \to K.$$

So we conclude

A3 $\{K\} \; Quicksort(m, v) \; \{K\}$

by the $\exists$-introduction rule and the consequence rule. Combining the correctness formulas **A1–A3** by the conjunction rule we get (1).

Proof of (2). In a similar way as above, we can prove the correctness formula

$$\{a = a_0\} \; Quicksort(w, n) \; \{perm(a, a_0, [w : n])\}.$$

By an application of the invariance rule we obtain

$$\{a = a_0 \land sorted(a_0[m : v]) \land v < w\}$$
$$Quicksort(w, n)$$
$$\{perm(a, a_0, [w : n]) \land sorted(a_0[m : v]) \land v < w\}.$$

Note now the following implications:

$$v < w \land sorted(a[m : v]) \to \exists a_0 : (a = a_0 \land sorted(a_0[m : v]) \land v < w),$$
$$(perm(a, a_0, [w : n]) \land sorted(a_0[m : v]) \land v < w) \to sorted(a[m : v]).$$

So we conclude

B1 $\{v < w \land sorted(a[m : v])\} \; Quicksort(w, n) \; \{sorted(a[m : v])\}$

by the $\exists$-introduction rule and the consequence rule. Further, by applying the instantiation rule to **Q3** we obtain

B2 $\{\mathbf{true}\} \; Quicksort(w, n) \; \{sorted(a[w : n])\}.$

Next, by the invariance axiom we obtain

B3 $\{J\}\ Quicksort(w, m)\ \{J\}$.

Further, using the implications

$$K \to \exists a_0 : (a = a_0 \wedge K[a := a_0]),$$
$$perm(a, a_0, [w : n]) \wedge K[a := a_0] \to K,$$

we can derive from **Q2**, in a similar manner as in the proof of **A3**,

B4 $\{K\}\ Quicksort(w, n)\ \{K\}$.

Combining the correctness formulas **B1–B4** by the conjunction rule and observing that $K \to v < w$ holds, we get (2).

The final application of the consequence rule in the proof outline given in Figure 2 is justified by the following crucial implication:

$$sorted(a[x : v]) \wedge sorted(a[w : y]) \wedge K[m, n := x, y] \to$$
$$sorted(a[x : y]).$$

Also note that $J \wedge m \geq n \to sorted(a[x : y])$, so the implicit **else** branch is properly taken care of. This concludes the proof of Claim 2.   □

We can now derive **Q3** by the recursion rule. In summary, we proved

$$\textbf{P3, Q2} \vdash \textbf{Q3}.$$

The proof of partial correctness of $Quicksort$ is now immediate: it suffices to combine **Q2** and **Q3** by the conjunction rule. Then after applying the substitution rule with the substitution $[x', y' := x, y]$ and the consequence rule we obtain **Q1**, or more precisely

$$\textbf{P1, P2, P3} \vdash \textbf{Q1}.$$

Total Correctness

To prove termination, by the decomposition rule discussed in Section 4, it suffices to establish

Q4 $\{\textbf{true}\}\ Quicksort(x, y)\ \{\textbf{true}\}$

in the sense of total correctness. In the proof we rely on the property **P4** of $Partition$:

$$\{m < n\}\ Partition(m, n)\ \{ri - m < n - m \wedge n - le < n - m\}.$$

The provability symbol $\vdash$ refers below to the proof system *TD* augmented with the block rule, the instantiation rule and the the auxiliary rules A2–A6. For the termination proof of the recursive procedure call $Quicksort(x, y)$ we use

$$t \equiv \max(y - x, 0)$$

as the bound function. Since $t \geq 0$ holds, the appropriate claim needed for the application of the recursion rule II is:

Claim 3

$$\text{\textbf{P4}, } \{t < z\} \, Quicksort(x, y) \, \{\textbf{true}\} \vdash$$
$$\{t = z\} \, \textbf{begin local } m, n := x, y; S_Q \textbf{ end } \{\textbf{true}\}.$$

Proof. In Figure 3 a proof outline for total correctness is given that uses as assumptions the correctness formulas **P4** and $\{t < z\} \, Quicksort(x, y) \, \{\textbf{true}\}$. In the following we

$\{t = z\}$
begin local
$\{\max(y - x, 0) = z\}$
$m, n := x, y;$
$\{\max(n - m, 0) = z\}$
if $n < k$ **then**
$\quad \{\max(n - m, 0) = z \wedge m < n\}$
$\quad \{n - m = z \wedge m < n\}$
$\quad Partition(m, n);$
$\quad \{n - m = z \wedge m < n \wedge ri - m < n - m \wedge n - le < n - m\}$
$\quad$ **begin local**
$\quad \{n - m = z \wedge m < n \wedge ri - m < n - m \wedge n - le < n - m\}$
$\quad v, w := ri, le;$
$\quad \{n - m = z \wedge m < n \wedge v - m < n - m \wedge n - w < n - m\}$
$\quad \{\max(v - m, 0) < z \wedge \max(n - w, 0) < z\}$
$\quad Quicksort(m, v);$
$\quad \{\max(n - w, 0) < z\}$
$\quad Quicksort(w, n)$
$\quad \{\textbf{true}\}$
$\quad$ **end**
$\quad \{\textbf{true}\}$
fi
$\{\textbf{true}\}$
end
$\{\textbf{true}\}$

Fig. 3. Proof outline establishing termination of the $Quicksort$ procedure

justify the correctness formulas about $Partition$ and the recursive calls of $Quicksort$ used in this proof outline. Since $m, n, z \notin change(D)$, we deduce from **P4** using the invariance rule the correctness formula

$$\{n - m = z \wedge m < n\}$$
$$Partition(m, n) \qquad\qquad (3)$$
$$\{n - m = z \wedge ri - m < n - m \wedge n - le < n - m\}.$$

Consider now the assumption

$$\{t < z\}\ Quicksort(x, y)\ \{\mathbf{true}\}.$$

Since $n, w, z \notin change(D)$, the instantiation rule and the invariance rule yield

$$\{\max(v - m, 0) < z \wedge \max(n - w, 0) < z\}$$
$$Quicksort(m, v)$$
$$\{\max(n - w, 0) < z\}$$

and

$$\{\max(n - w, 0) < z\}\ Quicksort(w, n)\ \{\mathbf{true}\}.$$

The application of the consequence rule preceding the first recursive call of $Quicksort$ is justified by the following two implications:

$$(n - m = z \wedge m < n \wedge v - m < n - m) \rightarrow \max(v - m, 0) < z,$$
$$(n - m = z \wedge m < n \wedge n - w < n - m) \rightarrow \max(n - w, 0) < z.$$

This completes the proof of Claim 3.   □

Applying now the simplified version of the recursion rule II we derive **Q4**. In summary, we proved

$$\mathbf{P4} \vdash \mathbf{Q4}.$$

6   Conclusions

The issue of modularity has been by now well-understood in the area of program construction. It also has been addressed in the program verification. Let us just mention two references, an early one and a recent one: [8] focused on modular verification of temporal properties of concurrent programs which were modelled as a set of modules that interact by means of procedure calls. In turn, [19] considered modular verification of heap manipulating programs, where the focus has been on the automatic extraction and verification specifications.

However, to our knowledge no approach has been proposed to deal with correctness of recursive programs in a modular fashion. When proving correctness of the $Quicksort$ program we found that the simple approach here proposed allowed us to structure the proof better by establishing the 'permutation property' first and then using it in the proof of the 'sorting property'.

So in our approach we propose modularity at the level of *proofs* and not at the level of *programs*. This should be of help when organizing a mechanically verified correctness proof, by expressing the proofs of the subsidiary properties as subsidiary lemmas. In general, modular correctness proofs of programs are proofs from assumptions about subprograms, which can be considered as 'black boxes' of the given programs. Zwiers [20] has investigated an appropriate notion of completeness for such proofs from assumptions about black boxes, called *modular completeness*.

The first proof of partial correctness of $Quicksort$ is given in [7]. That proof establishes the permutation and the sorting property simultaneously, in contrast to our approach. For dealing with recursive procedures, [7] use proof rules corresponding to our rules for blocks, instantiation, and recursion (for the case of one recursive procedure). They also use a so-called *adaptation rule* of [11] that allows one to adapt a given correctness formula about a program to other pre- and postconditions. In our approach we use several auxiliary rules which together have the same effect as the adaptation rule. The expressive power of the adaptation rule has been analyzed in [14]. No proof rule for the termination of recursive procedures is proposed in [7], only an informal argument is given why *Quicksort* terminates. An informal proof of total correctness of *Partition* is given in [12] as part of the program *Find* given in [9].

The recursion rule is modelled after the so-called Scott induction rule for fixed points that appeared first in the unpublished manuscript Scott and De Bakker [18]. Recursion rule II for total correctness is taken from America and De Boer [1], where also the completeness of a proof system similar to *TR* is established. The modularity rule corresponds to a theorem due to Bekić [5] which states that for systems of monotonic functions iterative fixed points coincide with simultaneous fixed points.

Acknowledgment

We thank the reviewer for helpful suggestions.

References

1. America, P., de Boer, F.S.: Proving total correctness of recursive procedures. Information and Computation 84(2), 129–162 (1990)
2. Apt, K.R.: Ten years of Hoare's logic, a survey, part I. ACM Transactions on Programming Languages and Systems 3, 431–483 (1981)
3. Apt, K.R., de Boer, F.S., Olderog, E.-R.: Verification of Sequential and Concurrent Programs, 3rd extended edn. Springer, New York (2009) (to appear)
4. Barendregt, H.P.: The Lambda Calculus. North Holland, Amsterdam (1984)
5. Bekić, H.: Definable operations in general algebras, and the theory of automata and flow charts. Technical report, IBM Laboratory, Vienna (1969); Typescript
6. Filliâtre, J.-C., Magaud, N.: Certification of sorting algorithms in the system Coq. In: Theorem Proving in Higher Order Logics: Emerging Trends (1999)
7. Foley, M., Hoare, C.A.R.: Proof of a recursive program: Quicksort. Computer Journal 14(4), 391–395 (1971)
8. Hailpern, B., Owicki, S.: Modular verification of concurrent programs. In: POPL 1982: Proceedings of the 9th ACM SIGPLAN-SIGACT symposium on Principles of programming languages, pp. 322–336. ACM, New York (1982)
9. Hoare, C.A.R.: Algorithm 65, Find. Communications of the ACM 4(7), 321 (1961)
10. Hoare, C.A.R.: Quicksort. Comput. J. 5(1), 10–15 (1962)
11. Hoare, C.A.R.: Procedures and parameters: an axiomatic approach. In: Engeler, E. (ed.) Proceedings of Symposium on the Semantics of Algorithmic Languages, New York. Lecture Notes in Mathematics, vol. 188, pp. 102–116. Springer, Heidelberg (1971)
12. Hoare, C.A.R.: Proof of a program: Find. Communications of the ACM 14(1), 39–45 (1971)
13. Kaldewaij, A.: Programming: The Derivation of Algorithms. Prentice-Hall, Englewood Cliffs (1990)

14. Olderog, E.-R.: On the notion of expressiveness and the rule of adaptation. Theoretical Computer Science 30, 337–347 (1983)
15. Owre, S., Shankar, N.: Writing PVS proof strategies. In: Archer, M., Di Vito, B., Muñoz, C. (eds.) Design and Application of Strategies/Tactics in Higher Order Logics (STRATA 2003), number CP-2003-212448 in NASA Conference Publication, Hampton, VA, September 2003, pp. 1–15. NASA Langley Research Center (2003)
16. Plotkin, G.D.: A structural approach to operational semantics. Technical Report DAIMI-FN 19, Department of Computer Science, Aarhus University (1981)
17. Plotkin, G.D.: A structural approach to operational semantics. J. of Logic and Algebraic Programming, 60–61, 17–139 (2004); Revised version of [16]
18. Scott, D., de Bakker, J.W.: A theory of programs. Notes of an IBM Vienna Seminar (1969)
19. Taghdiri, M.: Automating Modular Program Verification by Refining Specifications. Ph.D thesis. MIT, Cambridge, Mass (2008),
    http://alloy.mit.edu/community/files/mana_thesis.pdf
20. Zwiers, J.: Compositionality, Concurrency, and Partial Correctness. LNCS, vol. 321. Springer, Heidelberg (1989)

Appendix

We list here the used axioms and proof rules that were not defined earlier in the text. To establish correctness of **while** programs we rely on the following axioms and proof rules. In the proofs of partial correctness the loop rule is used, while in the proofs of total correctness the loop II rule is used.

AXIOM 1: SKIP
$$\{p\}\ skip\ \{p\}$$

AXIOM 2: ASSIGNMENT
$$\{p[u := t]\}\ u := t\ \{p\}$$

AXIOM 3: PARALLEL ASSIGNMENT
$$\{p[\bar{x} := \bar{t}]\}\ \bar{x} := \bar{t}\ \{p\}$$

RULE 4: COMPOSITION
$$\frac{\{p\}\ S_1\ \{r\}, \{r\}\ S_2\ \{q\}}{\{p\}\ S_1;\ S_2\ \{q\}}$$

RULE 5: CONDITIONAL
$$\frac{\{p \wedge B\}\ S_1\ \{q\}, \{p \wedge \neg B\}\ S_2\ \{q\}}{\{p\}\ \textbf{if}\ B\ \textbf{then}\ S_1\ \textbf{else}\ S_2\ \textbf{fi}\ \{q\}}$$

RULE 6: LOOP

$$\frac{\{p \wedge B\} \, S \, \{p\}}{\{p\} \ \textbf{while} \ B \ \textbf{do} \ S \ \textbf{od} \ \{p \wedge \neg B\}}$$

RULE 7: CONSEQUENCE

$$\frac{p \rightarrow p_1, \{p_1\} \, S \, \{q_1\}, q_1 \rightarrow q}{\{p\} \, S \, \{q\}}$$

RULE 8: LOOP II

$$\frac{\begin{array}{l} \{p \wedge B\} \, S \, \{p\}, \\ \{p \wedge B \wedge t = z\} \, S \, \{t < z\}, \\ p \rightarrow t \geq 0 \end{array}}{\{p\} \ \textbf{while} \ B \ \textbf{do} \ S \ \textbf{od} \ \{p \wedge \neg B\}}$$

where t is an integer expression and z is an integer variable that does not appear in p, B, t or S.

Additionally, we rely on the following auxiliary axioms and proof rules that occasionally refer to the assumed set of procedure declarations D.

AXIOM A1: INVARIANCE

$$\{p\} \, S \, \{p\}$$

where $free(p) \cap (change(D) \cup change(S)) = \emptyset$.

RULE A2: DISJUNCTION

$$\frac{\{p\} \, S \, \{q\}, \{r\} \, S \, \{q\}}{\{p \vee r\} \, S \, \{q\}}$$

RULE A3: CONJUNCTION

$$\frac{\{p_1\} \, S \, \{q_1\}, \{p_2\} \, S \, \{q_2\}}{\{p_1 \wedge p_2\} \, S \, \{q_1 \wedge q_2\}}$$

RULE A4: ∃-INTRODUCTION

$$\frac{\{p\} \, S \, \{q\}}{\{\exists x : p\} \, S \, \{q\}}$$

where $x \notin change(D) \cup change(S) \cup free(q)$.

RULE A5: INVARIANCE

$$\frac{\{r\} \, S \, \{q\}}{\{p \wedge r\} \, S \, \{p \wedge q\}}$$

where $free(p) \cap (change(D) \cup change(S)) = \emptyset$.

RULE A6: SUBSTITUTION

$$\frac{\{p\}\, S\, \{q\}}{\{p[\bar{z} := \bar{t}]\}\, S\, \{q[\bar{z} := \bar{t}]\}}$$

where $(var(\bar{z}) \cup var(\bar{t})) \cap (change(D) \cup change(S)) = \emptyset$.

Semi-formal Evaluation of Conversational Characters

Ron Artstein, Sudeep Gandhe, Jillian Gerten, Anton Leuski, and David Traum

Institute for Creative Technologies, University of Southern California, 13274 Fiji Way, Marina del Rey, CA 90292 USA

Abstract. Conversational dialogue systems cannot be evaluated in a fully formal manner, because dialogue is heavily dependent on context and current dialogue theory is not precise enough to specify a target output ahead of time. Instead, we evaluate dialogue systems in a semi-formal manner, using human judges to rate the coherence of a conversational character and correlating these judgments with measures extracted from within the system. We present a series of three evaluations of a single conversational character over the course of a year, demonstrating how this kind of evaluation helps bring about an improvement in overall dialogue coherence.

1   Introduction

In the past two decades, the field of Computational Linguistics has placed an increasing emphasis on formal evaluation of systems and system components; typically, this involves creating a target reference ("gold standard"), and measuring system performance against this reference. The availability of standard targets – most notably the Penn Treebank [1] – has greatly facilitated the use of machine learning for computational linguistic tasks, and formal system evaluation has opened up the field to competitions between systems working on a shared task.

While formal system evaluation is responsible for much of the progress made in Computational Linguistics in recent years, it has limitations. In order to conduct a formal evaluation, the desired target performance needs to be defined prior to the task, and this gets more difficult as we move from the surface of an utterance to more abstract levels of representation. Thus it is fairly straightforward to specify the desired output of a speech recognizer for a set of input utterances; specifying the desired part-of-speech labels or syntactic parse is somewhat more controversial, and defining a desired semantic representation or translation to a different language is even more difficult. The present study is concerned with dialogue systems, where the abstraction continues to climb with dialogue acts, and reaches the most abstract level with response selection – the decision *what* to say in response to a user utterance. For conversational characters this is essentially an open-ended problem.

Even in the absence of a predefined target, some dialogue systems can be evaluated formally. The performance of task-oriented systems can be measured

O. Grumberg et al. (Eds.): Francez Festschrift, LNCS 5533, pp. 22–35, 2009.

by the task success rate, length of the dialogues (assuming it is desirable to complete the task quickly), and other proxies for user satisfaction. Such measures are useful not only for evaluation and comparison but also for machine learning of dialogue strategies that optimize system performance according to the chosen criteria [2,3]. In contrast, conversational characters are not designed to help a user accomplish a specific goal; the criteria for successful dialogue are "soft" ones such as user satisfaction and tutoring outcomes, and at present we do not have proxies for these criteria that can be measured automatically.

We present a conversational character, Sergeant Star, who answers questions about the U.S. Army. He appears at conventions and conferences together with live exhibitors, and his purpose is to generate interest and engage the audience rather than to conduct efficient dialogues aimed at achieving a specific task. SGT Star's components include a number of statistical subsystems such as a speech recognizer and a response classifier; these components are formally trained and evaluated using large sets of data mapping inputs to outputs. But we have no way to formally evaluate the overall, end-to-end performance of SGT Star. Instead, we use what we call a "semi-formal" approach: we get the "soft" performance metrics by using human judges to rate the coherence of SGT Star's responses from actual field deployments, and then use these ratings together with measures taken from within the system in order to gain a better understanding of SGT Star's performance. The combination of ratings with system data allows us to see patterns in the overall behavior that would be difficult to detect in a detailed item-by-item qualitative analysis, and this influences the continued development of SGT Star.

We describe the SGT Star system in section 2 and our rating studies in section 3. Results and analysis, correlating the ratings with system data, are in sections 4 and 5. The conclusions in section 6 describe how the results are used in the authoring process to improve SGT Star's performance from one iteration to the next.

2   Sergeant Star

Sergeant Star is a virtual question-answering character developed for the U.S. Army Recruiting Command as a high-tech attraction and information source about the Army. He is a life-size character built for demos in mobile exhibits, who listens to human speech and responds with pre-recorded, animated voice answers (Figure 1). SGT Star is based on technology similar to that used in previous efforts [4,5], which treats question-answering as an information retrieval problem: given a natural-language question, the character should retrieve the most appropriate response from a predefined list. An Automatic Speech Recognition (ASR) module transforms the user's question to text, and then a statistical classifier trained on question-response pairs ranks possible responses according to their similarity to a language model derived from the user's utterance; the top-ranked response is SGT Star's best guess about the correct answer to the user's question. The size of SGT Star's domain is about 200 responses, and the training

Fig. 1. SGT Star

data contain a few hundred questions that link to the responses in a many-to-many mapping. The classifier has a few simple dialogue management capabilities, such as detecting when the best response has a low score (and thus might not be appropriate), avoiding repetitive responses, and prompting the user to ask a relevant question.

In a typical exhibit setting, SGT Star interacts with groups of attendees. Since SGT Star can only talk to one person at a time, conversation with the group is mediated by a human handler, who uses SGT Star to create a two-man show. There is a small group of handlers who demonstrate SGT Star at various shows, and acoustic models of the speech recognition component are tuned to their voices in order to get the best recognition in noisy convention environments.

Evaluation of SGT Star is based on performance in three actual field deployments: the National Future Farmers of America Convention (Indianapolis, October 2007 and 2008) and the National Leadership and Skills Conference (Kansas City, June 2008).[1] The main advantage of a field test is that the interactions being evaluated are real, rather than simulated interactions from the lab (Ai et al. [8] show that dialogues with real users have different characteristics from dialogues with lab subjects). The two main challenges presented by field evaluation are the lack of experimental controls and the demands of interacting with a live audience, which take precedence over experimental needs.

Initially, the field studies were intended as a general evaluation, but the results quickly turned these into a detailed study of SGT Star's "off-topic" responses. SGT Star is a simple question-answering character who does not have a dialogue manager to keep track of the state of the dialogue, the commitments and obligations of the various participants, his own goals and desires, and so on. Instead of a separate manager, dialogue management capabilities are incorporated

---

[1] Evaluations of the October 2007 and June 2008 deployments were reported in [6,7].

into the classifier. The most important capability is detecting when the best response is not good enough: if the score of the top-ranked classifier output falls below a specified threshold, SGT Star does not produce that output, but instead chooses among a set of predefined "off-topic" responses (e.g. "Sorry, I didn't catch that. Could you say it again?"). The threshold is set automatically during training in order to find an optimal balance between false positives (inappropriate responses above threshold) and false negatives (appropriate responses below threshold). We should note that the labels "on-topic" and "off-topic" characterize SGT Star's responses, not the user's questions: an in-domain question can receive an off-topic response (e.g., if it was not properly recognized), and such a response may well turn out to be coherent in the context of the dialogue; an out-of-domain question can also receive an on-topic response, though this usually indicates that SGT Star misunderstood the question and therefore the response is typically not appropriate.

The off-topic strategy for dealing with classification failures has been successful for other efforts such as SGT Blackwell, a general-domain question-answering character who interacts with visitors in a museum setting [4,9]. The environment in which SGT Star is deployed differs from that of SGT Blackwell in two important ways: speech input to SGT Star typically comes from trained handlers rather than from the general public, and the handlers try to engage SGT Star for a conversation consisting of a greeting phase, a few information exchanges, and a closing routine. Since handlers are trained, few user utterances are genuine out-of-domain questions, and most of SGT Star's classifier failures are caused by faulty speech recognition or insufficient training data. Since interactions are fairly long (compared to SGT Blackwell), random off-topic interruptions are very disruptive. Initial versions of SGT Star were very successful at providing on-topic responses, but rather poor when an off-topic response was called for: in the October 2007 study, the vast majority of the on-topic responses (80.7%) received the maximum coherence rating of 5, whereas the majority of off-topic responses (80.1%) were rated between 1 and 2. An individual analysis of the off-topic responses showed that requests for repetition were usually ranked as more coherent than other types of off-topic responses.

To improve the coherence of off-topic responses we re-authored many of the responses, and implemented a new off-topic selection policy. We were not able to use a separate classifier trained on out-of-domain questions [10], because very few of the questions SGT Star gets are truly outside his domain. Instead, we designed a strategy based on the knowledge that the vast majority of SGT Star's off-topic responses are triggered by speech recognition errors and classification failures. If SGT Star fails to find an answer, then in all likelihood he either misheard the user's utterance or misunderstood it. We therefore authored off-topic responses for SGT Star in the following four classes.

Clarify: Ask the user to repeat the question, for example:
    Could you throw that at me again?
    I didn't copy that. Could you repeat that?
    Sorry, I didn't catch that. Could you say it again?

Stall: Wait for user initiative, for example:
>   Aw this feels too much like school. I didn't study last night.
>   Sometimes I think you're just testing me.
>   You know, sometimes you ask a lot of questions.

Move on: Decline to answer, for example:
>   Can we talk about what I want to talk about?
>   Uh, next question.
>   You can get answers at GoArmy dot com.

Prompt: Direct the user to a new topic, for example:
>   Ask me how I can see in the dark.
>   Hey why don't you ask me about my badges?
>   Why don't you ask me about becoming an MP in the army?

The off-topic response classes were designed to fit into a simple strategy: the first off-topic response after a successful dialogue is always of the "clarify" type; if the following user utterance is also not understood, SGT Star produces a "stall" utterance, and if a third successive off-topic response is needed, SGT Star produces a "move on" utterance followed immediately by a prompt. The rationale behind this strategy is to ensure that the majority of SGT Star's off-topic responses are requests for clarification (which have been shown to be more coherent), while avoiding repetitiveness and progressively encouraging the user to move the conversation to a new and more fruitful direction. Due to an unrelated installation problem, we were not able to use the off-topic selection policy in the June 2008 study, and we had to revert to a version of the software which selects off-topic responses at random. This gave us an unintended experimental control, since we had the re-authored off-topic response classes, but without the policy. The October 2008 study used the same off-topic responses as the June 2008 study, with the full selection policy implemented.

3 Coherence Rating and Reliability

In the absence of a predetermined target performance, we used a post-hoc rating of SGT Star's responses. Raters were asked to rate the *coherence* (appropriateness) of SGT Star's responses rather than their correctness. The idea is that the more coherent a character is, the better he can engage the audience. An appropriate response to a question does not have to be a direct answer: a question or off-topic comment may sometimes be more appropriate, and SGT Star's off-topic responses were designed to allow him to hold a coherent conversation when he does not have a straight answer.

Following each field deployment, we transcribed all the user utterances from the recordings, and then created full transcripts of the dialogues from the transcribed user utterances and the character's responses obtained from system logs. These transcripts were presented as web pages on which judges rated each of SGT Star's responses, in the context of the immediately preceding dialogue, on a scale of 1 to 5 (Figure 2). In order to reduce the burden on the judges, we automatically identified the cases where the transcribed user utterance was identical

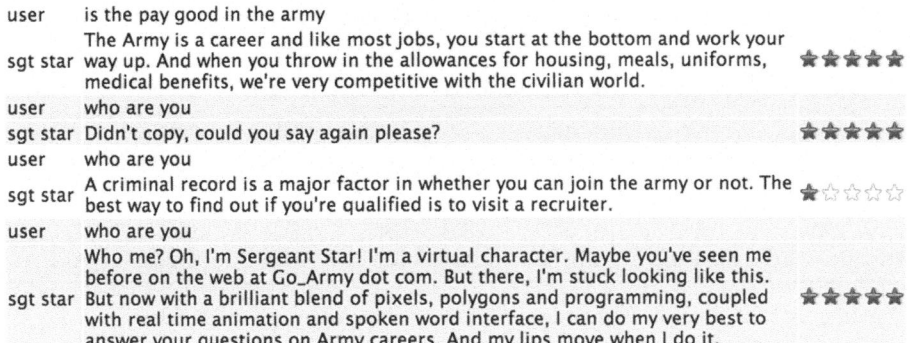

		user	is the pay good in the army
		sgt star	The Army is a career and like most jobs, you start at the bottom and work your way up. And when you throw in the allowances for housing, meals, uniforms, medical benefits, we're very competitive with the civilian world.
		user	who are you
		sgt star	Didn't copy, could you say again please?
		user	who are you
		sgt star	A criminal record is a major factor in whether you can join the army or not. The best way to find out if you're qualified is to visit a recruiter.
		user	who are you
		sgt star	Who me? Oh, I'm Sergeant Star! I'm a virtual character. Maybe you've seen me before on the web at Go_Army dot com. But there, I'm stuck looking like this. But now with a brilliant blend of pixels, polygons and programming, coupled with real time animation and spoken word interface, I can do my very best to answer your questions on Army careers. And my lips move when I do it.

Fig. 2. The rating interface

Table 1. Number of responses

	Total	Perfect match (On-topic)		Rated					
				On-topic		Off-topic		Prompt	
		N	%	N	%	N	%	N	%
October 2007	3216	703	22	1283	40	1230	38	—	
June 2008	2095	578	28	704	34	632	30	181	9
October 2008	1321	282	21	705	53	281	21	51	4

to a training question and the response was linked to that question in the training data, and these were automatically rated as 5; the remaining responses were rated by the judges. Table 1 gives the number of responses rated in each study.

To ensure the ratings were meaningful we calculated inter-rater reliability using Krippendorff's α [11].[2] Four raters participated in each reliability study: the first author participated in all studies, the third author in the October 2008 and November 2008 studies, and the remaining judges were student annotators (one of whom had also transcribed the utterances). In each study, at least 2 of the raters marked all of the responses while the remaining raters only marked a portion for calculating reliability. The results of the reliability studies are shown in Table 2.

---

[2] Krippendorff's α is a chance-corrected agreement coefficient, similar to the more familiar K statistic [12]. Like K, α ranges from -1 to 1, where 1 signifies perfect agreement, 0 obtains when agreement is at chance level, and negative values show systematic disagreement. The main difference between α and K is that α takes into account the magnitudes of the individual disagreements, whereas K treats all disagreements as equivalent; α is more appropriate for our study because the ratings are numerical, and the disagreement between ratings of 2 and 3, for example, is clearly lower than between 2 and 5. For additional background, definitions and discussion of agreement coefficients, see [13].

Table 2. Reliability of rater judgments (Krippendorff's α)

	All responses			On-topic	Off-topic	Prompt
	All Raters	Excluding Outlier	Range[a]	All Raters	All Raters	All Raters
October 2007	0.786	0.886	0.676–0.901	0.794	0.097	—
June 2008	0.583	0.655	0.351–0.680	0.842	0.017	0.080[b]
October 2008	0.699	0.757	0.614–0.763	0.841	0.219	0.155

[a] Reliability for the most discordant and most concordant pairs of coders.
[b] Value reported for the two main judges only, because of the small number of prompts rated by the control judges.

Reliability of ratings for on-topic responses was $\alpha = 0.8$ or above in all studies, demonstrating that the coders share an understanding of the task and are able to apply it consistently. In contrast, reliability for off-topic responses and prompts was essentially at chance level for the October 2007 and June 2008 studies, and only slightly above chance for the October 2008 study; this reflects the fact that evaluating the coherence of an off-topic response is much more difficult than evaluating the coherence of an on-topic response. The improvement in reliability for off-topic and prompt ratings in the October 2008 study is statistically significant, and may be attributed to improved instructions, or to improved system performance which makes the off-topic responses better overall, allowing for better discrimination among coherent and incoherent ones (in other words, the fact that the character's overall coherence has improved makes it easier to rate his coherence).

Overall reliability decreased from the October 2007 to the June 2008 study. The reason for the drop in reliability is the improvement in the actual ratings of off-topic responses. In both studies the raters showed little ability to agree with each other on the ratings of individual off-topic responses, but in the October 2007 study these ratings were all very low (80.1% of the off-topic responses were rated between 1 and 2) and thus had little effect on the overall reliability; in the June 2008 study, off-topic responses were not ranked so low, making them less distinct from the on-topic ratings and therefore reducing overall reliability. Overall reliability improved with the October 2008 study despite the fact that an additional improvement in the off-topic ratings made them even more similar to on-topic ratings. This improvement in overall reliability is probably due to the improvement in the reliability of off-topic ratings.

We also calculated confidence intervals for α following a bootstrapping method similar to that of [14]; however, we found that variation in reliability among different subsets of raters was typically larger than the 95% confidence interval for a specific set of raters. We therefore report in Table 2 only the variation among subsets of raters (the range of reliability scores displayed by different rater subsets). We also note that in each study, removing one outlying rater bumps up reliability by 6 to 10 percentage points (the outlier was not always the same person). Since overall reliability was close to acceptable in all studies, and since we do not have reason to believe that the outlying raters are less correct than the others, we continued our analysis using the mean rating for each response (mean of all available scores in October 2007 and October 2008, mean of the two main raters but not the controls in June 2008).

4   Response Ratings

A straightforward way to measure system performance is to look at the coherence ratings. The overall mean is not very telling, because the distribution is far from normal. Instead, Figure 3 shows histograms of all the response ratings (including those that were automatically rated as 5). The three studies show a similar

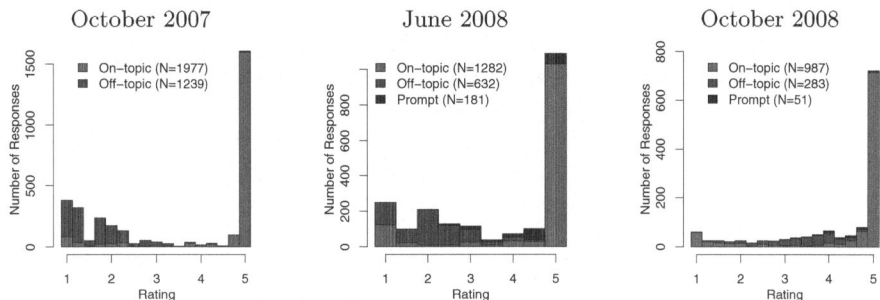

Fig. 3. Individual response ratings

pattern for on-topic response ratings. There is a very strong concentration of responses that are rated very high – in each of the studies, more than 80% of the on-topic responses received a mean rating of 4.5 or above; this means that when SGT Star's response score is above threshold, the response is usually very appropriate. There is also a discernible (though much smaller) bump at the lower end of the scale, which shows that when SGT Star chooses a wrong on-topic response, it is usually very inappropriate. In contrast to the stable on-topic rating pattern, the off-topic response ratings show a consistent improvement – the improvement from October 2007 to June 2008 is due to rewriting the responses in the three off-topic classes, and the improvement from June 2008 to October 2008 is due to the implementation of the off-topic selection policy.

We gain additional insight by looking at the individual response types, comparing their ratings with the frequency in which they occur in the dialogues (Figure 4). Again, the pattern for on-topic responses is the same in all studies: the frequent responses are more highly rated. The likely explanation, as we proposed in [6], is that the handlers are aware of which responses are easy to elicit, and target their questions to elicit these responses. The pattern thus demonstrates an interplay between the inherent capabilities of the system and the human handlers who are working to maximize its performance in a live show. The

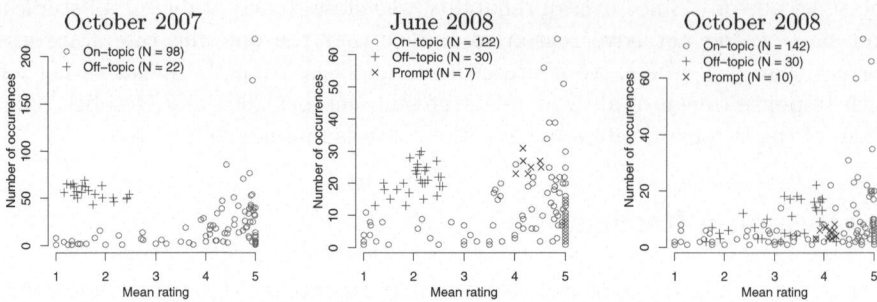

Fig. 4. Rating and frequency of response types

off-topic pattern, on the other hand, shows substantial variation. The October 2007 study shows a negative correlation between response rating and frequency ($r = -0.55, p < 0.01, df = 20$). The reason for this, as we argued in [6], is that some off-topic responses were linked to out-of-domain questions in the training data – for example, the question "so do you have a girlfriend?" was linked to the response "ha ha, you're a bad man". This boosted the frequency of the linked responses, but these turned out to be lower rated than clarification requests like "I didn't hear that, could you repeat the question?", which were typically not linked to any question. This observation led to reauthoring the off-topic responses, dividing them into classes, and unlinking all of them in the training data. The result in the June 2008 study is the absence of a significant correlation between rating and frequency for off-topics ($r = 0.35, p = 0.06, df = 28$), since they all appear with similar frequency. Finally, the implementation of the off-topic selection policy in the October 2008 study resulted in a positive correlation between rating and frequency ($r = 0.53, p < 0.005, df = 28$), reflecting the fact that the more coherent clarification requests are now also the most frequent, due to the fact that the off-topic policy chooses them as the first response. The last panel of Figure 4 shows that in the October 2008 study, the off-topic responses fall into the same pattern as on-topic responses with respect to the rating-frequency relation, and no longer stand out as a distinct cluster.

The question remains, whether the improvement in the scores of the frequent off-topic responses is due only to the fact that the new policy ensures that clarification requests are more frequent, or whether part of the improvement can be attributed to the actual sequencing. In other words: would placing clarification requests as the second and third responses in a sequence of off-topics improve coherence or degrade it? The data do not provide a clear answer. Figure 5 is a rating-frequency plot like Figure 4, highlighting the separate off-topic response classes. The different response classes have similar frequencies in the June 2008 study but different frequencies in the October 2008 study – this is a direct result of implementing the off-topic selection policy. In both studies, clarification questions are rated as the most coherent, though the difference between "clarify" and "move on" is not significant in the June 2008 study. Note that the October 2008 study had fewer data points overall, which makes reaching significance more

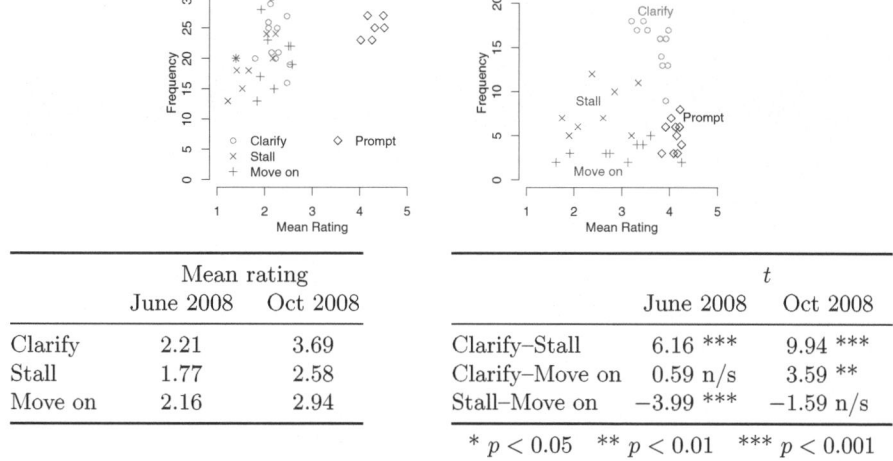

Fig. 5. Rating and frequency of off-topic responses

difficult, especially for the less frequent classes. We do not have an explanation for the overall increase in ratings for all classes of off-topics from June 2008 to October 2008; the responses were identical in the two studies, the only difference being the selection policy. It would be nice to attribute the increase in coherence to the implementation of the policy, though another possibility is that the raters may have interpreted the task differently, perhaps due to an emphasis shift in the instructions (these were revised between the two studies). The ratings for the October 2008 study are probably more trustworthy, since inter-rater reliability was somewhat higher.

5   Speech Recognition and Classifier Scores

Automatic speech recognition (ASR) affects performance [4]: if what SGT Star hears doesn't match what the user said, then SGT Star's response is more likely to be inappropriate. We computed the word error rate for each user utterance by comparing the ASR output with the transcribed speech.[3] Mean word error rate was 0.469 in October 2007, 0.365 in June 2008, and 0.428 in October 2008; the results are not directly comparable because the language models were re-trained for each study, and the acoustic environments differed. Figure 6 shows the distribution of utterance word error rates.

In all three studies we found a highly significant negative correlation (ranging from $r = -40$ to $r = -0.47$) between the rating of SGT Star's on-topic responses

---

[3] Word error rate is the number of substitutions, deletions and insertions needed to transform one string into the other, divided by the number of words in the actual (transcribed) speech; values above 1 were recorded as 1.

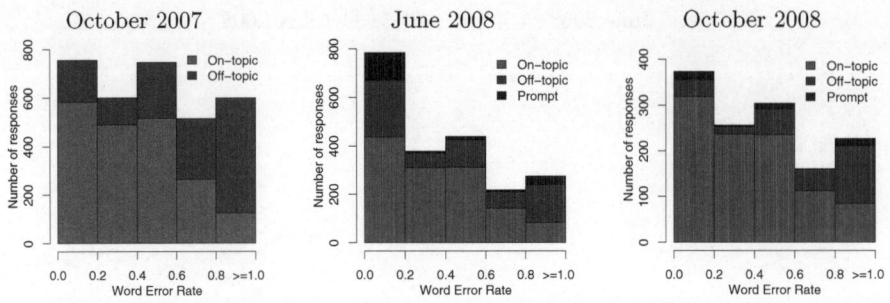

Fig. 6. Word error rates and the responses they triggered

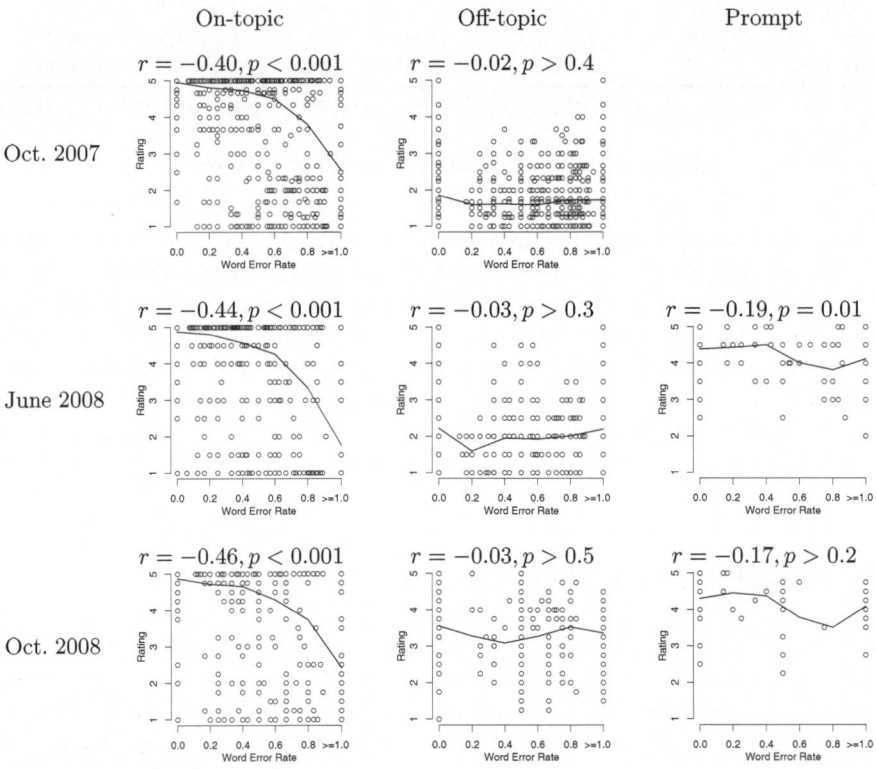

Fig. 7. Word error rates and ratings: the lines show the mean rating for each WER band

and the word error rate of the immediately preceding user utterances; off-topic responses and prompts typically did not exhibit such a correlation (Figure 7). The negative correlation between rating and word error rate for on-topic responses is expected: the less SGT Star understands the spoken utterance, the less likely he is to come up with a suitable on-topic response, so if an on-topic

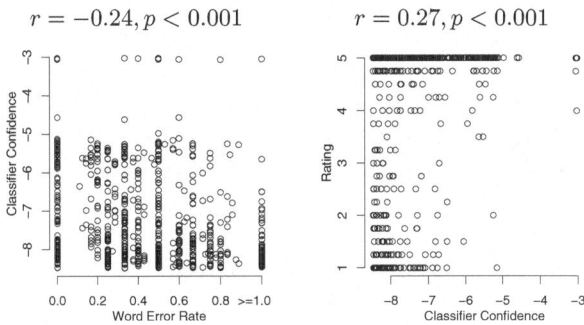

Fig. 8. Relation between speech recognition, classifier confidence and rating

response is selected it is more likely to be inappropriate. Off-topic responses and prompts are not expected to degrade with the mismatch between actual and recognized user utterance.

Our final measures concern the classifier confidence. We mentioned above that the decision whether to utter an on-topic or off-topic response depends on the classifier's confidence in the adequacy of the top-ranked response: if the confidence exceeds a specified threshold (determined during training), SGT Star utters that response, otherwise he gives an off-topic. With the rating study we can check how effective this strategy is. Figure 8 plots the word error rate of each user utterance against the classifier's confidence in the top-ranked response for that utterance, and the classifier's confidence against the rating (we only have data for the on-topic responses for the October 2008 study). The results are what we expect. Classifier confidence shows a negative correlation with word error rate, because noisy input is less similar to the input on which the classifier is trained. Confidence is positively correlated with coherence, meaning that the measure the classifier uses – similarity between language models – is similar to human judgment of the appropriateness of the responses. To evaluate the suitability of the threshold chosen by the system we will need to collect additional data, namely confidence and ratings for top-ranked responses that fall below the off-topic threshold.

6   Conclusion

We choose to call our method of evaluation "semi-formal" because it combines "hard" numbers taken from within the system with the "soft" numbers of the human rating study. The analysis is quantitative, but the conclusions are qualitative: The numbers are used to identify which classes of responses work in different dialogue contexts, and eventually which individual responses are good and which need to be improved upon. We believe that this sort of analysis allows better insight into the operation of SGT Star than a simple qualitative analysis of the dialogues, because the numbers reveal patterns such as the relative success of the various off-topic response classes, or the rating-frequency correlation for on-topic responses which exposes the effect that the handlers have on steering SGT Star towards his more popular (and more successful) responses. Additionally, the ratings can be used to evaluate aggregations of specific responses from different contexts, which is an important tool for improved authoring.

In a system where responses are selected from a fixed pool rather than generated on the fly, authoring plays an important role: the improvement of off-topic responses from October 2007 to June 2008 is due mainly to the re-authoring of these responses, since the selection policy was not implemented until October 2008. But authoring appropriate responses for an interactive system that can respond to any user input is a difficult task. SGT Star's ability to respond appropriately depends on his training data – a list of questions, a list of responses, and links between the two. Developing the training corpus is an iterative process, whereby new questions are added based on actual user data, responses get added as the character's domain expands and are modified based on their acceptance by the audience, and links keep getting tweaked in order to improve the character's performance. The information from the semi-formal analysis is helpful at the level of the individual response, but more importantly it exposes global patterns in SGT Star's performance which influence the directions of future character development.

Acknowledgments

SGT Star is loosely based on a character created by Next IT for the Army's recruiting web site, http://www.GoArmy.com/. SGT Star's body and animations were developed by the Mixed Reality team at ICT: Diane Piepol, Kip Haynes, Matt Liewer, Josh Williams, and Tae Yoon. Thanks to Jacob Cannon, Joe Henderer, Ryan Kim and Jordan Peterson for rating SGT Star's utterances.

The project or effort described here has been sponsored by the U.S. Army Research, Development, and Engineering Command (RDECOM). Statements and opinions expressed do not necessarily reflect the position or the policy of the United States Government, and no official endorsement should be inferred.

References

1. Marcus, M.P., Marcinkiewicz, M.A., Santorini, B.: Building a large annotated corpus of English: the Penn Treebank. Computational Linguistics 19(2), 313–330 (1993)
2. Levin, E., Pieraccini, R., Eckert, W.: A stochastic model of human–machine interaction for learning dialog strategies. IEEE Transactions on Speech and Audio Processing 8(1), 11–23 (2000)
3. Walker, M.A.: An application of reinforcement learning to dialogue strategy selection in a spoked dialogue system for email. Journal of Artificial Intelligence Research 12, 387–416 (2000)
4. Leuski, A., Patel, R., Traum, D., Kennedy, B.: Building effective question answering characters. In: Proceedings of the 7th SIGdial Workshop on Discourse and Dialogue, Sydney, Australia, Association for Computational Linguistics, July 2006, pp. 18–27 (2006)

5. Leuski, A., Traum, D.: A statistical approach for text processing in virtual humans. In: 26th Army Science Conference, Orlando, Florida (December 2008)
6. Artstein, R., Gandhe, S., Leuski, A., Traum, D.: Field testing of an interactive question-answering character. In: ELRA Workshop on Evaluation, Marrakech, Morocco, May 2008, pp. 36–40 (2008)
7. Artstein, R., Cannon, J., Gandhe, S., Gerten, J., Henderer, J., Leuski, A., Traum, D.: Coherence of off-topic responses for a virtual character. In: 26th Army Science Conference, Orlando, Florida (December 2008)
8. Ai, H., Raux, A., Bohus, D., Eskenazi, M., Litman, D.: Comparing spoken dialog corpora collected with recruited subjects versus real users. In: Keizer, S., Bunt, H., Paek, T. (eds.) Proceedings of the 8th SIGdial Workshop on Discourse and Dialogue, Antwerp, Belgium, September 2007, pp. 124–131. Association for Computational Linguistics (2007)
9. Robinson, S., Traum, D., Ittycheriah, M., Henderer, J.: What would you ask a conversational agent? Observations of human-agent dialogues in a museum setting. In: LREC 2008 Proceedings, Marrakech, Morocco (May 2008)
10. Patel, R., Leuski, A., Traum, D.: Dealing with out of domain questions in virtual characters. In: Gratch, J., Young, M., Aylett, R.S., Ballin, D., Olivier, P. (eds.) IVA 2006. LNCS, vol. 4133, pp. 121–131. Springer, Heidelberg (2006)
11. Krippendorff, K.: Content Analysis: An Introduction to Its Methodology, ch. 12, pp. 129–154. Sage, Beverly Hills (1980)
12. Siegel, S., Castellan Jr., N.J.: Nonparametric Statistics for the Behavioral Sciences, 2nd edn., ch. 9.8, pp. 284–291. McGraw-Hill, New York (1988)
13. Artstein, R., Poesio, M.: Inter-coder agreement for computational linguistics. Computational Linguistics 34(4), 555–596 (2008)
14. Hayes, A.F., Krippendorff, K.: Answering the call for a standard reliability measure for coding data. Communication Methods and Measures 1(1), 77–89 (2007)

Scope Dominance with Generalized Quantifiers

Gilad Ben-Avi[1] and Yoad Winter[2]

[1] Technion – Israel Institute of Technology
[2] Technion – Israel Institute of Technology and Utrecht University

Abstract. When two quantifiers Q_1 and Q_2 satisfy the scheme $Q_1 x\, Q_2 y\, \phi \to Q_2 y\, Q_1 x\, \phi$, we say that Q_1 is *scopally dominant* over Q_2. This relation is central in analyzing and computing entailment relations between different readings of ambiguous sentences in natural language. This paper reviews the known results on scope dominance and mentions some open problems.

1 Basic Definitions

An *arbitrary generalized quantifier* of signature $\langle n_1, ..., n_k \rangle$ over a non-empty domain E is a relation $f \subseteq \wp(E^{n_1}) \times ... \times \wp(E^{n_k})$, where $k \geq 1$, and $n_i \geq 1$ for all $i \leq k$ (e.g. Peters and Westerståahl , 2006, p.65). In short, we say that f is a *quantifier* when it is of signature $\langle 1 \rangle$, a *determiner* (relation) when it is of signature $\langle 1, 1 \rangle$, and a *dyadic quantifier* when it is of signature $\langle 2 \rangle$. When $\mathcal{R}$ is a binary relation over some domain $\mathcal{E}$ (not necessarily E), we denote for every $\mathcal{X}, \mathcal{Y} \in \mathcal{E}$:

(1) a. $\mathcal{R}_{\mathcal{X}} = \{\mathcal{Y} \in \mathcal{E} : \mathcal{R}(\mathcal{X}, \mathcal{Y})\}$
     b. $\mathcal{R}^{\mathcal{Y}} = \{\mathcal{X} \in \mathcal{E} : \mathcal{R}(\mathcal{X}, \mathcal{Y})\}$

In theories of natural language semantics, determiner relations are useful in describing the meaning of *determiner expressions* as in (2).

(2) *every:*   **every** $= \{\langle A, B \rangle \subseteq E^2 : A \subseteq B\}$

    some: **some** $= \{\langle A, B \rangle \subseteq E^2 : A \cap B \neq \emptyset\}$

    more than half: **mth** $= \{\langle A, B \rangle \subseteq E^2 : |A \cap B| > |A \cap \overline{B}|\}$

It is well-known (Peters and Westerståaahl , 2006, p.469) that meanings of natural language determiners – e.g. of the expression *more than half* – may be beyond what is expressible in first order logic.

We assume that nouns denote sets $A \subseteq E$. Noun phrase meanings are then described as in (3) using a quantifier D_A, where the noun denotation is the left argument of the determiner relation D (cf. (1a)).

(3) *every student:*   **every**$_S = \{B \subseteq E : S \subseteq B\}$

    some teacher: **some**$_T = \{B \subseteq E : T \cap B \neq \emptyset\}$

    more than half of the students: **mth**$_S = \{B \subseteq E : |S \cap B| > |S \cap \overline{B}|\}$

Truth values of simple sentences with intransitive verbs are derived as in (4), using the membership statement that the set denotation of the verb is in the

O. Grumberg et al. (Eds.): Francez Festschrift, LNCS 5533, pp. 36–44, 2009.
© Springer-Verlag Berlin Heidelberg 2009

quantifier denotation of the subject, or, equivalently, that the pair of sets denoted by the noun and the verb are in the determiner relation.

(4) *every student smiled:* $SM \in \mathbf{every}_S \Leftrightarrow \langle S, SM \rangle \in \mathbf{every} \Leftrightarrow$
$$S \subseteq SM$$

    some teacher cried: $C \in \mathbf{some}_T \Leftrightarrow \langle T, C \rangle \in \mathbf{some} \Leftrightarrow$
$$T \cap C \neq \emptyset$$

The *iteration* of two quantifiers Q_1 and Q_2 is the dyadic quantifier Q_1–Q_2 defined in (5). Iteration is used as in (6b) for describing meanings of simple sentences like (6a), with transitive verbs that denote binary relations.

(5) Q_1–$Q_2 \overset{def}{=} \{R \subseteq E^2 : \{x \in E : R_x \in Q_2\} \in Q_1\}$
(6) a. *some teacher praised every student*
    b. $P \in \mathbf{some}_T$–$\mathbf{every}_S \Leftrightarrow$
      $\{x \in E : P_x \in \mathbf{every}_S\} \in \mathbf{some}_T \Leftrightarrow$
      $T \cap \{x \in E : S \subseteq P_x\} \neq \emptyset$

The statement in (6b) is equivalent to the predicate calculus formula (7a). However, a well-known problem in linguistics (Ruys and Winter, 2008) is that transitive sentences like (6a) also have an "inverse scope" reading (7b).

(7) a. $\exists x[T(x) \land \forall y[S(y) \to P(x,y)]]$
    b. $\forall y[S(y) \to \exists x[T(x) \land P(x,y)]]$

A way to use (generalized) quantifiers Q_1 and Q_2 for deriving the meaning of formula (7b), is to define an operator '$\sim$' of *inverse iteration*. The dyadic quantifier $Q_1 \sim Q_2$ defined in (8) is used in (9b) for obtaining an alternative analysis of (9a) (=(6a)).

(8) $Q_1 \sim Q_2 \overset{def}{=} \{R^{-1} : R \in Q_2$–$Q_1\}$
    $= \{R \subseteq E^2 : \{y \in E : R^y \in Q_1\} \in Q_2\}$

(9) a. *some teacher praised every student*
    b. $P \in \mathbf{some}_T \sim \mathbf{every}_S \Leftrightarrow$
      $\{y \in E : P^y \in \mathbf{some}_T\} \in \mathbf{every}_S \Leftrightarrow$
      $S \subseteq \{y \in E : T \cap P^y \neq \emptyset\}$

A trivial fact of first order logic is the entailment (7a)$\Rightarrow$(7b). In natural language semantics, this is reflected in the logical relation between the two readings of sentences like (9a). To describe the general phenomenon, we define a notion of *scope dominance* (in short, "dominance") between quantifiers.

Definition 1. *A quantifier Q_1 is **scopally dominant** over a quantifier Q_2, if Q_1–$Q_2 \subseteq Q_1 \sim Q_2$.*

In cases of determiners D, D', where the quantifier D_A is dominant over D'_B for any two sets A and B, we say that D is dominant over D'. For example, the classical entailment (7a)⇒(7b) amounts to the fact that for all $S, T \subseteq E$, the quantifier $\textbf{some}_S$ is dominant over $\textbf{every}_T$. In short: **some** is dominant over **every**. Furthermore, as can be easily verified, **some** is also dominant over **mth**, where the latter quantifier is not first-order definable. The latter scope dominance is reflected in the relation between the two readings of sentence (10).

(10) *A guard is standing in front of more than half of the churches.*

The likely reading of sentence (10), in which more than half of the churches have a guard in front of them (potentially different guards), entails the less likely reading, in which more than half of the churches are associated with the same guard.

To describe such cases of scope dominance in natural languages, the general task is:

Characterize the quantifiers Q_1, Q_2 such that Q_1 is dominant over Q_2.

In this paper we give a review of previous results about this question, and point to some open problems.

2   Results on Scope Dominance

2.1   Scope Dominance and Duality

The *complement* $\overline{Q}$ of a quantifier Q is the set $\wp(E) \setminus Q$, whereas Q's *postcomplement* is the set $Q- \overset{def}{=} \{A \subseteq E : E \setminus A \in Q\}$. The *dual* Q^d of a quantifier Q is the complement of Q's postcomplement:

$$Q^d \overset{def}{=} \overline{(Q-)} = (\overline{Q})- = \{A \subseteq E : E \setminus A \notin Q\}.$$

Obviously, these three relations between quantifiers are symmetric, and $(\overline{Q})^d = Q-$. This is naturally described using figure 1, which generalizes the Aristotelian *square of opposition*.

Whenever two determiner relations D and D' form complement (postcomplement, dual) quantifiers D_A and D'_A on every set A, we say that the determiners are each other's complements (postcomplements, duals) and write that $D' = \overline{D}$ ($D' = D-$, $D' = D^d$, respectively). The classical square of opposition is between the four determiner expressions *every, some, no* and *not every*. More generally, the opposition holds between the following determiners, which form the classical square in the special case $n = 0$.

(11) $D\ \ = \textbf{all\_but\_at\_most\_}n\ \ \ \ \ = \{\langle A, B \rangle \subseteq E^2 : |A \setminus B| \leq n\}$

$\ \ \ \ \ \ D^d = \textbf{more\_than\_}n\ \ \ \ \ \ \ \ \ = \{\langle A, B \rangle \subseteq E^2 : |A \cap B| > n\}$

$\ \ \ \ \ \ D- = \textbf{at\_most\_}n\ \ \ \ \ \ \ \ \ \ \ = \{\langle A, B \rangle \subseteq E^2 : |A \cap B| \leq n\}$

$\ \ \ \ \ \ \overline{D}\ \ = \textbf{not(all\_but\_at\_most\_}n) = \{\langle A, B \rangle \subseteq E^2 : |A \setminus B| > n\}$

The relation between quantifier duality and scope dominance is described in the following fact (Westerståhl, 1986, p.278).

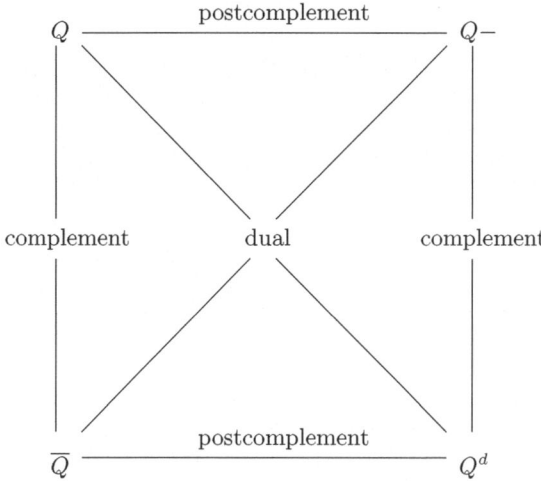

Fig. 1. The Square of Opposition

Fact 1 *For all quantifiers Q_1 and Q_2: Q_1 is scopally dominant over Q_2 iff Q_2^d is dominant over Q_1^d.*

For instance, just like **some** (=**more_than_0**) is dominant over **more_than_3**, so is **all_but_at_most_3** dominant over **every** (=**all_but_at_most_0**).

2.2   Special Cases of Scope Dominance

Two different special cases of scope dominance were studied by Zimmermann (1993) and Westerståhl (1996). Zimmermann characterizes the class of *scopeless* quantifiers: those quantifiers Q that satisfy for all $Q_1 \subseteq \wp(E)$: $Q\text{-}Q_1 = Q\sim Q_1$. He shows that the scopeless quantifiers over E are precisely the *principal ultrafilters* over E: the quantifiers $\{B \subseteq E : a \in B\}$ for some arbitrary $a \in E$.[1] Westerståhl (1996) characterizes the class of *self-commuting* quantifiers: those quantifiers Q that satisfy $Q\text{-}Q = Q\sim Q$. He shows that Q is self-commuting iff Q is either a union or an intersection of principal ultrafilters, or a finite symmetric difference of principal ultrafilters, or a complement of such a symmetric difference.

Clearly, the notion of scope dominance is more general than scopelessness or self-commutativity: a quantifier Q is scopeless iff Q and Q^d are both dominant over any quantifier, while Q is self-commuting iff it is dominant over itself. However, it should be noted that the results on scope dominance that we survey below do not fully subsume these results by Zimmermann and Westerståhl, which hold irrespectively of quantifier monotonicity and the cardinality of the domain.

---

[1]   Zimmermann characterizes scopelessness in a more general case, where Q and Q_1 are not necessarily defined over the same domain. The property we mention here is a direct result of his characterization.

2.3 Characterizing Scope Dominance

A quantifier Q over E is called *upward (downward) monotone* if $A \subseteq B \subseteq E$ implies that $B \in Q$ if (only if) $A \in Q$. A determiner D over E is *upward (downward) monotone* if D_A is upward (downward) monotone for all $A \subseteq E$.

Westerståhl (1986) characterizes the pairs D_1 and D_2 of upward monotone determiners for which D_1 is dominant over D_2. He shows that over finite domains these are precisely the cases where D_1 is **some** or D_2 is **every**. Westerståhl's characterization is stated in a more general form in the following theorem.[2]

Theorem 2. *Let Q_1 and Q_2 be two nontrivial upward monotone quantifiers over a finite domain E. Then Q_1 is scopally dominant over Q_2 iff $Q_1 = \textbf{some}_A$ or $Q_2 = \textbf{every}_A$ for some $A \subseteq E$.*

In general, however, this characterization is too narrow for *infinite* domains. For instance, the quantifier $Q_1 = \{X \subseteq \mathbb{N} : \{1,2\} \subseteq X\}$ is dominant over the quantifier $Q_2 = \{X \subseteq \mathbb{N} : |Y \setminus X| \in \mathbb{N}\}$. This is reflected in the logical relation between the two readings of sentence (12). But there is no $A \subseteq \mathbb{N}$ such that $Q_1 = \textbf{some}_A$ or $Q_2 = \textbf{every}_A$.

(12) *Both item 1 and item 2 cover all but finitely many cases.*

Altman *et al.* (2005) extend Westerståhl's result for upward monotone quantifiers over *countable* domains.[3] For the formulation of the characterization, they define a property of quantifiers called the *Descending Chain Condition* (DCC). A quantifier Q is said to satisfy (DCC) if for every descending sequence $A_1 \supseteq A_2 \supseteq \cdots A_n \supseteq \cdots$ in Q, the intersection $\bigcap_i A_i$ is in Q as well. For instance, any quantifier of the form $\textbf{every}_A$ satisfies (DCC), while a quantifier $\textbf{some}_A$ satisfies (DCC) if and only if A is finite. The following theorem by Altman et al. is a generalization of Theorem 2 to countable domains.

Theorem 3. *Let Q_1 and Q_2 be upward monotone quantifiers over a countable domain E. Then Q_1 is scopally dominant over Q_2 iff all of the following requirements hold:*

(i) Q_1^d or Q_2 are closed under finite intersections;
(ii) Q_1^d or Q_2 satisfy (DCC);
(iii) Q_1^d or Q_2 are not empty.

Note that, on finite domains, an upward monotone quantifier Q is closed under intersections if and only if it is of the form $\textbf{every}_A$. Furthermore, over finite domains, (ii) is a trivial consequence of (i). So this is indeed a generalization of Theorem 2.

---

[2] In fact, Westerståhl's result is restricted to "global" determiner *functors* (abstracting over the domain E), which are furthermore *logical*, i.e. satisfy the familiar restrictions of *conservativity, permutation invariance* and *extension*.

[3] See Altman *et al.* (2001) for an earlier, and more restricted, characterization of scope dominance for the same class of quantifiers.

If we consider not only upward but also *downward* monotone quantifiers, then also on finite domains there are instances of scope dominance that do not involve quantifiers of the form **every**$_A$ or **some**$_A$. For instance, the logical relation between the two readings of sentence (13) is a manifestation of the scope dominance relation between **mth** (cf. (2)) and **no** = **at_most_0** (cf. (11)).

(13) *More than half of the teachers praised no student.*

Theorem 4 below from Ben-Avi and Winter (2004), together with its dual, provide a characterization over finite domains of scope dominance between quantifiers of "opposite" (upward/downward) monotonicities. In this theorem we use the notion of a minimal set in a quantifier. Standardly, we say that a set $X \in Q$ is *minimal in Q* if $Y \subsetneq X$ implies $Y \notin Q$.

Theorem 4. *Let Q_1 and Q_2 be two (non-trivial) quantifiers over a finite domain E, s.t. Q_1 is upward monotone and Q_2 is downward monotone. Let*

$$n \overset{def}{=} \max\{|X| : X \text{ is minimal in } \overline{Q_2}\}$$

Then Q_1 is scopally dominant over Q_2 iff every $Q \subseteq Q_1$ with $|Q| \leq n+1$ has a nonempty intersection (i.e., $\bigcap Q \neq \emptyset$).

Theorem 4 captures the dominance in (13). For more examples of this sort see Ben-Avi and Winter (2004).

2.4   Inverse Linking Constructions

One of the puzzling structures for theories of quantifier scope in natural language involves sentences like the following, which are sometimes referred to as *inverse linking* constructions.

(14) *Some student from every city participated.*

In Predicate Calculus notation, the prominent reading of (14) is (15b), whereas a possible but less plausible reading is (15a).

(15)   a. $\exists x[S(x) \wedge \forall y[C(y) \to F(x,y)] \wedge P(x)]$
          ("there exists a student who is from every city, and that student participated")
       b. $\forall y[C(y) \to \exists x[S(x) \wedge F(x,y) \wedge P(x)]]$
          ("for every city y, there exists a student who is from y and who participated")

As a matter of syntactic structure, inverse linking sentences involve a noun phrase that appears within the *left* argument of a determiner expression, where this argument is further *restricted* by a noun. In the case of sentence (14), the noun phrase *every city* is in the left argument of the determiner expression *some*, restricted by the noun *student*. To capture the effect we call "restriction", we adopt the following notation for any determiner over E and a set $X \subseteq E$.

(16)   a. $D_{:X} = \{\langle A, B\rangle : \langle A \cap X, B\rangle \in D\}$
       b. $D^{:X} = \{\langle A, B\rangle : \langle A, B \cap X\rangle \in D\}$

The denotations in inverse linking sentences like (14) involve two determiners D and D', three sets A, B and P, and a binary relation R. In the example, the determiners correspond to *some* and *every*, the sets correspond to *student*, *city* and *participated*, and the binary relation corresponds to *from*. Using the "restriction" notation in (16), the two readings of inverse linking constructions are expressed as follows.

(17)   a. $R \in D^{P}_{:A} {-} D'_{B}$
       b. $R \in D^{P}_{:A} {\sim} D'_{B}$

It is easy to verify that the two predicate logic formulae in (15) are equivalent to the following claims.

(18)   a. $F \in \mathbf{some}^{P}_{:S} {-} \mathbf{every}_{C}$
       b. $F \in \mathbf{some}^{P}_{:S} {\sim} \mathbf{every}_{C}$

The intersectivity (cf. Peters and Westerståaahl , 2006, p.210) of the determiner **some** implies that $\mathbf{some}^{P}_{:S} = \mathbf{some}_{S \cap P}$. As a result, the familiar dominance of **some** over **every** accounts for the dominance of $\mathbf{some}^{P}_{:S}$ over $\mathbf{every}_{C}$, or the entailment from (18a) (=(15a)) to (18b) (=(15b)).

Matters get more involved when the first determiner in the inverse linking construction is not intersective. Consider sentence (19) and its two analyses in (19a-b).

(19) *Every student from some city participated.*
       a. $F \in \mathbf{every}^{P}_{:S} {-} \mathbf{some}_{C}$
       b. $F \in \mathbf{every}^{P}_{:S} {\sim} \mathbf{some}_{C}$

Two facts should be noted about the quantifier $Q = \mathbf{every}^{P}_{:S}$ in these analyses. First, the effect of supplying the *right* argument P of the determiner **every** is that in general Q is not an upward monotone quantifier. Second, for any sets S and P, the quantifier Q is furthermore downward monotone, as a result of the simple fact $Q = \mathbf{every}^{P}_{:A} = \mathbf{every}^{P \cup \overline{A}}$, and the downward monotonicity of **every** in its left argument.[4] Consequently, we can use Theorem 4 to show that $\mathbf{every}^{P}_{:S}$ is dominant over $\mathbf{some}_{C}$, unless $C = \emptyset$. In fact, the dual of $\mathbf{some}_{C}$, $\mathbf{every}_{C}$, is upward monotone, and $\bigcap \mathbf{every}_{C} \neq \emptyset$, unless $C = \emptyset$. By Theorem 4, $\mathbf{every}_{C}$ is dominant over any downward monotone quantifier;[5] specifically, it is dominant over the dual of $\mathbf{every}^{P}_{:S}$. By Fact 1, $\mathbf{every}^{P}_{:S}$ is dominant over $\mathbf{some}_{C}$.

---

[4] More generally, the presentation $D^{P}_{:A} = D^{P \cup \overline{A}}$ follows for any co-intersective determiner D. Co-intersectivity requires that if $A \backslash B = A' \backslash B'$, then $B \in D_A \Leftrightarrow B' \in D_{A'}$. (Keenan, 2006)

[5] It should be noted that Theorem 4 as stated here does not cover trivial quantifiers. However, $C \neq \emptyset$ implies that $\mathbf{every}_{C}$ is not trivial, in which case it is easy to verify that it is dominant over any trivial quantifier.

We conclude that the scheme in (17) allows us to use the theorems above, which are stated in terms of quantifiers, for characterizing at least some of the entailments between readings of inverse linking constructions, which are initially stated in terms of determiners. Further study of the way to characterize such entailments must wait for further research.

2.5   Open Questions

In addition to the characterization of entailments between readings, results about scope dominance itself can be extended and generalized in some directions. This includes a full characterization of scope dominance between arbitrary quantifiers, and over domains with arbitrary cardinality. At present there are no such general results known to us.

3   Discussion

Nissim Francez (p.c.) has expressed some doubts about the importance of scope dominance for theoretical and computational semantics of natural language. As always, we are happy to differ. Although it is a "purely" combinatorial problem, it seems clear that a full characterization of scope dominance would deepen our understanding of ambiguity in natural language. It is a fact that some sentences show intricate logical relations between readings that grammarians find necessary to assume. This fact has led to considerable confusion in the theoretical linguistic literature (Ruys, 2002). Having sound rules of thumb for the situations in which such relations may appear could undoubtedly help to prevent further descriptive inadequacies. Furthermore, some works (Chaves, 2003; Altman and Winter, 2005) have started to explore the possibilities of computing scope dominance with natural language sentences. As Francez has argued, at this point it is too early to know if such algorithms can be useful for reasoning under ambiguity (e.g. Reyle, 1995; Van Deemter, 1996; Van Eijck and Jaspars, 1996). However, this is one domain where non-trivial problems about inference in natural language seem tractable. We therefore believe that scope dominance introduces an interesting challenge for a realm in which Francez (Fyodorov et al., 2003; Zamansky et al., 2006; Francez and Dyckhoff, 2007) has continuously contributed.

References

Altman, A., Winter, Y.: Computing dominant readings with upward monotone quantifiers. Research on Language and Computation 3, 1–43 (2005)

Altman, A., Keenan, E., Winter, Y.: Monotonicity and relative scope relations. In: van Rooy, R., Stokhof, M. (eds.) Proceedings of the 13th Amsterdam Colloquium, pp. 25–30 (2001)

Altman, A., Peterzil, Y., Winter, Y.: Scope dominance with upward monotone quantifiers. Journal of Logic, Language and Information 14, 445–455 (2005)

Ben-Avi, G., Winter, Y.: Scope dominance with monotone quantifiers over finite domains. Journal of Logic, Language and Information 13, 385–402 (2004)

Chaves, R.P.: Non-redundant scope disambiguation in underspecified semantics. In: ten Cate, B. (ed.) Proceedings of the 8th ESSLLI student session, pp. 47–58 (2003)

Francez, N., Dyckhoff, R.: Proof-theoretic semantics for natural language. Mathematics of Language (MOL) 10 (2007)

Fyodorov, Y., Winter, Y., Francez, N.: Order-based inference in natural logic. Logic Journal of the IGPL 11, 385–417 (2003)

Keenan, E.L.: Quantifiers: Semantics. In: Brown, K. (ed.) Encyclopedia of Language and Linguistics, vol. 10. Elsevier, Oxford (2006)

Peters, S., Westerståhl, D.: Quantifiers in Language and Logic. Clarendon Press, Oxford (2006)

Reyle, U.: On reasoning with ambiguities. In: Proceedings of the EACL 1995 (1995)

Ruys, E.: Wide scope indefinites: genealogy of a mutant meme. Utrecht University (2002) (unpublished ms),
http://www.let.uu.nl/~Eddy.Ruys/personal/download/indefinites.pdf

Ruys, E., Winter, Y.: Quantifier scope in formal linguistics. In: Gabbay, D. (ed.) Handbook of Philosophical Logic, 2nd edn. (2008) (to appear)

Van Deemter, K.: Towards a logic of ambiguous expressions. In: van Deemter, K., Peters, S. (eds.) Semantic Ambiguity and Underspecification, pp. 203–237. Cambridge University Press, Cambridge (1996)

Van Eijck, D.J.N., Jaspars, J.O.M.: Ambiguity and Reasoning. Technical report, Computer Science / Department of Software Technology, Centrum voor Wiskunde en Informatica (CWI), CS-R9616 (1996)

Westerståhl, D.: On the order between quantifiers. In: Furberg, M., et al. (eds.) Acta Universitatis Gothoburgensis, pp. 273–285. Göteborg University (1986)

Westerståhl, D.: Self-commuting quantifiers. The Journal of Symbolic Logic 61, 212–224 (1996)

Zamansky, A., Winter, Y., Francez, N.: A 'natural logic' inference system using the lambek calculus. Journal of Logic, Language and Information 15, 273–295 (2006)

Zimmermann, T.E.: Scopeless quantifiers and operators. Journal of Philosophical Logic 22, 545–561 (1993)

Nonassociative Lambek Calculus with Additives and Context-Free Languages

Wojciech Buszkowski[1,2] and Maciej Farulewski[1]

[1]Faculty of Mathematics and Computer Science
Adam Mickiewicz University in Poznań
Chair of Logic and Computation
[2]University of Warmia and Mazury in Olsztyn
buszko@amu.edu.pl, maciejf@amu.edu.pl

Abstract. We study Nonassociative Lambek Calculus with additives $\wedge, \vee$, satisfying the distributive law (Distributive Full Nonassociative Lambek Calculus **DFNL**). We prove that categorial grammars based on **DFNL**, also enriched with assumptions, generate context-free languages. The proof uses proof-theoretic tools (interpolation) and a construction of a finite model, earlier employed in [11] in the proof of Finite Embeddability Property (FEP) of **DFNL**; our paper is self-contained, since we provide a simplified version of the latter proof. We obtain analogous results for different variants of **DFNL**, e.g. **BFNL**, which admits negation $\neg$ such that $\wedge, \vee, \neg$ satisfy the laws of boolean algebra, and **HFNL**, corresponding to Heyting algebras with an additional residuation structure. Our proof also yields Finite Embeddability Property of boolean-ordered and Heyting-ordered residuated groupoids. The paper joins proof-theoretic and model-theoretic techniques of modern logic with standard tools of mathematical linguistics.

1 Introduction

Nonassociative Lambek Calculus **NL** proves the order formulas $\alpha \leq \beta$, valid in *residuated groupoids*, i.e. ordered algebras $(M, \cdot, \backslash, /, \leq)$ such that $(M, \leq)$ is a poset, and $\cdot, \backslash, /$ are binary operations on M, satisfying the residuation law:

$$a \cdot b \leq c \text{ iff } b \leq a \backslash c \text{ iff } a \leq c/b, \tag{1}$$

for all $a, b, c \in M$. As an easy consequence of (1), we obtain:

$$a(a \backslash b) \leq b, \ (a/b)b \leq a, \tag{2}$$

if $a \leq b$ then $ca \leq cb, ac \leq bc, c \backslash a \leq c \backslash b, a/c \leq b/c, b \backslash c \leq a \backslash c, c/b \leq c/a,$ (3)

for all $a, b, c \in M$. Hence every residuated groupoid is a partially ordered groupoid, if one forgets residuals $\backslash, /$ (we refer to $\cdot$ as product).

NL was introduced by Lambek [19] as a variant of Syntactic Calculus [18], now called Associative Lambek Calculus **L**, which yields the order formulas

O. Grumberg et al. (Eds.): Francez Festschrift, LNCS 5533, pp. 45–58, 2009.
© Springer-Verlag Berlin Heidelberg 2009

valid in residuated semigroups ($\cdot$ is associative). Both are standard type logics for categorial grammars [3,8,21,22]. While $\mathbf{L}$ is appropriate for expressions in the form of strings, $\mathbf{NL}$ corresponds to tree structures. The cut-elimination theorem holds for $\mathbf{NL}$ and $\mathbf{L}$, and it yields the decidability of these systems [18,19].

$\mathbf{NL}$ and $\mathbf{L}$ are examples of *substructural logics*, i.e. non-classical logics whose sequent systems lack some structural rules (Weakening, Contraction, Exchange). Besides multiplicatives $\cdot, \backslash, /$, substructural logics usually admit additives $\wedge, \vee$. The related algebras are residuated lattices $(M, \wedge, \vee, \cdot, \backslash, /, 1)$: here $(M, \wedge, \vee)$ is a lattice, $(M, \cdot, 1)$ is a monoid (i.e. a semigroup with 1), and (1) holds. For the nonassociative case, monoids are replaced by groupoids or unital groupoids (i.e. groupoids with 1); the resulting algebras are called lattice-ordered residuated (unital) groupoids. An algebra of that kind is said to be distributive, if its lattice reduct is distributive. The complete logic for residuated lattices is Full Lambek Calculus $\mathbf{FL}$; it admits cut-elimination and is decidable [13]. $\mathbf{FL}$ amounts to the *-free fragment of Pratt's action logic [10].

Full Nonassociative Lambek Calculus $\mathbf{FNL}$ is the complete logic of lattice-ordered residuated groupoids. We present it in the form of a sequent system. The cut-elimination theorem holds for this system [13]. It is not useful for our purposes, since we consider the consequence relation of $\mathbf{FNL}$ which requires the cut rule to be complete with respect to algebraic models. Furthermore, our main issue is $\mathbf{DFNL}$ in which the distributive law is affixed as a new axiom; the cut rule is necessary in $\mathbf{DFNL}$.

Categorial grammars based on $\mathbf{NL}$ generate precisely the $\epsilon-$free context-free languages [7,17,15]. Pentus [24] proves the same for $\mathbf{L}$. Using $\mathbf{FL}$, even without $\vee$, one can generate languages which are not context-free, i.e. meets of two context-free languages [16]. This also holds for $\mathbf{FL}$ with distribution, since it is conservative over its $\vee-$free fragment. The provability problem for $\mathbf{L}$ is NP-complete [25]; for $\mathbf{FL}$, the upper bound is P-SPACE. Categorial grammars with (partial) commutation can generate non-context-free languages [6,12].

We prove that (in opposition to $\mathbf{FL}$) categorial grammars based on $\mathbf{DFNL}$ generate context-free languages, and it remains true if one adds an arbitrary finite set of assumptions to $\mathbf{DFNL}$. For $\mathbf{NL}$, an analogous result has been proved in [9]. The latter paper also proves the polynomial time decidability of the consequence relation of $\mathbf{NL}$ (extending a result of [14] for pure $\mathbf{NL}$), but it does not hold in the presence of additives. The consequence relation of $\mathbf{DFNL}$ is decidable; this follows from FEP. It is known that the consequence relation for $\mathbf{L}$ is undecidable [9].

The construction of a finite model, used in the proof of FEP for $\mathbf{DFNL}$ in [11], will also be employed here (in a modified form) in order to prove the subformula property (of logics with cut) and, consequently, an interpolation lemma in an unrestricted version (see section 3).

Our methods can be extended to multi-modal variants of $\mathbf{DFNL}$ which admit several 'product' operations (of arbitrary arity) and the corresponding residual operations. Without additives, this multi-modal framework was presented in [8,9]. It is also naturally related to multi-modal extensions of Lambek Calculus, studied in e.g. [21,22]. This leads us to the proof of FEP for **BFNL**, which is the complete logic of boolean-ordered residuated groupoids, and the context-freeness of the corresponding grammars. Similar results are obtained for **HFNL**, which is the complete logic of Heyting-ordered residuated groupoids. Our results cannot directly be adapted for systems describing non-distributive lattices.

(External) consequence relations for substructural logics have been studied in different contexts; see e.g. [1,5,13]. Put it differently, one studies logics enriched with (finitely many) assumptions. Assumptions are sequents (not closed under substitution) added to axioms of the system (with the cut rule). Categorial grammars are usually required to be lexical in the sense that the logic is common for all languages and all information on the particular language is contained in the type lexicon. But, there are approaches allowing non-lexical assumptions, which results in a more efficient description of the language and an increase of generative power [8,20]. Let us emphasize that our results on context-freeness are new even for pure logics **DFNL**, **BFNL** and **HFNL**, and assumptions do not change anything essential in proofs.

Categorial grammars with additives are not popular in the linguistic literature. There are, nonetheless, good reasons for studying them. As it has been mentioned above, additives are standard operations in linear logics and other substructural logics [13]. The syntactic category of type α is usually understood as the set of all strings (or: trees; this seems more natural for the nonassociative case) which are assigned type α by the grammar. Then, it is natural to consider basic boolean operations on sets and to explicitly represent them in the grammar formalism. In general, they can result in a refinement of language description.

Types with $\wedge$ were used already in Lambek [19] in order to change a finite type assignment $a \mapsto \alpha_i$, $i = 1, \ldots, n$, into a rigid type assignment $a \mapsto \alpha_1 \wedge \cdots \wedge \alpha_n$ in the type lexicon of a categorial grammar (see section 4 for a definition of a categorial grammar). Standard basic categories, e.g. 's' (sentence), 'np' (noun phrase), 'pro' (pronoun), are not sufficient for a really effective description of natural language. One can divide them in subcategories. Lambek [20] considers three types of sentence: s_1 (sentence in the present tense), s_2 (sentence in the past tense), and s (sentence when the tense is not relevant) and four types of personal pronouns: π_1, π_2, π_3 for first, second and third person pronoun, respectively, and π when the person is irrelevant. This naturally leads to non-lexical assumptions, like $s_i \Rightarrow s$, $\pi_i \Rightarrow \pi$, added to the basic logic. An alternative solution is to define $s = s_1 \vee s_2$, $\pi = \pi_1 \vee \pi_2 \vee \pi_3$ which introduces $\vee$ on the scene. Kanazawa [16] proposes feature decomposition of basic categories; 'walks' is assigned type (np $\wedge$ sing)\s, 'walk' type (np $\wedge$ pl)\s, 'walked' type np \s, 'John' type np $\wedge$ sing, 'the Beatles' type np $\wedge$ pl, and 'became' type (np \s)/(np$\vee$ ap), where 'ap' stands for 'adjective phrase'. Types with negation can be employed to represent a negative information; for instance, John $\mapsto \neg s$ means that 'John' is not a sentence. It is well-known that negative information makes learning algorithms for formal grammars more efficient. It opens an interesting area of application for grammars with negation.

2 Restricted Interpolation

We admit a denumerable set of variables $p, q, r, \ldots$. Formulas are built from variables by means of $\cdot, \backslash, /, \wedge, \vee$. Formula structures (shortly: structures) are built from formulas according to the rules: (1) every formula is a structure, (2) if X, Y are structures then (X, Y) is a structure. We denote arbitrary formulas by $\alpha, \beta, \gamma, \ldots$ and structures by X, Y, Z. A *context* is a structure $X[\circ]$ containing a single occurrence of a special substructure $\circ$ (a place for substitution); $X[Y]$ denotes the result of substitution of Y for $\circ$ in $X[\circ]$.

Sequents are of the form $X \Rightarrow \alpha$. **FNL** assumes the following axioms and inference rules:

$$(\text{Id}) \ \alpha \Rightarrow \alpha,$$

$$(\cdot\text{L}) \ \frac{X[(\alpha, \beta)] \Rightarrow \gamma}{X[\alpha \cdot \beta] \Rightarrow \gamma}, \ (\cdot\text{R}) \ \frac{X \Rightarrow \alpha; \ Y \Rightarrow \beta}{(X, Y) \Rightarrow \alpha \cdot \beta},$$

$$(\backslash\text{L}) \ \frac{X[\beta] \Rightarrow \gamma; \ Y \Rightarrow \alpha}{X[(Y, \alpha\backslash\beta)] \Rightarrow \gamma}, \ (\backslash\text{R}) \ \frac{(\alpha, X) \Rightarrow \beta}{X \Rightarrow \alpha\backslash\beta},$$

$$(/\text{L}) \ \frac{X[\beta] \Rightarrow \gamma; \ Y \Rightarrow \alpha}{X[(\beta/\alpha, Y)] \Rightarrow \gamma}, \ (/\text{R}) \ \frac{(X, \alpha) \Rightarrow \beta}{X \Rightarrow \beta/\alpha},$$

$$(\wedge\text{L}) \ \frac{X[\alpha_i] \Rightarrow \beta}{X[\alpha_1 \wedge \alpha_2] \Rightarrow \beta}, \ (\wedge\text{R}) \ \frac{X \Rightarrow \alpha; \ X \Rightarrow \beta}{X \Rightarrow \alpha \wedge \beta},$$

$$(\vee\text{L}) \ \frac{X[\alpha] \Rightarrow \gamma; \ X[\beta] \Rightarrow \gamma}{X[\alpha \vee \beta] \Rightarrow \gamma}, \ (\vee\text{R}) \ \frac{X \Rightarrow \alpha_i}{X \Rightarrow \alpha_1 \vee \alpha_2},$$

$$(\text{CUT}) \ \frac{X[\alpha] \Rightarrow \beta; \ Y \Rightarrow \alpha}{X[Y] \Rightarrow \beta}.$$

In $(\wedge\text{L})$ and $(\vee\text{R})$, the subscript i equals 1 or 2. The latter rules and $(\cdot\text{L})$, $(\backslash\text{R})$, $(/\text{R})$ have one premise; the remaining rules have two premises, separated by semicolon. **DFNL** admits the additional axiom scheme:

$$(\text{D}) \ \alpha \wedge (\beta \vee \gamma) \Rightarrow (\alpha \wedge \beta) \vee (\alpha \wedge \gamma).$$

Notice that the converse sequent is provable in **FNL**. (CUT) can be eliminated from **FNL** but not from **DFNL**. For instance, $\alpha \wedge (\beta \vee (\gamma \vee \delta)) \Rightarrow (\alpha \wedge \beta) \vee ((\alpha \wedge \gamma) \vee (\alpha \wedge \delta))$ cannot be proved without (CUT).

Let Φ be a set of sequents. We write $\Phi \vdash X \Rightarrow \alpha$ if $X \Rightarrow \alpha$ is deducible from Φ in **DFNL**. By $F(X)$ we denote the formula arising from X after one has replaced each comma by $\cdot$. By $(\cdot\text{L})$ and (Id), $(\cdot\text{R})$, (CUT), $X \Rightarrow \alpha$ and $F(X) \Rightarrow \alpha$ are mutually deducible. Consequently, without loss of generality we can assume that Φ consists of sequents of the form $\alpha \Rightarrow \beta$ (simple sequents). In models, $\Rightarrow$ is interpreted as $\leq$ and, by definition, an assignment f satisfies $X \Rightarrow \alpha$ iff $f(F(X)) \leq f(\alpha)$.

In what follows, we always assume that Φ is a finite set of simple sequents. T denotes a set of formulas. By a T−sequent we mean a sequent such that all formulas occurring in it belong to T. We write $X \Rightarrow_{\Phi,T} \alpha$ if $X \Rightarrow \alpha$ has a deduction from Φ in **DFNL** which consists of T−sequents only (then, $X \Rightarrow \alpha$ must be a T−sequent). In proofs we write $\Rightarrow_T$ for $\Rightarrow_{\Phi,T}$. The following lemma is proved for **DFNL** but the same proof works for **FNL**.

Lemma 1. *Let T be closed under $\wedge, \vee$. Let $X[Y] \Rightarrow_{\Phi,T} \gamma$. Then, there exists $\delta \in T$ such that $X[\delta] \Rightarrow_{\Phi,T} \gamma$ and $Y \Rightarrow_{\Phi,T} \delta$.*

Proof. δ is called an interpolant of Y in $X[Y] \Rightarrow \gamma$. The proof proceeds by induction on T−deductions of $X[Y] \Rightarrow \gamma$ from Φ.

The case of axioms and assumptions is easy; they are simple sequents $\alpha \Rightarrow \gamma$, so $Y = \alpha$ and $\delta = \alpha$.

Let $X[Y] \Rightarrow \gamma$ be the conclusion of a rule. (CUT) is easy. If Y comes from one premise of (CUT), then we take an interpolant from this premise. Otherwise Y must contain Z, where the premises are $X[\alpha] \Rightarrow \gamma$, $Z \Rightarrow \alpha$. So, $Y = U[Z]$, and it comes from $U[\alpha]$ in the first premise. Then, an interpolant δ of $U[\alpha]$ in this premise is also an interpolant of Y in the conclusion, by (CUT).

Let us consider other rules. First, we assume that Y does not contain the formula, introduced by the rule (the active formula). If Y comes from exactly one premise of the rule, then one takes an interpolant from this premise. Let us consider ($\wedge$R). The premises are $X[Y] \Rightarrow \alpha$, $X[Y] \Rightarrow \beta$, and the conclusion is $X[Y] \Rightarrow \alpha \wedge \beta$. By the induction hypothesis, there are interpolants δ of Y in the first premise and δ' of Y in the second one. We have $X[\delta] \Rightarrow_T \alpha$, $X[\delta'] \Rightarrow_T \beta$, $Y \Rightarrow_T \delta$, $Y \Rightarrow_T \delta'$. Then, $\delta \wedge \delta'$ is an interpolant of Y in the conclusion, by ($\wedge$L), ($\wedge$R). Let us consider ($\vee$L). The premises are $X[\alpha][Y] \Rightarrow \gamma$, $X[\beta][Y] \Rightarrow \gamma$, and the conclusion is $X[\alpha \vee \beta][Y] \Rightarrow \gamma$, where Y does not contain $\alpha \vee \beta$. As above, there are interpolants δ, δ' of Y in the premises. Again $\delta \wedge \delta'$ is an interpolant of Y in the conclusion, by ($\wedge$L), ($\vee$L) and ($\wedge$R). For ($\cdot$R) with premises $U \Rightarrow \alpha$, $V \Rightarrow \beta$ and conclusion $(U, V) \Rightarrow \alpha \cdot \beta$, if $Y = (U, V)$, then we take $\delta = \alpha \cdot \beta$.

Second, we assume that Y contains the active formula (so, the rule must be an L-rule). If Y is a single formula, then we take $\delta = Y$. Assume that Y is not a formula. For ($\cdot$L), ($\wedge$L), we take an interpolant of Y' in the premise, where Y' is the natural source of Y. For ($\backslash$L) with premises $X[\beta] \Rightarrow \gamma$, $Z \Rightarrow \alpha$ and conclusion $X[(Z, \alpha\backslash\beta)] \Rightarrow \gamma$, we consider the source Y' of Y (Y' occurs in $X[\beta]$ and contains β). Then, Y arises from Y' by substituting $(Z, \alpha\backslash\beta)$ for β. Hence, an interpolant of Y' in the first premise is also an interpolant of Y in the conclusion, by ($\backslash$L). The case of (/L) is similar. The final case is ($\vee$L) with premises $Z[U[\alpha]] \Rightarrow \gamma$, $Z[U[\beta]] \Rightarrow \gamma$ and conclusion $Z[U[\alpha \vee \beta]] \Rightarrow \gamma$, where $Y = U[\alpha \vee \beta]$. Let δ be an interpolant of $U[\alpha]$ in the first premise and δ' be an interpolant of $U[\beta]$ in the second premise. Then, $\delta \vee \delta'$ is an interpolant of Y in the conclusion, by ($\vee$L),($\vee$R). $\qquad\square$

3   Finite Models and Interpolation

We prove an (extended) subformula property and an interpolation lemma for the deducibility relation $\vdash$ in **DFNL**. We need some constructions of lattice-ordered residuated groupoids.

Let $(M, \cdot)$ be a groupoid. On the powerset $P(M)$ one defines operations: $U \cdot V = \{ab : a \in U, b \in V\}$, $U \backslash V = \{c \in M : U \cdot \{c\} \subseteq V\}$, $U/V = \{c \in M : \{c\} \cdot V \subseteq U\}$, $U \vee V = U \cup V$, $U \wedge V = U \cap V$. $P(M)$ with these operations is a distributive lattice-ordered groupoid (it is a complete lattice). The order is $\subseteq$.

An operator $C : P(M) \mapsto P(M)$ is called *a closure operator* (or: a nucleus) on $(M, \cdot)$, if it satisfies the following conditions: (C1) $U \subseteq C(U)$, (C2) if $U \subseteq V$ then $C(U) \subseteq C(V)$, (C3) $C(C(U)) \subseteq C(U)$, (C4) $C(U) \cdot C(V) \subseteq C(U \cdot V)$, for all $U, V \subseteq M$ [13]. A set $U \subseteq M$ is said to be closed, if $C(U) = U$. By $C(M, \cdot)$ we denote the family of all closed subsets of M. Operations on $C(M, \cdot)$ are defined as follows: $U \otimes V = C(U \cdot V)$, $U \backslash V$, U/V and $U \wedge V$ as above, $U \vee V = C(U \cup V)$. (The product operation in $C(M, \cdot)$ is denoted by $\otimes$ to avoid collision with $\cdot$ in $P(M)$.) It is known that $C(M, \cdot)$ with these operations is a complete lattice-ordered residuated groupoid [13]; it need not be distributive. The order is $\subseteq$.

(C4) is essential in the proof that $C(M, \cdot)$ is closed under $\backslash, /$. Actually, if U is closed, then $V \backslash U$ and U/V are closed, for any $V \subseteq M$. Let us consider U/V. Since (2) hold in $P(M)$, then $(U/V) \cdot V \subseteq U$. We get $C(U/V) \cdot V \subseteq C(U/V) \cdot C(V) \subseteq C((U/V) \cdot V) \subseteq C(U) = U$, and consequently, $C(U/V) \subseteq U/V$, by (1) for $P(M)$. The reader is invited to prove that (1) holds in $C(M, \cdot)$ and $C(U \cup V)$ is the join of U, V in $C(M, \cdot)$.

Let T be a nonempty set of formulas. By T^* we denote the set of all structures formed out of formulas from T. $T^*[\circ]$ denotes the set of all contexts $X[\circ]$ whose all atomic substructures different from $\circ$ belong to T.

T^* is a (free) groupoid with the operation $X \cdot Y = (X, Y)$. Hence $P(T^*)$ is a lattice-ordered residuated groupoid with operations defined as above. For $Z[\circ] \in T^*[\circ]$ and $\alpha \in T$, we define a set:

$$[Z[\circ], \alpha] = \{X \in T^* : Z[X] \Rightarrow_{\Phi, T} \alpha\}. \tag{4}$$

The family of all sets $[Z[\circ], \alpha]$, defined in this way, is denoted $B(T)$. An operator $C_T : P(T^*) \mapsto P(T^*)$ is defined as follows:

$$C_T(U) = \bigcap \{[Z[\circ], \alpha] \in B(T) : U \subseteq [Z[\circ], \alpha]\}, \tag{5}$$

for $U \subseteq T^*$. It is easy to see that C_T satisfies (C1), (C2), (C3). We prove (C4). Let $U, V \subseteq T^*$ and $X \in C_T(U)$, $Y \in C_T(V)$. We show $(X, Y) \in C_T(U \cdot V)$. Let $[Z[\circ], \alpha] \in B(T)$ be such that $U \cdot V \subseteq [Z[\circ], \alpha]$. For any $X' \in U$, $Y' \in V$, $(X', Y') \in [Z[\circ], \alpha]$, whence $Z[(X', Y')] \Rightarrow_T \alpha$. So, $U \subseteq [Z[(\circ, Y')], \alpha]$, whence $C_T(U) \subseteq [Z[(\circ, Y')], \alpha]$, by (5), and the latter holds for any $Y' \in V$. Then, $Z[(X, Y')] \Rightarrow_T \alpha$, for any $Y' \in V$. We get $V \subseteq [Z[(X, \circ)], \alpha]$, which yields $C_T(V) \subseteq [Z[(X, \circ)], \alpha]$, by (5). Consequently, $Z[(X, Y)] \Rightarrow_T \alpha$, whence $(X, Y) \in [Z[\circ], \alpha]$ (see [23,2,13,11] for similar arguments).

We have shown that C_T is a closure operator on $(T^*, \cdot)$. We consider the algebra $C_T(T^*, \cdot)$, further denoted by $M(T, \Phi)$. Clearly, all sets in $B(T)$ are closed under C_T. We define:

$$[\alpha] = [\circ, \alpha] = \{X \in T^*; X \Rightarrow_{\Phi, T} \alpha\}. \tag{6}$$

For $\alpha \in T$, $[\alpha] \in B(T)$. The following equations are true in $M(T, \Phi)$ provided that all formulas appearing in them belong to T.

$$[\alpha] \otimes [\beta] = [\alpha \cdot \beta], \ [\alpha]\backslash[\beta] = [\alpha\backslash\beta], \ [\alpha]/[\beta] = [\alpha/\beta], \tag{7}$$

$$[\alpha] \vee [\beta] = [\alpha \vee \beta], \ [\alpha] \wedge [\beta] = [\alpha \wedge \beta]. \tag{8}$$

We prove the first equation (7). If $X \Rightarrow_T \alpha$ and $Y \Rightarrow_T \beta$ then $(X, Y) \Rightarrow_T \alpha \cdot \beta$, by ($\cdot$R). Consequently, $[\alpha] \cdot [\beta] \subseteq [\alpha \cdot \beta]$. Then $[\alpha] \otimes [\beta] = C_T([\alpha] \cdot [\beta]) \subseteq [\alpha \cdot \beta]$, by (C2), (C3). We prove the converse inclusion. Let $[Z[\circ], \gamma] \in B(T)$ be such that $[\alpha] \cdot [\beta] \subseteq [Z[\circ], \gamma]$. By (Id), $\alpha \in [\alpha]$, $\beta \in [\beta]$, whence $Z[(\alpha, \beta)] \Rightarrow_T \gamma$. Then, $Z[\alpha \cdot \beta] \Rightarrow_T \gamma$, by ($\cdot$R). Hence, if $X \in [\alpha \cdot \beta]$ then $Z[X] \Rightarrow_T \gamma$, by (CUT), which yields $X \in [Z[\circ], \gamma]$. We have shown $[\alpha \cdot \beta] \subseteq C_T([\alpha] \cdot [\beta])$.

We prove the second equation (7). Let $X \in [\alpha]\backslash[\beta]$. Since $\alpha \in [\alpha]$, then $(\alpha, X) \in [\beta]$. Hence $(\alpha, X) \Rightarrow_T \beta$, which yields $X \Rightarrow_T \alpha\backslash\beta$, by ($\backslash$R). We have shown $\subseteq$. To prove the converse inclusion it suffices to show $[\alpha] \cdot [\alpha\backslash\beta] \subseteq [\beta]$. If $X \Rightarrow_T \alpha$ and $Y \Rightarrow_T \alpha\backslash\beta$, then $(X, Y) \Rightarrow_T \beta$, since $(\alpha, \alpha\backslash\beta) \Rightarrow_T \beta$, by (Id), ($\backslash$L), and one applies (CUT). The proof of the third equation (7) is similar. Proofs of (8) are left to the reader.

We say that formulas $\alpha, \beta \in T$ are T−equivalent, if $\alpha \Rightarrow_T \beta$ and $\beta \Rightarrow_T \alpha$. By (Id), (CUT), T−equivalence is an equivalence relation. By $\overline{T}$ we denote the smallest set of formulas which contains all formulas from T and is closed under subformulas and $\wedge, \vee$. If T is closed under subformulas, then $\overline{T}$ is the closure of T under $\wedge, \vee$. The following lemma is a syntactic variant of the well-known fact that distributive lattices are locally finite (it means: every finitely generated distributive lattice is finite).

Lemma 2. *If T is a finite set of formulas, then $\overline{T}$ is finite up to $\overline{T}$−equivalence.*

Proof. If T is finite, then the set T' of subformulas of formulas from T is also finite. $\overline{T}$ is the closure of T' under $\wedge, \vee$. The converse of (D) $(\alpha \wedge \beta) \vee (\alpha \wedge \gamma) \Rightarrow \alpha \wedge (\beta \vee \gamma)$ is provable in **FNL** (it is valid in all lattices); if $\alpha, \beta, \gamma \in \overline{T}$, then the proof uses $\overline{T}$−sequents only. Consequently, for $\alpha, \beta, \gamma \in \overline{T}$, both sides of (D) are $\overline{T}$−equivalent. It follows that every formula from $\overline{T}$ is $\overline{T}$−equivalent to a finite disjunction of finite conjunctions of formulas from T'. If one omits repetitions, then there are only finitely many formulas of the latter form. □

Recall that an assignment in a model M is a homomorphism from the formula algebra into M.

Lemma 3. *Let T be a nonempty, finite set of formulas. Then, $M(\overline{T}, \Phi)$ is a finite distributive lattice-ordered residuated groupoid. For any assignment f in $M(\overline{T}, \Phi)$ such that $f(p) = [p]$, for any $p \in \overline{T}$, and any $\overline{T}$−sequent $X \Rightarrow \alpha$, f satisfies $X \Rightarrow \alpha$ in $M(\overline{T}, \Phi)$ if and only if $X \Rightarrow_{\overline{T}} \alpha$.*

Proof. As shown above, $M(\overline{T}, \Phi)$ is a lattice-ordered residuated groupoid. We prove the second part of the lemma. Let f satisfy $f(p) = [p]$, for any variable $p \in \overline{T}$. Using (7), (8), one proves $f(\alpha) = [\alpha]$, for all $\alpha \in \overline{T}$, by easy formula induction.

Assume that f satisfies the $\overline{T}$–sequent $X \Rightarrow \alpha$. For any formula β appearing in X, we have $\beta \in [\beta] = f(\beta)$, whence $X \in f(F(X))$. Since $f(F(X)) \subseteq f(\alpha)$, then $X \in f(\alpha) = [\alpha]$. Thus $X \Rightarrow_{\overline{T}} \alpha$. Assume $X \Rightarrow_{\overline{T}} \alpha$. Then, there exists a $\overline{T}$–deduction of $X \Rightarrow \alpha$ from Φ in **DFNL**. By induction on this deduction, we prove that f satisfies $X \Rightarrow \alpha$ in $M(\overline{T}, \Phi)$. f obviously satisfies axioms (Id). Assumptions from Φ and instances of (D), restricted to $\overline{T}$–sequents, are of the form $\beta \Rightarrow \gamma$, where $\beta, \gamma \in \overline{T}$. Since $\beta \Rightarrow_{\overline{T}} \gamma$, then $[\beta] \subseteq [\gamma]$, by (CUT), which yields $f(\beta) \subseteq f(\gamma)$. The rules of **FNL** are sound for any assignment in a lattice-ordered residuated groupoid, which finishes this part of proof.

Let R be a selector of the family of equivalence classes of $\overline{T}$–equivalence (R chooses one formula from each equivalence class). By Lemma 2, R is a nonempty finite subset of $\overline{T}$. We show that every nontrivial (i.e. nonempty and not total) closed subset of $\overline{T}^*$ equals $[\alpha]$, for some $\alpha \in R$. Let U be nontrivial and closed.. Let $X \in U$. There exists a set $[Z[\circ], \beta] \in B(\overline{T})$ such that $U \subseteq [Z[\circ], \beta]$. So, $Z[X] \Rightarrow_{\overline{T}} \beta$. By Lemma 1, there exists $\delta \in \overline{T}$ such that $Z[\delta] \Rightarrow_{\overline{T}} \beta$ and $X \Rightarrow_{\overline{T}} \delta$. We get $X \in [\delta]$ and $[\delta] \subseteq [Z[\circ], \beta]$, by (CUT). Clearly we can take $\delta \in R$. We can find such a formula $\delta \in R$, for any set $[Z[\circ], \beta] \in B(\overline{T})$ such that $U \subseteq [Z[\circ], \beta]$. Let γ_X be the conjunction of all formulas δ, fulfilling the above. By (8), (C3) and (6), $X \in [\gamma_X]$ and $[\gamma_X] \subseteq U$. We can replace γ_X by a $\overline{T}$–equivalent formula from R. So, we stipulate $\gamma_X \in R$. Let α be the disjunction of all formulas γ_X, for $X \in U$. By (8), $[\alpha] \subseteq U$ and, evidently, $U \subseteq [\alpha]$. We can assume $\alpha \in R$.

It follows that $M(\overline{T}, \Phi)$ is finite. We prove that it is distributive. It suffices to prove $U \wedge (V \vee W) \subseteq (U \wedge V) \vee (U \wedge W)$, for any closed sets U, V, W. This inclusion is true, if at least one of the sets U, V, W is empty or total, since $M(\overline{T}, \Phi)$ is a lattice. So, assume U, V, W be nontrivial. By the above paragraph, $U = [\alpha]$, $V = [\beta]$, $W = [\gamma]$, for some $\alpha, \beta, \gamma \in R$. Then, the inclusion follows from (8) and the fact that $[\alpha \wedge (\beta \vee \gamma)] \subseteq [(\alpha \wedge \beta) \vee (\alpha \wedge \gamma)]$.   □

We are ready to prove an extended subformula property and an interpolation lemma for **DFNL**.

Lemma 4. *Let T be a finite set of formulas, containing all formulas appearing in $X \Rightarrow \alpha$ and Φ. If $\Phi \vdash X \Rightarrow \alpha$ then $X \Rightarrow_{\Phi, \overline{T}} \alpha$.*

Proof. Let f be an assignment in $M(\overline{T}, \Phi)$, satisfying $f(p) = [p]$, for any variable $p \in \overline{T}$. Let $\beta \Rightarrow \gamma$ be a sequent from Φ. Then $\beta \Rightarrow_{\overline{T}} \gamma$, which yields $f(\beta) \subseteq f(\gamma)$, by Lemma 3. So, f satisfies all sequents from Φ.

Assume $\Phi \vdash X \Rightarrow \alpha$. Since **DFNL** is strongly sound with respect to distributive lattice-ordered residuated groupoids, then f satisfies $X \Rightarrow \alpha$. Consequently, $X \Rightarrow_{\overline{T}} \alpha$, by Lemma 3.   □

Lemma 5. *Let T be a finite set of formulas, containing all formulas appearing in $X[Y] \Rightarrow \alpha$ and Φ. If $\Phi \vdash X[Y] \Rightarrow \alpha$ then there exists $\delta \in \overline{T}$ such that $\Phi \vdash X[\delta] \Rightarrow \alpha$ and $\Phi \vdash Y \Rightarrow \delta$.*

Proof. Assume $\Phi \vdash X[Y] \Rightarrow \alpha$. By Lemma 4, $X[Y] \Rightarrow_{\overline{T}} \alpha$. Apply Lemma 1. □

Lemma 3 except for the finiteness and distributivity of $M(\overline{T}, \Phi)$ and Lemmas 4 and 5 also hold for **FNL**. For **NL**, Lemma 4 and Lemma 5 hold with $\overline{T}$ defined as the closure of T under subformulas [9] (for pure **NL**, a weaker form of the latter lemma was proved in [15]).

This yields Strong Finite Model Property (SFMP) of **DFNL**: if $\Phi \vdash X \Rightarrow \alpha$ does not hold in **DFNL**, then there exist a finite distributive lattice-ordered residuated groupoid $\mathcal{M}$ and an assignment f such that all sequents from Φ are true under f, but $X \Rightarrow \alpha$ is not (take T as in Lemma 4 and use Lemma 3). Equivalently, every Horn formula which is not valid in distributive lattice-ordered residuated groupoids is falsified in a finite model. If a class of algebras $\mathcal{K}$ is closed under (finite) products, then SFMP for $\mathcal{K}$ is equivalent to FEP for $\mathcal{K}$: every finite partial subalgebra of an algebra from $\mathcal{K}$ is embeddable into a finite algebra from $\mathcal{K}$ [13]. (The latter is equivalent to FMP of the universal theory of $\mathcal{K}$.) Then, FEP holds for the class of distributive lattice-ordered residuated groupoids (proved in [11] in collaboration with the first author; the present proof is simplified).

4   Categorial Grammars Based on DFNL

A categorial grammar based on a logic $\mathcal{L}$ (presented as a sequent system) is defined as a tuple $G = (\Sigma, I, \alpha_0, \Phi)$ such that Σ is a finite alphabet, I is a nonempty finite relation between elements of Σ and formulas of $\mathcal{L}$, α_0 is a formula of $\mathcal{L}$, and Φ is a finite set of sequents of $\mathcal{L}$. Elements of Σ are usually interpreted as words from the lexicon of a language and strings on Σ as phrases. Formulas of $\mathcal{L}$ are called types. I assigns finitely many types to each word from Σ. If I is fixed, then one often writes $a \mapsto \beta$ for $(a, \beta) \in I$. α_0 is a designated type; often one takes a designated variable s (the type of sentences). $\mathcal{L}$ is the logic of type change and composition. Φ is a finite set of assumptions added to $\mathcal{L}$.

Our logic $\mathcal{L}$ is **DFNL**. Let $G = (\Sigma, I, \alpha_0, \Phi)$ be a categorial grammar. By $T(G)$ we denote the set of all types appearing in the range of I. Let T be the smallest set containing $T(G)$, all types from Φ and α_0. For any type β, we define $L(G, \beta) = \{X \in \overline{T}^* : \Phi \vdash X \Rightarrow \beta\}$. Elements of $\overline{T}^*$ can be seen as finite binary trees whose leaves are labeled by types from $\overline{T}$. The tree language of G, denoted by $L_t(G)$, consists of all trees which can be obtained from trees in $L(G, \alpha_0) \cap T(G)^*$ by replacing each type γ by some $a \in \Sigma$ such that $(a, \gamma) \in I$. The language of G, denoted by $L(G)$, is the yield of $L_t(G)$.

Theorem 1. $L(G)$ *is a context-free language, for any categorial grammar G based on* **DFNL**.

Proof. Fix $G = (\Sigma, I, \alpha_0, \Phi)$. We define a context-free grammar G' such that $L(G') = L(G)$. Let T be defined as above. By Lemma 2, $\overline{T}$ is finite up to $\overline{T}$−equivalence. We choose a set $R \subseteq \overline{T}$ which contains one formula from each equivalence class. For $\beta \in \overline{T}$, by $r(\beta)$ we denote the unique type from R which is $\overline{T}$−equivalent to β.

G' is defined as follows. The terminal alphabet is Σ. The nonterminal alphabet is R. Production rules are: (R1) $\beta \mapsto \gamma$, for $\beta, \gamma \in R$ such that $\Phi \vdash \gamma \Rightarrow \beta$, (R2) $\beta \mapsto \gamma\delta$, for $\beta, \gamma, \delta \in R$ such that $\Phi \vdash (\gamma, \delta) \Rightarrow \beta$, (R3) $r(\beta) \mapsto a$, for $\beta \in T(G)$, $a \in \Sigma$ such that $(a, \beta) \in I$. The initial symbol is $r(\alpha_0)$.

Every derivation tree in G' can be treated as a deduction from Φ in **DFNL** which is based on deducible sequents appearing in (R1), (R2) and (CUT). Then, $L(G') \subseteq L(G)$. The converse inclusion follows from Lemma 5. Let $x \in L(G)$. Then, x is the yield of some $Y \in L_t(G)$. There exists $X \in L(G, \alpha_0)$ such that Y is obtained from X in the way described above. Let $r(X)$ denote the tree resulting from X after one has replaced each type β by $r(\beta)$. Clearly, if $X \in L(G, \gamma)$, then $r(X) \in L(G, r(\gamma))$. It suffices to prove that, for any $\gamma \in \overline{T}$ and any $X \in L(G, \gamma)$, there exists a derivation of $r(X)$ from $r(\gamma)$ (as a derivation tree) in G'. We proceed by induction on the number of commas in X. Let $X \in L(G, \gamma)$ be a single type, say, $X = \beta$. Then, $\Phi \vdash \beta \Rightarrow \gamma$, whence $\Phi \vdash r(\beta) \Rightarrow r(\gamma)$. Then, $r(X) = r(\beta)$ is derivable from $r(\gamma)$, by (R1). Let $X \in L(G, \gamma)$ contain a comma. Then, X must contain a substructure of the form (δ_1, δ_2), where $\delta_i \in \overline{T}$. We write $X = Z[(\delta_1, \delta_2)]$. By Lemma 5, there exists $\delta \in \overline{T}$ such that $\Phi \vdash Z[\delta] \Rightarrow \gamma$ and $(\delta_1, \delta_2) \Rightarrow \delta$. By the induction hypothesis, $r(Z[\delta])$ can be derived from $r(\gamma)$ in G'. Then, $r(X)$ can be derived from $r(\gamma)$, by (R2). $\qquad\square$

It has been shown in [7,17] that every ϵ−free context-free language can be generated by a categorial grammar based on **NL** which uses very restricted types only: $p, p\backslash q, p\backslash(q\backslash r)$, where p, q, r are variables; the designated type is also a variable s. Now, we use the fact that **DFNL** is conservative over **NL**, since every residuated groupoid can be embedded into a powerset algebra over a groupoid [8]. Accordingly, every ϵ−free context-free language is generated by some categorial grammar based on **DFNL**.

The tree language $L_t(G)$ is regular, for any grammar G based on **DFNL**. It follows from the above theorem (and its proof), since $L_t(G)$ equals the tree language determined by derivation trees of G'. A direct proof uses the well-known fact that a tree language $L \subseteq T^*$ is regular if and only if there exists a congruence of finite index on T^* which is compatible with L (it means: L is the join of some family of equivalence classes of $\sim$). For $X, Y \in T(G)^*$, define $X \sim Y$ iff, for all $\alpha \in \overline{T}$, $\Phi \vdash X \Rightarrow \alpha$ iff $\Phi \vdash Y \Rightarrow \alpha$. One can replace 'for all $\alpha \in \overline{T}$' by 'for all $\alpha \in R$', and consequently, there are at most 2^n equivalence classes of $\sim$, where n is the cardinality of R. So, $\sim$ is of finite index. Clearly $\sim$ is compatible with $L(G, \alpha_0)$. We prove that $\sim$ is a congruence on $T(G)^*$. Assume $X \sim Y$. We show $(Z, X) \sim (Z, Y)$, for any $Z \in T(G)^*$. Assume $\Phi \vdash (Z, X) \Rightarrow \alpha$, $\alpha \in \overline{T}$. By Lemma 5, there exists $\delta \in \overline{T}$ such that $\Phi \vdash (Z, \delta) \Rightarrow \alpha$ and $\Phi \vdash X \Rightarrow \delta$. Then, $\Phi \vdash Y \Rightarrow \delta$, whence $\Phi \vdash (Z, Y) \Rightarrow \alpha$, by (CUT). We have shown that $\Phi \vdash (Z, X) \Rightarrow \alpha$ implies $\Phi \vdash (Z, Y) \Rightarrow \alpha$. The converse implication has a similar proof. Therefore $(Z, X) \sim (Z, Y)$. Similarly $(X, Z) \Rightarrow (Y, Z)$. Consequently $L(G, \alpha_0) \cap T(G)^*$ is regular. By the construction of $L_t(G)$, the latter tree language is also regular.

The proof of Theorem 1 provides a construction of a context-free grammar equivalent to a given categorial grammar based on **DFNL**, since $\vdash$ for **DFNL** is decidable (it follows from FEP), and R can effectively be constructed. Actually we can construct a set R' such that $R \subseteq R'$. For T, defined as above, let T' denote the set of all subformulas of formulas from T. Then, $\overline{T}$ is the closure of T' under $\wedge, \vee$. Let $\alpha_1, \ldots, \alpha_k$ be all formulas from T'. We form all finite conjunctions $\alpha_{i_1} \wedge \cdots \wedge \alpha_{i_l}$ such that $1 \leq i_1 < \cdots < i_l \leq k$ and arrange them in a list $\beta_1, \ldots, \beta_{2^k-1}$. We define R' as the set of all finite disjunctions $\beta_{i_1} \vee \cdots \vee \beta_{i_l}$ such that $1 \leq i_1 < \cdots < i_l \leq 2^k - 1$. Clearly every formula from $\overline{T}$ is equivalent to some formula from R'. Using a decision method for **DFNL**, one can extract R from R'.

A syntactic decision method for **DFNL** can be designed like the one for **NL** in [9]. One can assume that R' is closed under $\wedge, \vee$ up to $\overline{T}-$equivalence. More precisely, for $\alpha, \beta \in R'$, one defines $\alpha \wedge' \beta$ as the formula from R' which results from a natural reduction of $\alpha \wedge \beta$ to a formula from R'. $\alpha \vee' \beta$ is defined in a similar way. Sequents of the form $\alpha \Rightarrow \beta$ and $(\alpha, \beta) \Rightarrow \gamma$ such that $\alpha, \beta, \gamma \in R'$ are called basic sequents. As in [9], one proves that a basic sequent is deducible from Φ if and only if there exists a deduction of this sequent from Φ which consists of basic sequents only (in rules for $\wedge, \vee$, these connectives are replaced by $\wedge'$ and $\vee'$, respectively). This yields a doubly exponential decision method for **DFNL** and a construction of G' of the same complexity. (In [9], it yields a polynomial time decision procedure for $\vdash$ in **NL**.)

Analogous results can be obtained for the $\vee-$free fragment of **FNL** (with assumptions). In the proof of Lemma 3, $\vee$ cannot be used; the finiteness of the model follows from the fact that every closed set is the join of a family of equivalence classes of $\sim$ (see above). The construction of G' requires exponential time.

5   Variants

The methods of this paper cannot be applied to associative systems **L**, **FL**, **DFL**. Consequence relations for these systems are undecidable; see [9,13]. Analogues of Theorem 1 are false (see Introduction).

They can be applied to several other nonassociative systems. The first example is **DFNL**$_e$, i.e. **DFNL** with the exchange rule:

$$(\text{EXC}) \quad \frac{X[(Y, Z)] \Rightarrow \alpha}{X[(Z, Y)] \Rightarrow \alpha}.$$

DFNL$_e$ is complete with respect to distributive lattice-ordered commutative residuated groupoids ($ab = ba$, for all elements a, b). Then, $a \backslash b = b/a$, and one considers one residual only, denoted $a \to b$. All results from sections 2, 3 and 4 can be proved for **DFNL**$_e$, and proofs are similar as above. Exception: not every $\epsilon-$free context-free language can be generated by a categorial grammar based on **DFNL**$_e$.

One can add the multiplicative constant 1, interpreted as the unit in unital groupoids. We need the axiom (1R): $\Rightarrow 1$, and the rules:

$$(\text{1Ll}) \quad \frac{X[Y] \Rightarrow \alpha}{X[(1, Y)] \Rightarrow \alpha}, \quad (\text{1Lr}) \quad \frac{X[Y] \Rightarrow \alpha}{X[(Y, 1)] \Rightarrow \alpha}.$$

The empty antecedent is understood as the empty structure Λ, and one admits $(\Lambda, X) = X$, $(X, \Lambda) = X$ in metatheory. Again, there are no problems with adapting our results for **DFNL** with 1 and **DFNL**$_e$ with 1. Caution: $\overline{T}$ contains 1, for any set T, and T^* contains Λ.

Additive constants $\perp$ and $\top$ can also be added, with axioms:

$$(\perp \text{L})\ X[\perp] \Rightarrow \alpha,\ (\top \text{R})\ X \Rightarrow \top.$$

They are interpreted as the lower bound and the upper bound, respectively, of the lattice. $\overline{T}$ must contain these constants. In the proof of (an analogue of) Lemma 1, one must consider new cases: $X[Y] \Rightarrow \alpha$ is an axiom $(\perp \text{L})$ or $(\top \text{R})$. For the first case, if Y contains $\perp$ then $\perp$ is an interpolant of Y; otherwise, $\top$ is an interpolant of Y. For the second case, $\top$ is an interpolant of Y. In the proof of Lemma 3, $M(\overline{T}, \Phi)$ interprets $\perp$ as $C_{\overline{T}}(\emptyset)$ and $\top$ as $\overline{T}^*$. Now, every closed set equals α, for some $\alpha \in \overline{T}$.

Instead of one binary product $\cdot$ one may admit a finite number of operations $o, o', \ldots$ of arbitrary arity: unary, binary, ternary and so on. Each n-ary operation o is associated with n residual operations o^i, for $i = 1, \ldots, n$. In models, one assumes the (generalized) residuation law:

$$o(a_1, \ldots, a_n) \leq b \text{ iff } a_i \leq o^i(a_1, \ldots, b, \ldots, a_n), \tag{9}$$

for all $i = 1, \ldots, n$ (on the right-hand side, b is the i-th argument of o^i). The corresponding formal system, called Generalized Lambek Calculus **GLC**, was presented in [8,9]. To each n-ary operation o one attributes a structure constructor $(X_1, \ldots, X_n)_o$, and formula structures can contain different structure constructors. Unary operations can be identified with (different) unary modalities. **GLC** represents a multi-modal variant of **NL**. The introduction rules for o and o^i look as follows:

$$\frac{X[(\alpha_1, \ldots, \alpha_n)_o] \Rightarrow \beta}{X[o(\alpha_1, \ldots, \alpha_n)] \Rightarrow \beta},\ \frac{X_1 \Rightarrow \alpha_1; \ldots; X_n \Rightarrow \alpha_n}{(X_1, \ldots, X_n)_o \Rightarrow o(\alpha_1, \ldots, \alpha_n)},$$

$$\frac{X[\alpha_i] \Rightarrow \beta;\ Y_j \Rightarrow \alpha_j, \text{ for all } j \neq i}{X[(Y_1, \ldots, o^i(\alpha_1, \ldots, \alpha_n), \ldots, Y_n)_o] \Rightarrow \beta},\ \frac{(\alpha_1, \ldots, X, \ldots, \alpha_n)_o \Rightarrow \alpha_i}{X \Rightarrow o^i(\alpha_1, \ldots, \alpha_n)}.$$

The consequence relation of **GLC** is polytime, and the corresponding categorial grammars generate ϵ-free context-free languages [9]. All results of this paper can easily be adapted for **GLC** with $\wedge$, $\vee$ and distribution (and, possibly, (EXC) for some binary operations, multiplicative units for them, and $\perp$, $\top$). In powerset algebras, one defines $o(U_1, \ldots, U_n)$ as the set of all elements $o(a_1, \ldots, a_n)$ such that $a_i \in U_i$, for $i = 1, \ldots, n$; o admits n residual operations o^i, $i = 1, \ldots, n$: $o^i(U_1, \ldots, U_n) = \{a \in M : o(U_1, \ldots, \{a\}, \ldots, U_n) \subseteq U_i\}$. (C4) takes the form: $o(C(U_1), \ldots, C(U_n)) \subseteq C(o(U_1, \ldots, U_n))$. Each operation o in the powerset algebra induces an operation on closed sets: $o_C(U_1, \ldots, U_n) = C(o(U_1, \ldots, U_n))$.

At the end, we consider **DFNL** with $\perp$, $\top$ and negation $\neg$, satisfying the laws of boolean algebra. To axioms for $\perp$ and $\top$ we add:

$$(\neg 1)\ \alpha \wedge \neg \alpha \Rightarrow \perp,\ (\neg 2)\ \top \Rightarrow \alpha \vee \neg \alpha.$$

The resulting system, denoted **BFNL**, is strongly complete with respect to lattice-ordered residuated groupoids whose underlying lattice is a boolean algebra.

All results of this paper can be proved for **BFNL**. We consider an auxiliary system **S** which is **DFNL** with $\top, \bot$ and an additional product connective & with residual $\rightarrow$ (so, it is a **GLC**-like system). We assume (EXC) for the structure constructor of &, whence & is commutative. We define $\neg\alpha = \alpha \rightarrow \bot$ and admit axioms ($\neg1$), ($\neg2$). It is easy to show that **S** is a conservative extension of **BFNL** (every model of the latter can be expanded to a model of the former, by identifying & with $\wedge$ and $\rightarrow$ with boolean implication). Now, for **S** we proceed as for **GLC** with additives. $\overline{T}$ must contain $\top, \bot$ and be closed under $\wedge, \vee, \neg$; then, Lemma 2 remains true, since boolean algebras are locally finite.

If we replace in **S** axioms ($\neg1$), ($\neg2$) by axioms $\alpha \wedge \beta \Rightarrow \alpha \& \beta$ and $\alpha \& \beta \Rightarrow \alpha \wedge \beta$, then we obtain **HFNL**: a complete logic of Heyting-ordered residuated groupoids. Now, $\overline{T}$ must be closed under $\wedge, \vee, \&$, but Lemma 2 is still true. All results of this paper can be proved for **HFNL** as well.

Theorem 2. *Categorial grammars based on* **BFNL**, **HFNL** *generate context-free languages. FEP holds for the classes of boolean-ordered and Heyting-ordered residuated groupoids (algebras).*

References

1. Avron, A.: The Semantics and Proof Theory of Linear Logic. Theoretical Computer Science 57, 161–184 (1988)
2. Belardinelli, F., Jipsen, P., Ono, H.: Algebraic Aspects of Cut Elimination. Studia Logica 77, 209–240 (2004)
3. van Benthem, J.: Language in Action: Categories, Lambdas and Dynamic Logic. North-Holland, Amsterdam (1991)
4. van Benthem, J., ter Meulen, A. (eds.): Handbook of Logic and Language. Elsevier, Amsterdam (1997)
5. Blok, W.J., van Alten, C.J.: On the finite embeddability property for residuated ordered groupoids. Transactions of AMS 357(10), 4141–4157 (2005)
6. Buszkowski, W.: A note on the Lambek - van Benthem calculus. Bulletin of The Section of Logic 131, 31–37 (1984)
7. Buszkowski, W.: Generative Capacity of Nonassociative Lambek Calculus. Bull. Polish Acad. Scie. Math. 34, 507–516 (1986)
8. Buszkowski, W.: Mathematical Linguistics and Proof Theory. In: [4], pp. 683–736
9. Buszkowski, W.: Lambek Calculus with Nonlogical Axioms. In: Casadio, C., Scott, P.J., Seely, R. (eds.) Language and Grammar. Studies in Mathematical Linguistics and Natural Language, pp. 77–93. CSLI Publications, Stanford (2005)
10. Buszkowski, W.: On Action Logic: Equational Theories of Action Algebras. Journal of Logic and Computation 17(1), 199–217 (2007)
11. Farulewski, M.: Finite Embeddabilty Property for Residuated Groupoids. Reports on Mathematical Logic 43, 25–42 (2008)
12. Francez, N., Kaminski, M.: Commutation-Augmented Pregroup Grammars and Mildly Context-Sensitive Languages. Studia Logica 87(2-3), 295–321 (2007)

13. Galatos, N., Jipsen, P., Kowalski, T., Ono, H.: Residuated Lattices: An Algebraic Glimpse at Substructural Logics. Elsevier, Amsterdam (2007)
14. de Groote, P., Lamarche, F.: Classical Nonassociative Lambek Calculus. Studia Logica 71(2), 355–388 (2002)
15. Jäger, G.: Residuation, structural rules and context-freeness. Journal of Logic, Language and Information 13, 47–59 (2004)
16. Kanazawa, M.: The Lambek Calculus Enriched with Additional Connectives. Journal of Logic, Language and Information 1(2), 141–171 (1992)
17. Kandulski, M.: The equivalence of nonassociative Lambek categorial grammars and context-free grammars. Zeitschrift für mathematische Logik und Grundlagen der Mathematik 34, 41–52 (1988)
18. Lambek, J.: The mathematics of sentence structure. American Mathematical Monthly 65, 154–170 (1958)
19. Lambek, J.: On the calculus of syntactic types. In: Jakobson, R. (ed.) Structure of Language and Its Mathematical Aspects, pp. 166–178. AMS, Providence (1961)
20. Lambek, J.: Type Grammars Revisited. In: Lecomte, A., Perrier, G., Lamarche, F. (eds.) LACL 1997. LNCS (LNAI), vol. 1582, pp. 1–27. Springer, Heidelberg (1999)
21. Moortgat, M.: Categorial Type Logic. In: [4], pp. 93–177
22. Morrill, G.: Type Logical Grammar. Categorial Logic of Signs. Kluwer, Dordrecht (1994)
23. Okada, M., Terui, K.: The finite model property for various fragments of intuitionistic linear logic. Journal of Symbolic Logic 64, 790–802 (1999)
24. Pentus, M.: Lambek Grammars are Context-Free. In: Proc. 8th IEEE Symp. Logic in Computer Sci., pp. 429–433 (1993)
25. Pentus, M.: Lambek calculus is NP-complete. Theoretical Computer Science 357, 186–201 (2006)

On Lazy Commutation

Nachum Dershowitz[*]

School of Computer Science, Tel Aviv University, Ramat Aviv, Israel

For Nissim, distinguished scholar
and longtime friend.

Abstract. We investigate combinatorial commutation properties for re-ordering a sequence of two kinds of steps, and for separating well-foundedness of unions of relations. To that end, we develop the notion of a constricting sequence. These results can be applied, for example, to generic path orderings used in termination proofs.

> *"Loop" as a train destination means that*
> *the train enters the Loop elevated structure in downtown Chicago,*
> *does a complete circle, and then returns the way it came.*
> *Red Line and Blue Line trains serve the Loop area of Chicago....*
>
> – Paul Inast (wikipedia.org)

1   Introduction

Imagine a city that has two coöperating bus companies, a Red line and a Blue line. An unusual condition of Doornbos, Backhouse, and van der Woude [13] guarantees that if a circular excursion is possible, then there is one traveling solely on one of the lines. This condition, denominated here "lazy commutation", states that if you can get somewhere by first taking a Red bus and then transferring to a Blue one, then – unless the Red line can take you directly to your destination – the Blue line will be able take you someplace whence you can also get to your destination by some combination of segments.

We demonstrate that lazy commutation implies that there is no need for the Red line to ever issue transfers to Blue routes, provided there are no circular trips that are purely Red or purely Blue. Bear in mind that the fact that one bus line operates from point X to point Y does not mean that it – or its competitor – necessarily operates in the opposite direction, from Y to X.

One of the earlier works on using commutation properties in termination proofs, especially of complicated forms of rewriting, was by Porat and Francez [29]. That same avenue is explored here.

Lazy commutation turns out to be a very meaningful concept. It has three important consequences regarding sequences of Red and Blue steps:

---

[*] Research supported in part by the Israel Science Foundation (grant no. 254/01).

O. Grumberg et al. (Eds.): Francez Festschrift, LNCS 5533, pp. 59–82, 2009.

1. It implies "finite separation", in the above sense, namely that Blue steps followed by Red steps will get one to any destination – unless there is an infinite monochromatic (single-step or multiple-step) excursion.
2. It implies "infinite separation", in that if there is an infinite combined Red-Blue excursion, then there must be an infinite monochromatic excursion as well, as was shown in [13].
3. If there is an endless combined excursion, but no endless purely Blue ones, then there is also an endless Red excursion from which a transfer at any point to the Blue line precludes wandering the lines forever.

After fixing notations and some terminology in the next section, we collect some new and old observations on separability in Sect. 3. Then, in Sect. 4, we consider conditions that ensure that a positive number of steps separates into a likewise nonempty sequence. This is followed by a definition, in Sect. 5, of "constricting" sequences, expanding on a notion of Plaisted's [28].

A weaker notion than laziness, dubbed "selection" and defined in Sect. 6, suffices to prove the main separation results in Sects. 7 and 8. Section 9 discusses the "lifting" of termination of one color to that of the other in the presence of separation.

We close with some thoughts on the implications that lazy commutation has for proofs of well-foundedness of orderings, such as the path orderings commonly used in proofs of termination (strong normalization) of rewriting.

2   Notions and Notations

All relations in this paper are binary. We use juxtaposition for composition of relations, represent union by $+$, and denote the inverse, n-fold composition, transitive closure, reflexive closure and reflexive-transitive closure of relation E by E^-, E^n, E^+, E^ε and E^*, respectively. Let I be the identity relation.

A relation E on a finite or infinite set of elements V may be viewed as a directed graph with vertices V and edges E. So, there is a (directed) path from s to t, for $s, t \in V$, if, and only if, $s \, E^* \, t$.

The *immortal* elements of a set V equipped with a relation E are those $t \in V$ that initiate infinite E-chains of (not necessarily distinct) elements of V,

$$t \, E \, t' \, E \, t'' \, E \cdots .$$

We will also say that a step $t \, E \, t'$ is *immortalizing* if t' is immortal. Define

$$E^\infty = \{\langle u, v \rangle \mid u, v \in V, \ u \text{ is immortal for } E\} \, ,$$

relating immortal elements to all vertices, as is commonly done in relational program semantics (e.g. [13]). Thus,

$$E^\infty C \subseteq E^\infty \, , \tag{1}$$

for any relation C.

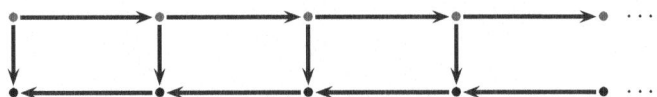

Fig. 1. Mortal (black) nodes on bottom and immortal (green) nodes on top

Fig. 2. Mortal in each alone (dashed Azure or solid Bordeaux), but immortal in their union

In Fig. 1, all the nodes on the upper row are immortal, while all those on the lower row are mortal. Taking a "wrong" turn can precipitate mortality, but, in contrast, once mortal, always mortal.

We compose relations, using notation like $s\,BA\,t$ to mean that there is two-step path from s to t, taking B followed by A. Similarly, $s \in AB^\infty$ means that s has an A-neighbor that is immortal for B. We say that a relation E is *well-founded* (regardless of whether E is transitive) if it admits no cases of immortality, that is, if $E^\infty = \varnothing$.

This paper explores properties of the (non-disjoint) union $E = A + B$ of two relations, A and B. Throughout, we will regularly use E, without elaboration, to denote the union $A + B$.

In Fig. 2, the individual relations A (think Azure in the illustrations) and B (Bordeaux) do not engender immortality, but their union does.

The following two main properties will claim our attention:

Definition 1 (Finite Separation). *Two relations A and B are finitely separable if*

$$(A + B)^* = A^* B^* \,.$$

Definition 2 (Infinite Separation). *Two relations A and B are infinitely separable if*

$$(A + B)^\infty = (A + B)^* (A^\infty + B^\infty) \,. \tag{2}$$

Nota bene. Finite separation is not a symmetric property: we want the A's before the B's. Infinite separation, on the other hand, is symmetric.

Any two points in the graph depicted in Fig. 3 can be connected by A's followed by B's, so the graph is finitely separable. Furthermore, every node is immortal in B alone, so it is infinitely separable, as well.

In terms of the bus routes (for which there are no simple [non-looping] infinite paths), infinite separability means that at least one company must provide a circular excursion, whenever it is possible to go in circles using both companies.

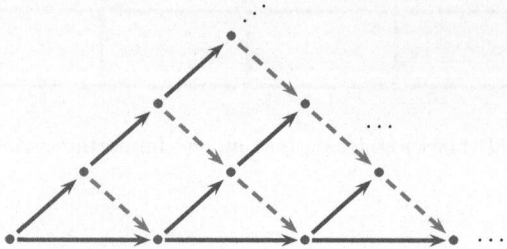

Fig. 3. The two relations are both finitely and infinitely separable

It is easy to see [33, Ex. 1.3.5(ii)] that finite separability (called "postponement" in [33]) is equivalent to the (global) "commutation" property

$$B^*A^* \subseteq A^*B^* . \tag{3}$$

Note 3. This is analogous to the equivalence of the Church-Rosser and global confluence properties of abstract rewriting [27].

Another trivial induction also shows that (3) is equivalent to an even simpler property:

Proposition 4 ([30]). *Two relations A and B are finitely separable if, and only if,*

$$B^+A \subseteq A^*B^* .$$

The point of infinite separability is that it ensures that the union $E = A + B$ is well-founded if each of A and B is, even without finite separability.

Example 5. An easy example of finite, but not infinite, separability is $s\,A\,t\,B\,s$.

Example 6. A simple example of infinite, but not finite, separability is $s\,A\,s\,BA$ t.

Definition 7 (Full Separation). *Two relations A and B are* fully (infinitely) separable *if*

$$
\begin{aligned}
(A + B)^\infty &= A^*B^*(A^\infty + B^\infty) \\
&= A^*B^*A^\infty + A^*B^\infty .
\end{aligned}
$$

Table 1. Types of separation ("coefficients" of right-hand side terms)

	$\subseteq$	A^*B^*	B^*	A^∞	B^∞
Weak (Def. 44)	E^*	I		A^*B^*	A^*
Finite (Def. 1)	E^*	I			
Productive (Def. 28)	E^+	A	B		
Infinite (Def. 2)	E^∞			E^*	E^*
Nice (Def. 9)	E^∞			E^*	
Full (Def. 7)	E^∞			A^*B^*	A^*
Neat (Def. 8)	E^∞			I	

With finite separability, infinite separability (2) is equivalent to full separability.

A pleasant special case is the following:

Definition 8 (Neat Separation). *Two relations A and B are neatly (infinitely) separable if*

$$(A + B)^\infty = A^\infty + A^*B^\infty .$$

A nice in-between (asymmetric) notion of separability is the following:

Definition 9 (Nice Separation). *Two relations A and B are nicely (infinitely) separable if*

$$(A + B)^\infty = (A + B)^*A^\infty + A^*B^\infty .$$

Clearly, then, If relations A and B are nicely, or fully, separable, and A is well-founded, then

$$E^\infty = A^*B^\infty .$$

Table 1 summarizes the various types of separability we will be using, from weaker down to stronger. For example, the penultimate row defines a relatively strong form of infinite separability (see Definition 7) and should be read as

$$E^\infty \subseteq A^*B^*A^\infty + A^*B^\infty .$$

It implies nice separation, and, a fortiori, (plain) infinite separation, but not neat separation.

3   Warmup

We begin with relatively simple cases of separability. We are looking for *local* conditions on double-steps BA that help establish separability, finite or infinite. The point is that these local conditions suggest how to eliminate BA patterns, which are exactly what are prohibited in separated sequences.

3.1 Finite Separation

Note that $CB^\varepsilon = CB + C$, and recall the following:

Proposition 10 ([20]). *If, for relations A and B,*

$$BA \subseteq A^* B^\varepsilon \qquad (4)$$

then A and B are finitely separable.

Note 11. When B is the inverse A^- of A, Eq. (4) becomes (one half of) Huet's [18] strong-confluence condition.

Proof. Repeatedly rewriting the last occurrence of BA in an E-chain with instances of $A^* B^\varepsilon$ must terminate in a sequence without any instance of BA, since either the number of B's decreases, or else the rightmost B moves rightward. □

By symmetry (of A and B and left and right):

Proposition 12. *If, for relations A and B,*

$$BA \subseteq A^\varepsilon B^* , \qquad (5)$$

then A and B are finitely separable.

Proof. An alternate proof to that of the previous proposition proceeds as follows: First, an easy induction shows that (5) implies that $B^n A \subseteq A^\varepsilon B^*$, for all n:

$$B^n BA \subseteq B^n A^\varepsilon B^* = B^n B^* + B^n AB^* = B^* + A^\varepsilon B^* = A^\varepsilon B^* .$$

Then, an application of Proposition 4, on account of $B^* A \subseteq A^\varepsilon B^*$, gives finite separation. □

As an aside, in direct analogy to Newman's Lemma [27], we have the following:

Proposition 13. *If, for relations A and B, $A + B^-$ is well-founded and*

$$BA \subseteq A^* B^* , \qquad (6)$$

then A and B are finitely separable.

This means that *local (weak) commutation* (6) implies separability, but only with a strong side condition of well-foundedness.

Proposition 14 ([17,5]). *If relation A is well-founded and*

$$BA \subseteq A^+ B^* , \qquad (7)$$

for relation B, then A and B are finitely separable.

Note 15. The necessity of either well-foundedness of $A + B^-$ or the limiting right-hand sides of inclusions (4) and (5) can be seen from the following example (adapted from [27]): $s\, B\, t\, B\, u\, A\, v$, and $t\, A\, u$. Pictorially:

Note that the edge $t\,E\,u$ in this graph belongs to both relations, A and B.

More generally, we have the following definition and result:

Definition 16 (Quasi-commutation [2]). *Relation A quasi-commutes over relation B when*

$$BA \subseteq A(A+B)^* . \qquad (8)$$

The right-hand side $A(A+B)^*$ is equivalent to $(AB^*)^+$.

Obviously (by induction):

Lemma 17. *Relation A quasi-commutes over relation B if, and only if,*

$$B^+A \subseteq A(A+B)^* .$$

Theorem 18. *If relation A quasi-commutes over relation B, then*

$$(A+B)^* \subseteq A^*B^* + A^\infty . \qquad (9)$$

(Compare Definition 44 below.)

Proof. Consider any sequence $E^* = A^*(B^*A)^*$. Using the above lemma, repeatedly replace the first occurrence of B^+A with AE^*. If this process continues forever, then the initial A-segment grows ever larger. Otherwise, we are left with a sequence A^*B^* sans occurrences of BA. □

Corollary 19. *If relation A is well-founded, then quasi-commutation of A over a relation B implies their finite separability.*

Since A is well-founded, one may assume that a property in question holds for every t such that $s\,A\,t$, in an inductive proof that it holds for s. This suggests the following alternative proof of this corollary:

Proof. We show that every sequence E^* that is not in the separated form A^*B^* can be rearranged to be "more separated". By quasi-commutation (Eq. 8),

$$E^* \setminus A^*B^* = A^*BAE^* \subseteq A^*AE^* ,$$

and by induction on A

$$A^*AE^* \subseteq A^+B^* .$$

So, in all cases, $E^* \subseteq A^*B^*$. □

Note 20. Quasi-commutation applies to combinatory logic and to orthogonal (left-linear non-overlapping) term-rewriting systems, where A is a leftmost (outermost) step and B is anything but. This means that leftmost steps may precede all non-leftmost ones. Combined with the fact that $A^-B \subseteq B^*A^-$, this gives *standardization* (leftmost rewriting suffices for computing normal forms). See [23].

3.2   Infinite Separation

We start with some obvious aspects of infinite chains:

Proposition 21. *For all relations A and B,*

$$(A + B)^\infty = (B^*A)^*B^\infty + (B^*A)^\infty \tag{10}$$
$$= (A^*B)^*A^\infty + (A^*B)^\infty .$$

It follows from (10) that if infinitely many interspersed A-steps are precluded, then $A + B$ is well-founded if, and only if, B is:

Corollary 22. *If relation B is well-founded, then*

$$(A + B)^\infty = (B^*A)^\infty . \tag{11}$$

for any relation A.

In other words, in any E-chain, there are either infinitely many A's, in which case E^∞ must be an infinite sequence of segments of the form B^*A, or else from some point on, there are only B's, and B is not well-founded. Cf. [2, Lemma 1] and [6, Lemma 12B].

The simplest local condition guaranteeing infinite separability is quasi-commutation (8). We have the following analogue of Theorem 18, based on the proof of infinite separation in [2]:

Theorem 23. *If relation A quasi-commutes over relation B, that is, if*

$$BA \subseteq A(A + B)^* ,$$

then A and B are neatly separable, that is,

$$(A + B)^\infty = A^\infty + A^*B^\infty .$$

The proof is also analogous to that of that earlier proposition:

Proof. Consider any E-chain, $E^\infty = E^*B^\infty + A^*(B^*A)^\infty$, as decomposed in (10). In the first instance (E^*B^∞), apply Theorem 18 to E^* and get $E^\infty \subseteq E^*B^\infty \subseteq (A^*B^* + A^\infty)B^\infty \subseteq A^*B^\infty + A^\infty$. Otherwise, repeatedly replace the first occurrence of B^*A in $A^n(B^*A)^\infty$ with AE^*, giving $A^{n+1}E^*(B^*A)^\infty = A^{n+1}(B^*A)^\infty$. This process continues forever, there being an infinite supply of A's, and the initial A-segment grows ever larger, yielding A^∞.   □

Note 24. This form of separability was used to show that "forward closure" termination suffices for orthogonal term rewriting, where B are "residual" steps (at redexes already appearing *below the top* in the initial term) and A are at "created" redexes (generated by earlier rewrites). See [7].

Theorem 25. *If, for relations A and B,*

$$BA \subseteq A^*B, \tag{12}$$

then A and B are fully separable, that is,

$$(A + B)^\infty = A^*B^*A^\infty + A^*B^\infty.$$

In what follows (Theorem 72), this condition, equivalent to $BA \subseteq A^+B+B$, will be weakened to laziness, the condition that $BA \subseteq AE^* + B$ (Definition 52), to give separation. That way, each occurrence of BA need not always leave a B, as in this theorem, nor always an A, as in the previous, but might sometimes leave one and sometimes the other.

Proof. If the number of A's or B's in an infinite E-chain is finite, then by virtue of Proposition 10, we have either $A^*B^*A^\infty$ or A^*B^∞. In any case, easy inductions show that $BA^* \subseteq A^*B$ and that $B^nA^* \subseteq A^*B^n$, for all n. Thus, a sequence $A^mB^nA^*BE^\infty$ turns into either $A^{m+k}B^{n+1}E^\infty$, $k > 0$, in which case the number of initial A's increases, or else into $A^mB^{n+1}E^\infty$, in which case the number of following B's increases. In the final analysis, one gets either A^∞, or else A^*B^∞. □

Note 26. This form of separability was used to show that "forward closure" termination suffices for right-linear term rewriting, where B are "created" steps and A are "residual" ones. See [7].

Note 27. When B is the subterm relation, condition (12) is a form of quasi-monotonicity (weak monotonicity) of A. Compare [8,1].

4   Productive Separation

With finite separability, a non-empty looping sequence (as in Example 5) can be "reordered" to give an empty one. To preclude an empty reordering, one can use the following notions:

Definition 28 (Productive Separation). *Relations A and B are* (finitely) *productively separable if*

$$(A + B)^+ = A^+B^* + B^+,$$

or, expressed more symmetrically, if

$$(A + B)^+ = A^+B^* + A^*B^+.$$

Note 29. If, for example, there is no "mixed" loop, that is, if

$$(A + B)^+ \cap I \subseteq A^+ + B^+,$$

then finite separation implies productive separation.

Theorem 30. *If relation A quasi-commutes over relation B and A is well-founded, then A and B are productively separable.*

Proof. By Corollary 19, the relations are finitely separable. So, we have:

$$E^+ = AE^* + BE^* \subseteq AE^* + BA^*B^* = AE^* + B^+ + BAE^*$$
$$\subseteq AE^* + B^+ + AE^*E^* \subseteq A^+B^* + B^+ \ . \qquad \square$$

Definition 31 (Promotion). *Relation B promotes relation A if*

$$BA \subseteq AB^* + B^+ \ , \tag{13}$$

or, expressed more symmetrically, if

$$BA \subseteq (A + B)B^* \ .$$

Note 32. One can have productivity without promotion: $s \ A \ t \ A \ u$ with $s \ B \ t$.

An easy induction shows the following:

Lemma 33. *Relation B promotes relation A if, and only if,*

$$B^+A \subseteq AB^* + B^+ \ .$$

Theorem 34. *If relation B promotes relation A, then A and B are productively separable.*

Note 35. Obviously, promotion (13) cannot be weakened to allow the erasure of both the A and B (cf. 5), as can easily be seen from $t \ BA \ t$ (Example 5), which is finitely, but not productively, separable.

Proof. By Proposition 12, promotion (13) gives finite separability, $E^* = A^*B^*$. By the previous lemma, $B^*A \subseteq AB^* + B^+$, whence the proposition follows:

$$E^+ = E^*(A + B) = A^*B^*A + A^*B^*B = A^+B^* + A^*B^+ \ . \qquad \square$$

In symmetry with promotion, we also have the following:

Proposition 36. *If, for relations A and B,*

$$BA \subseteq A^*B + A^+ \ ,$$

or, equivalently,

$$BA^* \subseteq A^*B + A^+ \ ,$$

then A and B are productively separable.

5   Constriction

Let V and E be the vertices and edges of a graph, respectively, and let $U = V^2$ be all pairs of vertices, so $E \subseteq U$. For any relation $R \subseteq U$ and property $P \subseteq V$, $RP = \{v \in V \mid v \, R \, w, \text{ for some } w \in P\}$ are all the vertices from which R takes one to a vertex satisfying P. The kinds of properties P we are interested are:

- a specific goal vertex;
- a set of vertices along some path; and
- the set of immortal vertices.

The point is to consider the impact of a preference for edges $A \subseteq E$ over others en route to P.

Given some property P, it will be convenient throughout this section to restrict E to only those edges allowing one to get to P. Let $\underline{V} = E^*P$ be those vertices from which P is reachable, let $\underline{U} = \underline{V}^2$, and let $\underline{R} = R \cap \underline{U}$ be the restriction of any relation R to vertices leading to P.

For any $B \subseteq E$, let

$$B_\sharp = \underline{B} \setminus \underline{AU} \, ,$$

that is, a useful B-step at a point such that no useful A-step can lead to P. Such $B_\sharp$-steps will be called *constricting*.

Example 37. Fair termination of $E = A + B$ with respect to A, à la the work of Porat and Francez [29], means that the union E is well-founded, except for the possibility of an infinite sequence that "unfairly" avoids taking an A-step infinitely often. So, the only way E can diverge is if from some point on A-steps were possible infinitely many times, but were never taken. This property can be expressed as

$$(B^*A)^\infty = B_\sharp^\infty = \varnothing \, ,$$

where the property P that defines $B_\sharp$ is always satisfied. It means that there are no infinite sequences with infinitely many A's (which certainly are fair), nor are there infinite sequences of B's, at every point of which no A-step was possible (these are the $B_\sharp$).

Accordingly:

Definition 38 (Constricting Chains). *An E-chain $t_0 \, E \, t_1 \, E \, \cdots$ is constricting in A with respect to P if its non-A steps are constricting, in the sense that there is no case of $t_i \, AE^*P$, for any $t_i \, (E \setminus A) \, t_{i+1}$.*

This means that the steps are all in $A + B_\sharp$. Our notion generalizes the "constricting" sequences of [28].

For any property, one can always build a constricting sequence. The idea is to ignore B-steps whenever an A-step is available:

Theorem 39. *For any relations A and B and property P,*

$$(\underline{A} + \underline{B})^* \subseteq (\underline{A} + B_\sharp)^* + (\underline{A} + B_\sharp)^\infty$$
$$= (\underline{A} + B_\sharp)^* + (\underline{A} + B_\sharp)^* \underline{A}^\infty + (\underline{A}^* B_\sharp)^\infty ,$$

where $\underline{A}$ and $\underline{B}$ are A and B steps, respectively, from which P is reachable and $B_\sharp$ is a $\underline{B}$-step from whose origin there is no outgoing $\underline{A}$.

Proof. Just take an A-step whenever possible, that is, whenever there is one that leads to P. Only take B if there is no choice, so the B-steps are constricting. Then, every A-step is actually $\underline{A}$ and every B-step is $B_\sharp$ (as well as $\underline{B}$). This process either terminates, or else goes on forever, with either infinitely many B steps, or finitely many. □

Note 40. If there is an infinite A-path, then preferring A can be a bad idea. Witness: $s\ A\ s\ B\ t$, that is,

Note 41. Even if the sole infinite path involves both relations, A's can preclude reaching the goal, as in $t\ A\ s\ B\ t\ B\ u$:

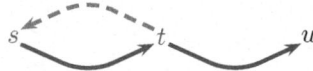

6   The Selection Property

In this section we deal with non-local properties for which constriction yields separation.

Definition 42 (Selection). *We say that relation B selects relation A if*

$$BA^+ \subseteq A(A+B)^* + B^+ . \tag{14}$$

Definition 43 (Weak Selection). *We say that relation B weakly selects relation A if*

$$BA^+ \subseteq A(A+B)^* + B^* + A^* B^*(A^\infty + B^\infty) . \tag{15}$$

Clearly, weak selection is weaker than selection.

Quasi-commutation gave a special case (Eq. 9) of the following property:

Definition 44 (Weak Separation).   *Two relations A and B are weakly (finitely) separable if*

$$(A+B)^* \subseteq A^* B^* + A^* B^* A^\infty + A^* B^\infty$$
$$= A^* B^*(I + A^\infty + B^\infty) .$$

It is straightforward to see that

Proposition 45. *Infinite separability plus weak separability give full separability.*

Restricting attention to constricting sequences teaches the following:

Theorem 46. *If relation B weakly selects relation A, then A and B are weakly separable.*

Proof. Suppose $s \, E^* \, t$ and let $P(x)$ hold only for $x = t$. The premise (15) means that

$$B_\sharp \underline{A}^* \subseteq B^* + B^*(A^\infty + B^\infty) = B^*(I + A^\infty + B^\infty) \,,$$

since there can be no (relevant) outgoing A-step where there is a constricting B-step. Theorem 39 asserts the (algebraic) equivalent of

$$\underline{E}^* \subseteq \underline{A}^*(B_\sharp \underline{A}^*)^* + \underline{A}^*(B_\sharp \underline{A}^*)^\infty + \underline{A}^*(B_\sharp \underline{A}^*)A^\infty \,,$$

and, replacing $B_\sharp \underline{A}^*$ as reasoned above,

$$E^* \subseteq A^* B^*[(I + A^\infty + B^\infty) + (I + A^\infty + B^\infty)^\infty + (I + A^\infty + B^\infty)A^\infty]$$
$$\subseteq A^* B^*(I + A^\infty + B^\infty) \,. \qquad \square$$

Note 47. An alternative weak version of selection,

$$BA^+ \subseteq A(A + B)^* + B^* + (A + B)^*(A^\infty + B^\infty) \,,$$

does not yield weak separability. For example, $s \, BA \, t \, A \, t$ is not weakly separable, though $BA^+ \subseteq B^* A^\infty$:

Theorem 48. *If relation B selects relation A and both are well-founded, then A and B are productively separable.*

Proof. Selection implies weak selection, which, per the previous theorem, gives weak separation, which, with well-foundedness, is finite separation. So selection, itself, becomes

$$BA^+ \subseteq AE^* + B^+ \subseteq A^+ B^* + B^+ \,,$$

as required by Definition 28. $\qquad \square$

Note 49. Without well-foundedness of A, one does not have separation, as may be seen from the following selecting, but unseparable example: $s \, A \, s \, BA \, t$. That is,

Note 50. Well-foundedness of B is also needed. To wit:

A somewhat analogous way of obtaining a "productive" version of weak separability, similar in flavor to promotion as used in Theorem 34, is the following:

Theorem 51. *If, for relations A and B,*

$$B^+ A \subseteq A(A + B)^* + B^+ + A^* B^* A^\infty , \tag{16}$$

then

$$(A + B)^+ \subseteq A^+ + A^* B^+ + A^* B^* A^\infty .$$

Proof. If a nonempty E-chain is not of the form $A^* B^*$, then it must be of the form $A^m B^+ A E^n$ $(m, n \geq 0)$. The premise (16) is then used to replace the segment $B^+ A$. The situation devolves into one of three possibilities:

1. We get $A^m A E^*$, in which case the initial run of A's grows by at least one, and the process may be repeated.
2. We get $A^m B^+ E^n$, in which case the $A E^*$-tail shrinks to length less than $n + 1$, and the process is repeated.
3. We get $A^m A^* B^* A^\infty$, in which case we are done.

So, either the process continues ad infinitum and infinitely many A's get pushed frontwards, giving A^∞, or else eventually the process ends happily. It can end either in immortality, when the third case transpires, or else in separation – with no subsequences $B^+ A$ to power the process. □

7   Lazy Commutation

In [13,14], the following local property is defined and examined:

Definition 52 (Lazy Commutation). *Relation A commutes lazily over relation B if*

$$BA \subseteq A(A + B)^* + B . \tag{17}$$

Figure 3 has lazily commuting relations; Fig. 2 does not.

Back to busses: Lazy commutation means that a Red-Blue trip can be substituted by a single ride on Red or by a junket beginning with a Blue segment.

Example 53. The relations $s\,B\,t\,A\,t$ commute lazily:

Note 54. Lazy commutation is noticeably weaker than quasi-commutation (8), which was shown in [2] to separate well-foundedness of the union into well-foundedness of each. (See [31] for an automated proof using Kleene algebras.) Corollary 76 below is stronger.

Note 55. As pointed out in [13], lazy commutation is also much weaker than transitivity of the union, which – by a direct invocation of a very simple case of the infinite version of Ramsey's Theorem – gives separation. (See [17,4] for some of the history of this idea.)

Note 56. Note that replacing BA with $AABB$ is a process that can continue unabated: BBA, $BAABB$, $AABBABB$, $AABAABBBB$, ..., with infinitely many paths through the graph.

The following is known:

Proposition 57 (**[14, Eq. 4.5]**). *If A commutes lazily over B, then*

$$BA^* \subseteq A(A+B)^* + B\,, \tag{18}$$

and – in particular – B selects A.

In short, laziness is stronger than selection.

Note 58. An example of selection without laziness is $s\,B\,t\,B\,u$ with $t\,A\,u$, the graph of which looks like this:

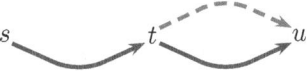

It is convenient to make use of the following notion:

Definition 59. *Relation X absorbs relation A if*

$$XA \subseteq X\,.$$

For example, A^*, E^*, B^∞, and $\varnothing$ all absorb A. (Recall Eq. 1.) Obviously, if X absorbs A, then it absorbs any number of A's, and CX absorbs A whenever X does.

Proposition 60. *If*
$$CA \subseteq C + X$$
for a relation X that absorbs A, then
$$CA^* \subseteq C + X \ .$$

Proof. By induction on n,
$$CA^n \subseteq C + X \ ,$$
since, trivially, $C \subseteq C + X$, while, by the inductive hypothesis and absorption,
$$CAA^n \subseteq (C + X)A^n = CA^n + XA^n \subseteq C + X + X = C + X \ . \qquad \square$$

Proposition 57 is a corollary, with $C = B$ and $X = AE^*$:
$$BA^+ \subseteq B + AE^* \subseteq B^+ + AE^* \ .$$

A weaker version of laziness, which allows for infinite exceptions, is the following:

Definition 61 (Lackadaisical Commutation). *Relation A commutes lackadaisically over relation B if*
$$BA \subseteq A(A + B)^* + B + A^*B^*(A^\infty + B^\infty) \ . \tag{19}$$

Theorem 62. *If relation A commutes lackadaisically over relation B, then B selects A weakly.*

Proof. Applying the previous proposition (E^*, A^∞, and B^∞ each absorb A), the premise (19) implies
$$BA^+ \subseteq AE^* + B + A^*B^*(A^\infty + B^\infty) \ ,$$
which is more than weak selection. $\qquad \square$

Now, by Theorem 46:

Theorem 63. *If relation A commutes lackadaisically over relation B, and – in particular – if A commutes lazily over B, then A and B are weakly separable.*

Note 64. That an infinite A-chain cannot be ruled out can be seen from $s\,B\,t\,A$ $t\,BA\,u$:

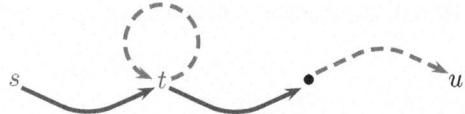

So lackadaisicalness gives only weak separability, rather than finite separability.

With well-foundedness, weak separation turns into finite separation and lackadaisical commutation into lazy commutation.

Corollary 65. *If relation A commutes lazily over relation B, and both are well-founded, then in fact*

$$BA \subseteq A^+ B^* + B .$$

Proof. Since laziness with well-foundedness gives finite separation,

$$BA \subseteq AE^* + B \subseteq AA^* B^* + B . \qquad \square$$

Moreover, we have productive separability (Definition 28):

Theorem 66. *If relation A commutes lazily over relation B, and both are well-founded, then A and B are productively separable.*

Proof. A simple induction – using the previous lemma – yields

$$BA^+ \subseteq A^+ B^* + B .$$

Thus, using finite separation,

$$E^+ = EA^* B^* = A^+ B^* + BA^+ B^* + B^+ \subseteq A^+ B^* + B^+ . \qquad \square$$

Note 67. Well-foundedness of A and of B are necessary, even in the presence of lazy commutation. Refer to the examples in Notes 49 and 50.

Note 68. Lazy commutation cannot be weakened to

$$BA \subseteq AE^* + B^+ .$$

Even

$$BA \subseteq AA + BB$$

causes trouble, as can be seen from the non-separable example in Note 15.

Note 69. Lazy commutation also cannot be weakened to

$$BA \subseteq AE^* + B^\varepsilon ,$$

as can be seen from the following non-separable graph:

8 Endless Commutation

We will see below that selection (Definition 42) suffices for the main consequences of lazy commutation, namely finite and infinite separation. But, first, observe the following:

Lemma 70. *If*

$$BA^+ \subseteq A(A+B)^* + B^+ + (A+B)^*(A^\infty + B^\infty) \,, \tag{20}$$

for relations A and B, then they are infinitely separable.

This will be our main tool for separation.

Note 71. The condition (20) clearly does not suffice for full separation. See the example in Note 47.

Proof. We use constriction (Sect. 5), letting $P(x)$ hold for immortal x. The premise (20) implies that

$$B_\sharp \underline{A}^* = B^+ + E^*(A^\infty + B^\infty) \,,$$

since $B_\sharp$ precludes there being an initial immortalizing A-step. By Theorem 39:

$$E^* \subseteq A^*(B_\sharp \underline{A}^*)^* + A^*(B_\sharp \underline{A}^*)^\infty + A^*(B_\sharp \underline{A}^*)^* A^\infty \,.$$

Infinite composition of this means that either the first possibility, $A^*(B_\sharp \underline{A}^*)^*$, repeats forever, or else it repeats finitely often and then one of the two infinite cases takes over:

$$
\begin{aligned}
E^\infty &= (A^*(B_\sharp \underline{A}^*)^*)^\infty \\
&\quad + (A^*(B_\sharp \underline{A}^*)^*)^* A^*(B_\sharp \underline{A}^*)^\infty + (A^*(B_\sharp \underline{A}^*)^*)^* A^*(B_\sharp \underline{A}^*)^* A^\infty \\
&= A^*(B_\sharp \underline{A}^*)^\infty + A^*(B_\sharp \underline{A}^*)^\infty + A^*(B_\sharp \underline{A}^*)^* A^\infty \,.
\end{aligned}
$$

By the premise, this is

$$A^* B^\infty + A^* B^* A^\infty + E^*(A^\infty + B^\infty) = E^*(A^\infty + B^\infty) \,,$$

as desired.   □

The net result is that

Theorem 72. *If relation B selects relation A, and – in particular – if A commutes lazily over B, then A and B are fully separable.*

Proof. Selection (14) implies the condition (20) of the previous lemma, giving infinite separation. Theorem 46 gives weak separation; Proposition 45 gives full separation; Proposition 57 gives the particular, lazy case.   □

The fact that laziness implies infinite separability is the main result of [13].

Example 73. The graph $s\ AB\ t\ A\ t$ is lazily commuting, since $BA \subseteq B$, and in fact $s\ ABA^\infty$, as can be seen in the following diagram:

So, the relations are fully separable, but not neatly separable.

Note 74. Obviously, sans lazy commutation, immortality cannot necessarily be separated. For example, with $s\ AB\ s$, only in the union is s immortal. See Example 5.

Note 75. Lazy commutation cannot be weakened to include $BA \subseteq AA + BB$. To see that, wrap the counterexample of Note 68 around itself, as follows:

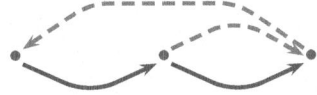

Combining this theorem with Theorem 48, we can summarize by saying the following:

Corollary 76. *If relations A and B are both well-founded, and, furthermore, B selects A, or – in particular – A commutes lazily over B, then the union $A + B$ is also well-founded and A and B are (finitely) productively separable.*

When the relations are finite, as for the bus companies of the introduction, any infinite tours are circular.

Corollary 77. *If relation B promotes relation A, then A and B are fully (infinitely) separable, as well as (finitely) productively separable.*

Proof. By Lemma 33, $B^+A \subseteq AB^* + B^+$, so A commutes lazily over B^+. Clearly, $E^\infty = (A + B^+)^\infty$. The result follows from Theorems 72 and 34.   □

Note 78. Promotion (Eq. 13) cannot be weakened to allow the erasure of both the A and B, as can be seen from $t\ BA\ t$, which only cycles in the union. Compare Note 35.

Actually:

Theorem 79. *Whenever relations A and B are (finitely) productively separable, they are also fully (infinitely) separable.*

Proof. Productive separability ($E^+ \subseteq A^+B^* + B^+$) implies that A^+ commutes lazily over B^+ (i.e. that $B^+A^+ \subseteq A^+E^* + B^+$) which means, thanks to Theorem 72, that A^+ and B^+ are fully separable, which is the same as A and B themselves being fully separable, since

$$E^\infty = (A^+ + B^+)^\infty = (A^+)^*(B^+)^*(A^+)^\infty + (A^+)^*(B^+)^\infty = A^*B^*A^\infty + A^*B^\infty .$$

□

Alternatively, one could have agreed directly from constriction.

9 Lifting Relations

For various applications of commutation arguments, the well-foundedness of one relation is dependent on that of the other.

First, note the following:

Proposition 80. *If, for relations A, B, and C, one has*

$$B \subseteq A^+ B + C \, ,$$

then

$$B \subseteq A^\infty + A^* C \, .$$

Proof. By unending application of the premise, we get (the greatest pre-fixpoint) $A^\infty + A^* C$ (for B).   □

In particular, if $B^\infty \subseteq A^+ B^\infty + A^\infty$, then $B^\infty \subseteq A^\infty$, meaning that B is well-founded if A is.

Theorem 81. *If relations A and B are infinitely separable and if*

$$B^\infty \subseteq A^+ B^\infty + (A + B)^* A^\infty \, ,$$

then

$$(A + B)^\infty = (A + B)^* A^\infty \, .$$

Proof. By the premise and the above proposition,

$$B^\infty \subseteq E^* A^\infty \, .$$

By infinite separation,

$$E^\infty = E^* A^\infty + E^* B^\infty \subseteq E^* A^\infty + E^* E^* A^\infty = E^* A^\infty \, .$$ □

Definition 82 (Lifting). *Relation A lifts to relation B if*

$$B^\infty \subseteq A(A + B)^\infty \, . \tag{21}$$

Regarding lifting (21), see [22,16,19]. It implies that if there is an infinite E-chain, then there is one with infinitely many interspersed A-steps. That is:

Lemma 83. *If relation A lifts to relation B, then*

$$(A + B)^\infty = (B^* A)^\infty \, .$$

Compare Eq. (11).

Proof. By Proposition 21 and lifting,

$$E^\infty = (B^*A)^* B^\infty + (B^*A)^\infty$$
$$\subseteq (B^*A)^* A E^\infty + (B^*A)^\infty = (B^*A)^+ E^\infty + (B^*A)^\infty .$$

By the previous proposition, we obtain the following:

$$E^\infty \subseteq (B^*A)^\infty + (B^*A)^* (B^*A)^\infty = (B^*A)^\infty . \qquad \square$$

Moreover, B steps need only be taken when no A step leads to immortality.

Lemma 84. *If relations A and B are nicely separable and A lifts to B, then*

$$(A+B)^\infty = (A+B)^* A^\infty .$$

Proof. Combining niceness and lifting, we have that

$$E^\infty \subseteq E^* A^\infty + A^* B^\infty \subseteq A^+ E^\infty + E^* A^\infty .$$

By Proposition 80, E^∞ gives either A^∞ or else $A^* E^* A^\infty$. $\qquad \square$

In other words:

Theorem 85. *If relations A and B are nicely separable and A lifts to B, then $A + B$ is well-founded if (and only if) A is.*

Finally, we have the following:

Theorem 86. *If relation A commutes lazily over relation B and A lifts to B, then $A + B$ is well-founded if, and only if, A is.*

Proof. Use Theorems 72 and 85, bearing in mind that full separation is more than just nice separation. $\qquad \square$

10 Discussion

Table 2 summarizes many of the more significant dependencies derived in the preceding sections.

The claims of Sects. 3 and 4 are amenable to automated proofs [32]. More work is needed, however, to automate the more advanced results of Sects. 6–8. This interest of ours in mechanization is why we have indulged in alternative proofs of some results.

In future work, we plan to use the ideas developed herein to show how laziness and constriction can contribute to proving well-foundedness of an abstract path ordering – in the style of [19,6] – which includes the nested multiset ordering [12], multiset path ordering [8], lexicographic path ordering [22], and recursive path ordering [24,9] as special cases. We seem to need a weaker alternative to lifting [16,19], in which lifting need only take place eventually.

Table 2. Implication graph (Legend: **Definition**; **Note**; **Proposition**; **Theorem**)

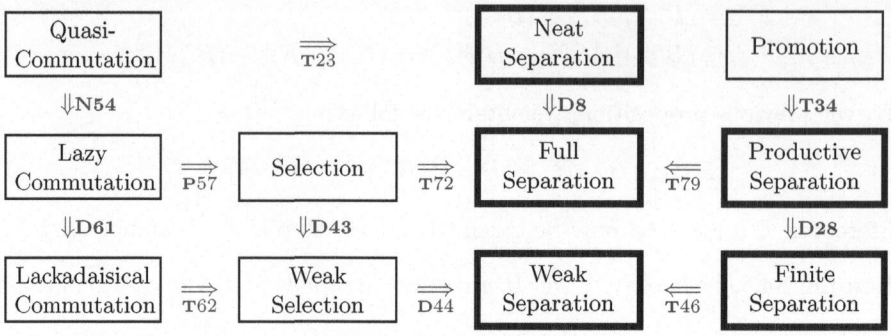

As pointed out in [10], there is an analogy between the use of reducibility predicates and the use of constricting derivations in proofs of well-foundedness. We are optimistic that the commutation-based approach taken here will likewise help for advanced path orderings, like the general path ordering [11] and higher-order recursive-path-ordering [15,21,3], without recourse to reducibility/computability predicates.

We plan to analyze minimal bad sequence arguments for well-quasi-orderings (pioneered in [26]) in a similar fashion. See [25]; compare [19]. We also hope to apply the results of the previous sections to analyze the dependency-pair method of proving termination. See [1]; compare [10].

Acknowledgements

Thank you, Frédéric Blanqui and Georg Struth.

References

1. Arts, T., Giesl, J.: Termination of term rewriting using dependency pairs. Theoretical Computer Science, vol. 236, pp. 133–178 (2000)
2. Bachmair, L., Dershowitz, N.: Commutation, transformation, and termination. In: Siekmann, J.H. (ed.) CADE 1986. LNCS, vol. 230, pp. 5–20. Springer, Heidelberg (1986),
http://www.cs.tau.ac.il/~nachum/papers/CommutationTermination.pdf
3. Blanqui, F., Jouannaud, J.-P., Rubio, A.: HORPO with computability closure: A reconstruction. In: Dershowitz, N., Voronkov, A. (eds.) LPAR 2007. LNCS, vol. 4790, pp. 138–150. Springer, Heidelberg (2007)
4. Blass, A., Gurevich, Y.: Program termination and well partial orderings. ACM Transactions on Computational Logic (TOCL) 9(3), Article No. 18 (June 2008),
http://research.microsoft.com/en-us/um/people/gurevich/opera/178.pdf
5. Di Cosmo, R., Piperno, A.: Expanding extensional polymorphism. In: Dezani-Ciancaglini, M., Plotkin, G.D. (eds.) TLCA 1995. LNCS, vol. 902, pp. 139–153. Springer, Heidelberg (1995)

6. Dawson, J.E., Gore, R.: Termination of abstract reduction systems. In: Proceedings of Computing: The Australasian Theory Symposium (CATS 2007), Ballarat, Australia, pp. 35–43 (2007)
7. Dershowitz, N.: Termination of linear rewriting systems (Preliminary version). In: Even, S., Kariv, O. (eds.) ICALP 1981. LNCS, vol. 115, pp. 448–458. Springer, Heidelberg (1981), http://www.cs.tau.ac.il/~nachum/papers/Linear.pdf
8. Dershowitz, N.: Orderings for term-rewriting systems. Theoretical Computer Science 17(3), 279–301 (1982)
9. Dershowitz, N.: Termination of rewriting. J. of Symbolic Computation 3, 69–116 (1987)
10. Dershowitz, N.: Termination by abstraction. In: Demoen, B., Lifschitz, V. (eds.) ICLP 2004. LNCS, vol. 3132, pp. 1–18. Springer, Heidelberg (2004), http://www.cs.tau.ac.il/~nachum/papers/TerminationByAbstraction.pdf
11. Dershowitz, N., Hoot, C.: Natural termination. Theoretical Computer Science 142(2), 179–207 (1995)
12. Dershowitz, N., Manna, Z.: Proving termination with multiset orderings. Communications of the ACM 22(8), 465–476 (1979)
13. Doornbos, H., Backhouse, R., van der Woude, J.: A calculational approach to mathematical induction. Theoretical Computer Science 179, 103–135 (1997)
14. Doornbos, H., von Karger, B.: On the union of well-founded relations. Logic Journal of the IGPL 6(2), 195–201 (1998)
15. Fernández, M., Jouannaud, J.-P.: Modular termination of term rewriting systems revisited. In: Astesiano, E., Reggio, G., Tarlecki, A. (eds.) Recent Trends in Data Type Specification. LNCS, vol. 906, pp. 255–272. Springer, Heidelberg (1995)
16. Ferreira, M.C.F., Zantema, H.: Well-foundedness of term orderings. In: Dershowitz, N. (ed.) CTRS 1994. LNCS, vol. 968, pp. 106–123. Springer, Heidelberg (1995)
17. Geser, A.: Relative Termination, Ph.D. dissertation, Fakultät für Mathematik und Informatik, Universität Passau, Germany (1990); also available as: Report 91-03, Ulmer Informatik-Berichte, Universität Ulm (1991), http://ginevras.pil.fbeit.htwk-leipzig.de/pil-website/public_html/geser/diss_geser.ps.gz
18. Huet, G.: Confluent reductions: Abstract properties and applications to term rewriting systems. J. of the Association for Computing Machinery 27(4), 797–821 (1980)
19. Goubault-Larrecq, J.: Well-founded recursive relations. In: Fribourg, L. (ed.) CSL 2001 and EACSL 2001. LNCS, vol. 2142, pp. 484–497. Springer, Heidelberg (2001)
20. Hindley, J.R.: The Church-Rosser Property and a Result in Combinatory Logic, Ph.D. thesis, University of Newcastle upon Tyne (1964)
21. Jouannaud, J.-P., Rubio, A.: Higher-order recursive path orderings. In: Proceedings 14th Annual IEEE Symposium on Logic in Computer Science (LICS), Trento, Italy, pp. 402–411 (1999)
22. Kamin, S., Lévy, J.-J.: Attempts for generalising the recursive path orderings, unpublished note, Department of Computer Science, University of Illinois, Urbana, IL (February 1980), http://pauillac.inria.fr/~levy/pubs/80kamin.pdf
23. Klop, J.W.: Combinatory Reduction Systems, Mathematical Centre Tracts 127, CWI, Amsterdam, The Netherlands (1980)
24. Lescanne, P.: On the recursive decomposition ordering with lexicographical status and other related orderings. J. Automated Reasoning 6(1), 39–49 (1990)
25. Melliès, P.-A.: On a duality between Kruskal and Dershowitz theorems. In: Larsen, K.G., Skyum, S., Winskel, G. (eds.) ICALP 1998. LNCS, vol. 1443, pp. 518–529. Springer, Heidelberg (1998)

26. Nash-Williams, C.St.J.A.: On well-quasi-ordering finite trees. Proceedings Cambridge Phil. Soc. 59, 833–835 (1963)
27. Newman, M.H.A.: On theories with a combinatorial definition of "equivalence". Annals of Mathematics 43(2), 223–243 (1942)
28. Plaisted, D.A.: Semantic confluence tests and completion methods. Information and Control 65(2/3), 182–215 (1985)
29. Porat, S., Francez, N.: Full-commutation and fair-termination in equational (and combined) term-rewriting systems. In: Siekmann, J.H. (ed.) CADE 1986. LNCS, vol. 230, pp. 21–41. Springer, Heidelberg (1986)
30. Staples, J.: Church-Rosser theorems for replacement systems. In: Crosley, J. (ed.) Algebra and Logic. Lecture Notes in Mathematics, vol. 450, pp. 291–307. Springer, Heidelberg (1975)
31. Struth, G.: Reasoning automatically about termination and refinement (Abstract). In: International Workshop on First-Order Theorem Proving (FTP 2007), Liverpool, UK (September 2007). Full Version at:
    http://www.dcs.shef.ac.uk/intranet/research/resmes/CS0710.pdf
32. Struth, G.: Personal communication (November 2007)
33. "Terese" (Bezem, M., Klop, J.W., de Vrijer, R., eds.): Term Rewriting Systems. Cambridge University Press, Cambridge (2002)

Aspect Oriented Approach for Capturing and Verifying Distributed Properties

Tzilla Elrad

Illinois Institute of Technology
elrad@iit.edu
www.iit.edu/~concur

Abstract. Given an initial decomposition of a distributed program into processes we are faced with two classes of properties that are tightly bounded. One
is the class of intra- process properties; these are relationships and requirements
over a local state of a process. The seγcond class is inter- processes properties;
these are relationships and requirements over different processes local states
that depict the consistencies and cooperation among a collection of cooperative
processes that might be tightly bounded to achieve a certain collective goal.
Both classes of system properties are crucial for system development and verification. The intra-process properties are relatively easier to define and carry on
through system development life cycle. This paper concentrates on the challenges of expressing and imposing inter-processes properties during a distributed execution.

The approach taken here is based on two completely independent areas of
research The first is the CCL – Communication Closed Layers; a formal system
for developing, maintaining, and verifying distributed programs [EF82] and the
second is the aspect oriented software development [AOSD, CACM] that enables simplification and automation of CCL implementations. Aspect-oriented
approach provides a natural framework to address inter-processes cooperation.
The essence of aspect oriented software developments is the localization of
crosscutting concerns. Under CCL constrains cooperative global properties can
be considered as a special case of crosscutting concerns. (Note that this is not
true in general; layers closeness, as defined in the following, is a necessary condition for the validity of cooperative global virtual assertions).

Both CCL and AOSD are software development methodologies; their synergy for distributed systems development and runtime verification of system
virtual global assertions is a promising match to handle the complexities of
global system cooperation and consistencies. Both are centered on the notion of
crosscutting concerns; CCL captures the semantics whereas AOSD provides the
expressive tools. Together, it enables a design-by-contract discipline [M93] to
be applicable to a wider range of distributed programs.

There is still ongoing research to use the WEAVR , an aspect oriented model
driven engineering [tool developed by Motorola, to automate code generation
from precise behavioral models to express layers and global assertions. [CE1,
CE2, CVE1, CVE2, CVE3].

O. Grumberg et al. (Eds.): Francez Festschrift, LNCS 5533, pp. 83–96, 2009.

1 Introduction

The Tyranny of Distributed Programs Processes Composition

Decomposition of a distributed system into processes provides only a syntactic representation of one dimension; the vertical dimension, no syntactic representation is available for the decomposition of the systems into its logical phases in terms of system's goals. For example, it is not syntactically visible when the system as a whole has achieved its first sub goal and is ready to launch into the second sub goal. At any given time different processes might be executing at different sub goals. The tyranny of system decomposition into processes overlooks the logical structure of the system as a whole.

A first attempt to break this tyranny of process decomposition is to impose a complementary horizontal decomposition. Artificial barriers may be introduced to gather and hold processes together at a certain point before allowing them to continue execution. When all processes have reached their respective local halting points the whole system halts, a global relationship among the different processes states is verified for consistency. Only if the system is on "the correct" milestone step it allows to continue. Artificial barriers allow both the sequential decomposition and the distributed composition to be syntactically visible. Global invariants could be inserted and the complementary structure of the system as a sequence of sub goals is observable. The huge benefit from such capabilities can be majored by their use in sequential, non-distributed systems, pre/post conditions and use of assertions and invariants during sequential life cycle process.

Obviously, this approach is far from practical. Artificial barriers result with over synchronization which slows the system performance more then necessary. In many cases just the process of detecting that distributed possesses have all reached a halting point is, in best case, a hazard and in many cases just impossible.

Breaking the tyranny of processes composition requires a more sophisticated approach. The idea behind the CCL is to find a reasonable set of restrictions under which processes do not have to "really" halt at their local halting points and yet the semantics of the program as a whole is as if they do. The horizontal decomposition of the system into its logical phases is observable and yet non-imposing in terms of execution. The proof of such semantic equivalence is given in [EF82, J94]

In [EBN96] we presented a synergy of object-oriented distributed programming and CCL called CHESSBOARD. CESSBOARD enables " design by contract" for distributed applications where processes are tightly bounded to achieve a unique common goal. The idea is that object orientation provides encapsulation of the communication among processes and hence enables a syntactic identification of a local process unit that collaborates with similar units in other processes. CCL enables the syntactic identification of such cooperation. The actual implementation of layers boundaries is inserted into each process code. This result with the well known tangling code phenomena of crosscutting concerns and here is where aspect orientation can provide a natural solution.

The integration between CCL and aspect orientation for distributed software development is simple; local halting points for each process would be realized by aspect oriented joint points. Pointcut designators will filter out a subset of all the join points in different processes for the identification of a virtual global state that represents the logic behind the horizontal structure of the program.

The Communication Closed Layer is a language independent methodology, it is represented only in terms of the semantics constraints among its components. Similarly, we choose to present CCL realization independent of any particular aspect oriented technology hence, we only use the common terminology of aspect orientation like cut point, pointcut, advice, aspects etc.

The paper is organized as follows. Section 2 is an overview of CCL – communication Closed Layers. Section 3 is an overview of basic concepts and terminology of AOSD – Aspect Oriented Software Development. Section 4 explains how to express layers boundaries using aspects. Section 5 illustrates the approach through a well known example of Two-Phase Commit Protocol by using aspects to capture the assertions at layers boundaries. Section 6 lists potential applications of AOSD and CCL synergy. The conclusion summarizes and evaluates the synergy and points to future work.

2   CCL - Communication Closed Layered

The notion of CCL is language independent. A complete formal definition and proof system can be found in [EF82, J94]. More work on CCL can be found in [EBN96, E84, EK90, EK91, EK93, PFZ93, GS96, JZ92, JPXZ94, PZ92, SdR87, JZ93, ZHLdR95, ZJ93, Moo]. Here we present only a general, more intuitive definition using CSP notation.

Let $[P1 \| P2 \| \dots Pn]$ be a distributed program $\mathbf{P}$ composed of n processes P1 to Pn.

Now assume each of these processes is decomposed into its logical local segments. Each Pi is refined into: begin Si1;Si2 …Sik end. Sij is the j's segment of process i.
The distributed program can be expressed as:

begin S11;S12 …S1k end
$\quad \|$
begin S21;S22 …S2k end
$\quad \|$

begin Sn1;Sn2 …Snk end

Not that this representation reflects the tyranny of processes decomposition. Now assume that all the j segments are communicating to achieve the overall system j's sub goal. We would like this fact to be syntactically represented.
$[S1j \| S2j \| \dots Snj]$ is called the j's layer of the system. Lj :: $[S1j \| S2j \| \dots Snj]$
Now we would like to use layers representation to compose the whole system back. There are two different composition rules.

The SCR – Sequential Composition Rule
Let L1 and L2 be two distributed program layers. The composition L1+L2 is defined as the distributed program $[S11 \| S21]$; $[S21 \| S22]$. This is, basically, the semantics

of our first attempt to make the logical structure of the whole distributed program syntactically visible. All processes must halt at layer boundaries and only when everyone has reached this synchronization point the second segment may start.

The DCR- Distributed Composition Rule

Let L1 and L2 be two distributed program layers. The composition L1*L2 is defined as the distributed program [S11;S12 || S21;S22] . This is, basically, the semantics which result from ignoring layer boundaries at execution time. DCR and SCR for more than two layers are defined inductively.

With respect to the example above, the Distributed Composition Rule; L1*L2* ...*Lk yields the distributed program in figure A1 Whereas the Sequential Composition Rule;

L1+L2+ ...+Lk, yields the distributed program in figure A2.

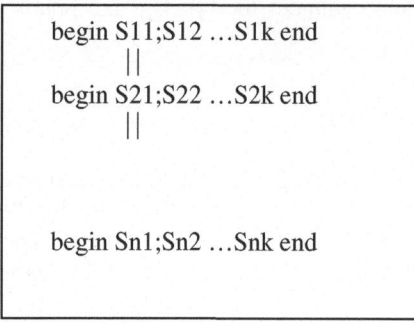

begin S11;S12 ...S1k end	[S11		S21		...Sn1]
			;
begin S21;S22 ...S2k end	[S12		S22	...Sn2]
			;
	;
begin Sn1;Sn2 ...Snk end	[S1k		S2k	...Snk]

Fig. A1. The DCR   **Fig. A2.** The SCR

Distributed Composition Rule   **Sequential Composition Rule**

Ideally, we would like to use the SCR though the software life cycle but use the DCR at runtime. The problem is that these two compositions, in general, do NOT result with the same semantics. The SCR program exhibits only a subset of all possible computation paths that can occur in the DCR program. This means that, in general, the transformation from SCR to DCR at runtime does not necessarily preserve the program semantics.

Communication- Closed Layer: A distributed layer Lj is called communication closed if for every communication both parties are taken from the j's segments. In other words, it is semantically impossible that a communication occurs across layers.

The Communication- Closed Layers Safety Theorem: Let L1 and L2 be two communication closed layers then the distributed programs L1+L2 and L1*L2 are semantically equivalent. A complete proof of the theorem can be found in [EF82] and also in [J94].

Using the CCL safety theorem we can safely use the SCR during software life cycle. Since under layer closeness transformation to the DCR preserves the distributed program semantics, the executable distributed program can be the DCR one. Layer closeness must be either verified or imposed. In the application section of this paper we will show how layer closeness could be imposed using aspect orientation. The DCR emphasizes processes composition whereas the SCR emphasizes the composition of logical sub goals.

3   Aspect Oriented Software Development

The core issue behind AOP, Aspect Oriented Programming, is to expand our ability to capture each concern as a modular software unit. The better we keep the representation of concerns separated throughout the software life cycle the more flexible and reusable our software becomes. AOSD, Aspect Oriented Software Development, refers to the refinement of software according to the aspect orientation process.

SoC, Separation of Concerns, is a core principle in software engineering with well established benefits. Yet the problem of applying this principle is at the heart of the ongoing software crisis. Ideally, we would first like to have heuristics criteria on how to separate concerns and second, software mechanisms to capture these concerns as modular units throughout the software refinement process. AOSD is a step in achieving these goals.

The object oriented approach focuses first on the data items that are being manipulated. The data items are characterized as active entities, objects which perform operations on and for them. System behavior is implemented through the interaction of objects. Such objects abstract together behavior and data into a single conceptual and physical entity. This methodology works well for software systems where most of the concerns could be modularized using the dominant decomposition of object orientation. Experience has showed that for more complex systems, especially concurrent and distributed systems, preservation of modularity is not possible for all concerns.

Crosscutting concerns - One special case where a program can fail to preserve modularity is when the implementation of a specific concern cuts across the modular representation of other concerns. A good example of such a concern is logging. A program code to implement logging may be scattered across different, otherwise modular components. Synchronization, scheduling, fault tolerance and security are concerns that are notoriously crosscutting. Under a dominant decomposition of core functionality, implementation of these concerns can not be modularized but rather must be scattered across various program components. To better understand crosscutting concerns we use the example taken from the introduction to the AspectJ language at aspectj.org. (Also appears in "Discussing Aspects of AOP" in the special section on aspect oriented programming in the Communication of the ACM October 2001 issue [CACM]).

Consider the UML for a simple figure editor, in which there are two concrete classes of figure elements, points and lines. These classes manifest good modularity, in that the source code in each class is closely related (cohesion) and each class has a clear and well-defined interface. But consider the concern that the screen manager should be notified whenever a figure element moves. This requires every method that moves a figure element to do the notification.

The Move Tracking box in the figure is drawn around every method that must implement this concern, just as the Point and Line boxes are drawn around every method that implements those concerns. Notice that the box for Move Tracking fits neither inside of nor around the other boxes in the picture - instead it cuts across the other boxes in the picture.

aspects <u>crosscut</u> classes

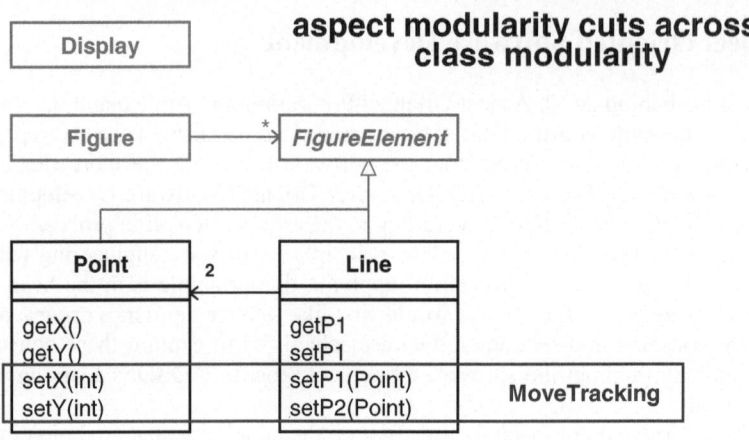

aspect modularity cuts across class modularity

Code tangling – An implementation of crosscutting concerns, using current programming languages, results in what is called code tangling. Ideally, software engineering principles instruct us to modularize our system software in such a way that (1) each module is cohesive in terms of the concerns it implements and (2) interface between modules is simple. Complying with these principles contributes to software that has many desirable outcomes; it is easier to produce, it can be naturally distributed among different programmers, it is easier to verify and to test, it is easier to maintain and reuse and it is more agreeable with the need for future adaptability. Now consider the impact of implementing a crosscutting concern, like logging, whose implementation is not modularized. Logging code is inserted into a variety of modules that each implements an activity that needs to be logged. (1) Cohesiveness is lost and (2) module interface is complicated. Violating the above software engineering principles bears the natural consequences; it is harder to produce, it can no longer be naturally distributed among different programmers, it is harder to verify and to test, it is harder to maintain and reuse and it is less agreeable with needs for future adaptability. Now continue to add on other crosscutting concerns such as security, scheduling, synchronization, fault tolerance etc. and eventually a software crisis is inevitable.

Progress in programming languages and design methods have always been driven by the discovery of a new design structure which can provide modularity where existing approaches fail. How does AOP fit within this paradigm? AOP introduces a new linguistic mechanism to modularize the implementation of crosscutting concerns. For example, logging code would no longer be scattered among different modules interfering with their cohesiveness but rather localized in one place called the logging aspect. This solves the problems above but now it introduces a new challenge; how to put Humpty Dumpy back together again? The logging code still needs to be inserted in the right places at runtime. Here is where different AOP approaches provide different solutions. To explain the different methods to achieve composition of core functionality and aspects we will first introduce AOP terminology.

Weaving - The process of composing core functionality modules with aspects is called weaving. Weaving can be done statically, at compile time or dynamically at runtime.

Join Points – are well-defined points, hooks in the execution flow of the program. A Join Point indicates where an aspect code can interface. A join point model provides the common frame of reference to enable the definition of the structure of aspects. Common join points are method calls. Each method call is one join point, the point in the flow when that method is called and when that method call returns.

Pointcut – A set of join points described by a pointcut designator. A pointcut expression filters out a subset of join points. This is an important feature of AOP because it provides a quantification mechanism. A programmer may designate all the join points in a program where, for example, a security code should be invoked. This eliminates the need to refer to each join point explicitly and hence reduces the likelihood that any aspect code would be incorrectly invoked. A Pointcut might also be specified in terms of properties of methods rather than their names.

Advice – advice declarations are used to define the aspect code that runs at join points. This, for example, might be the security code itself that runs at every join point that is filtered out by the security pointcut expression. For a method call-join points there are three kinds of advice: a before advice - advice code which runs at the moment before the method begins running. An after advice- advice code which runs at the moment control returns after the method is completed. An around advice - advice code which runs when the join point is reached and it can check for conditions that may control the selection of advice code. Advice is a method-like mechanism except that there is no explicit call to it at the point where it is executed.

Aspects – an aspect is a modular unit designed to implement crosscutting concerns. An aspect definition may contain some advice and the instructions on where, when and how to invoke them. It can be defined like a class and it can contain all intelligent information regarding aspect deployment. An aspect declaration may name the aspect, use pointcut designators to quantify all the possible places where the aspect advices should be invoked and defines the advice that should run at the designated join points.

4   Aspect Orientation for Expressing Communication-Closed Layers

We use the familiar aspect oriented semantics [KHHKPG] to provide a common frame of reference that makes it possible to define CCL join points. CCL pointcut designators, CCL advice and CC aspect.

CCL local join points: Join points are certain well-defined points in the execution flow of a program. *CCL joint points* are defined at the beginning and end of each program segment Sij.

CCL distributed join points: The collection of all CCL local joint points at the beginning and end of each layer Lj.

CCL distributed Pointcut designators: Pointcut designators identify particular join points by filtering out a subset of the entire join points in the program flow. *CCL pointcut designators* are filtering out the joint points at each layer boundaries. Pointcut layer-j filters out all CCL join points Sij for i=1..n. CCL distributed pointcut designator is a mapping into CCL distributed join points. CCL pointcut designators allows us to define distributed virtual global states

CCL distributed advice: Advice declarations are used to define additional code that runs at CCL distributed join points.

CCL distributed Aspect: An aspect is a modular unit of crosscutting implementation. *CCL aspects* are assertions over virtual global states that are verified at runtime. Since a global state is a collection of local states its realization is a crosscutting concern.

For sequential programs current research has already established the use of aspects in verifying and imposing pre-and post-conditions and use of "design by contract" [M93]. Aspects make it possible to implement pre- and post-conditions in a modular form. Also a consistent behavior across a large number of operations could be implemented in a much simpler way because of the localization of crosscutting concerns. The contribution of this paper is the extension of the class of properties that can be verified and imposed using aspect orientation approach. The "virtual global state" as defined in [EBN96] is the distributed programming equivalence to the simple state in sequential programming over which assertions are defined.

To best explain the use of aspect orientation in breaking the tyranny of process composition in distributed programs we use the well known two-phase commit protocol. A complete formal CCL development of this protocol is given in [J94].

5   An Aspect Oriented implementation of Two-Phase Commit Protocol

The two-phase commit protocol is an example used in distributed databases to guarantee consistency of the database. A coordinator process receives a request to initiate a voting, it should return, "COMMIT" if all processes participating in the voting process vote "yes" and "FAIL" otherwise. The voting process is a distributed program called COORDINATION (we are using uppercase for distributed programs and lowercase for sequential programs). CORDINATION can be farther decomposed into four layers that reflect the logical structure of the program into its sequential sub goals.

REQUEST – A vote-req message is passed between the coordinator and each of the participants.

VOTE – Each process, based on its local state deliberates a yes/no reply.

DECIDE – The coordinator collects the votes and computes the collective consensus.

EFFECTUATE – The final decision is passed back to all participants that act accordingly.

Note that this decomposition is orthogonal to the processes decomposition.

The process decomposition is assigning every voting participant process(i) a roll in each of these layers.

$\forall i \in (1..n)$ Process(i) :: BEGIN request-i; vote-i; decide-i; effectuate-i END

The two-phase commit protocol decomposition into layers is given in Figure B.

REQUEST layer is [request-1|| request-2|| ...|| request-n]
VOTE layer is [vote-1|| vote-2|| ... || vote-n]
DECIDE layer is [decide-1|| decide-2|| ...|| decide-n]
EFFECTUATE layer is [effectuate-1|| effectuate-2|| ...|| effectuate-n]

Fig. B. SCR of the two phases commit protocol

$\{ \forall i \in (1..n) \ (\text{vote-i} = \text{NONE} \land \text{decision-i} = \text{NONE}) \}$
REQUEST
$\{ \forall i \in (1..n) \ (\text{vote-i} = \text{NONE} \land \text{decision-i} = \text{NONE}) \land \text{message} = \text{VOTE\_REQ} \}$
VOTE
$\{ \forall i \in (1..n) \ ((\text{vote-i} = \text{YES}) \Leftrightarrow \text{stable-i} \land (\text{vote-i} = \text{NO}) \Leftrightarrow \text{not(stable-i)}) \}$
DECIDE
$\{ \forall i \in (1..n) \ ((\text{vote-i} = \text{YES}) \Leftrightarrow \text{stable-i} \land (\text{vote-i} = \text{NO}) \Leftrightarrow \text{not(stable-i)})$
$\land (\ (\text{decision} = \text{COMMIT} \land \forall i (\text{vote-i} = \text{YES}))$

$\lor (\text{decision} = \text{FAIL} \land \exists i \ (\text{vote-i} = \text{NO}) \ ) \}$
EFFECTUATE
$\{ \ (\ (\text{decision} = \text{COMMIT} \land \forall i (\text{vote-i} = \text{YES}) \lor (\text{decision} = \text{FAIL} \land \exists i)(\text{vote-i} = \text{NO}))$
$\land \forall i \ (\text{decision-i} = \text{decision}) \}$

Fig. C. Virtual global assertions for the two phases commit protocol

The communication-closed layers safety theorem applied to this example states that if each of the four layers is communication-closed - communications are allowed only between request segments, between vote segments, between decide segments and between effectuate segments but never between a request segment and a vote segment - then the two compositions are semantically equivalent. This means that we can use the SCR during software life cycle and the DCR for the actual execution.

The following is a formal specification in terms of pre-and post- conditions that reflects the "design by contract" of the two-phase commit protocol SCR. What we like to emphasize here are not so much the details of the specifications, but rather the nature of the global assertions that reflect distributed cooperation.

These assertions are called virtual global assertions because there might not be any real time at which any one of them holds. Each process reaches its own layer boundaries at a different time.

A Formal Distributed Aspect Oriented – CCL

We can use joint points and pointcut designators to define virtual global time and virtual global state.

Let [P1|| P2|| ...Pn] be a distributed program **P** composed of n processes P1 to Pn.

And let BEGIN LAYER-1;LAYER-2; ...; LAYER-k END be the complementary program composition into k layers.

$\forall i \in$ (1..n), process Pi is BEGIN layer-i1;layer-i2; ...; layer-ik END

$\forall i \in (1..n)$, $\forall j \in (1..k)$ let time-layer-ij, be the time stamp when process(i) reaches its joint point layer-ij. Clearly each process reaches this point at a different time.

$\forall j \in$ (1..k) A distributed virtual global time for the LAYER-j sub goal is defined as the vector: (time-layer-1j, time-layer-2j, ... , time-layer-nj) This is a collection of the different times each process has reached its own local segment referring to the collective j sub goal. The collection is identifiable by LAYER-j pointcut designator.

$\forall i \in$ (1..n),$\forall j \in$ (1..k) let state-layer-ij, be a snapshot of process(i) local state at time-layer-ij. *$\forall j \in (1..k)$ A virtual global state for LAYER-j sub goal* is defined as the vector: (state-layer-1j, state-layer-2j, ..., state-layer-kj). This is a collection of local states of each of the processes taken each at a different time by the advice code that runs on these joint points.

$\forall j \in (1..k)$ A distributed Virtual global assertion TEAMWORK-j: is defined as an assertion over the virtual global state of LAYER-j. TEAMWORK assertions depict the logical structure of a distributed program.

Regarding the two phases commit protocol, the executable program is the one obtained by applying the distributed composition. The program that we prove correctness is one obtained by applying the sequential composition. Since under layers closeness the two programs are semantically equivalent the proof holds for the distributed composition as well.

There are four classes of CCL local join points for every process(i): around request-i, around vote-i, around decide-i and around effectuate-i.

There are four CCL distributed pointcut designators: $\forall i \in$ (1..n) request-i, $\forall i \in$ (1..n) vote-i, $\forall i \in$ (1..n) decide-i, and $\forall i \in$ (1..n) effectuate-i.

Virtual global times that we would like to express are: The distributed virtual time when the system achieved its REQUEST sub goal, the distributed virtual time when the system achieved its VOTE sub goal, the distributed virtual time when the system achieved its DECIDE sub goal and the distributed virtual time when the system achieved its EFFECTUATE sub goal.

At each distributed virtual time we have the associated virtual global state and the distributed virtual global assertions as given in figure C.

The advice code that runs at joint points takes a snapshot of the process local state (or just an appropriate subset of all the variables that appear in a global invariant) and copies it into a pool of all such snapshots. When a pool is full; all processes have passed their appropriate layer boundaries, the verification of the global assertion can be evaluated. An intelligent decision could be made based on this evaluation.

The roles played by states, assertions and invariants in sequential programming design by contract discipline - can be played by distributed virtual global state, distributed virtual assertions and distributed virtual invariants in distributed programming design by contract discipline. The effectiveness of this approach increases with the degree of logical cooperation and the degree of communication between the processes. AOSD principles support the CCL distributed software development. CCL practical implementation relies on an effective handling of crosscutting concerns.

6   Potential Applications of AOSD and CCL Synergy

One of the principles in software development is the visibility rule: a significant concern should be syntactically visible. Aspect orientation strength is mainly due to elevating crosscutting concerns to be syntactically visible. CCL strength is mainly due to elevating the cooperative structure of distributed software to be syntactically visible. In the past we had mostly applications where processes, for the most part, did not interfere with each other. Resources management enforced sharing. Now, we see more applications where there is a higher degree of processes cooperation, the processes do not merely share resources, but actually have common goals. Such a distributed program common goals are significant concerns yet these concerns are not syntactically visible. Given a distributed program it is impossible to decompose it back to its logical structure in terms of common sub goals. These types of applications can benefit from an aspect orientation realization of CCL development.

The following are examples of the benefits:

Testing – Virtual global assertions and program's global invariants could be tested during run time or just at production phase. This is important because each program run might produce a different sequence of events.

Verification – Syntactic visibility of sub goals is a necessary condition for the ability to prove their correctness. Formal verification of a distributed program is a very complex task hence its decomposition into layers makes it more manageable.

Intelligent decision – Evaluation of a virtual global state allows for the insertion of intelligence into global system behavior. A system may choose to proceed differently based on certain relationship among local processes states.

Imposing layers closeness – Layer closeness is a necessary condition for the semantics equivalence between the DCR and the SCR. Proving layers closeness is a difficult task. Since aspect orientation enables localization of layers boundaries it is possible to verify that two communicating parties belong to the same layer. This can be done at run time or during software production phase only.

Layers Adaptability –Aspect orientation allows for aspects adaptability, aspects with different policy can replace one another. CCL allows for layers adaptability, layers complying with the same input/output global assertions can replace one another.

Performance – Overall system performance depends on all system processes proceeding at the same phase. Otherwise, a process that is legging behind might cause suspension in other processes. For example, if one process is executing at layer-1 and all the rest are already at layer-2 any cooperation with the legging process concerning the second sub goal need to be put on suspension A smart scheduling can prevent this by always preferring a process that is executing in a lower layer over one that is executing in a higher one.

Real-time applications – CCL provides the virtual global time vector. The vector component with the highest value can be considered as the "real-time" at which a sub goal has been achieved. Different applications might need to eliminate execution of a non-crucial layer in case of time constrains. When a real-time computation cannot be completed, at least we get an approximation by considering the latest assertion evaluated.

A partial list of more applications of CCL can be found in [EBN96, E84, EK90, EK91, EK93, PFZ93, GS96, JZ92, JPXZ94, PZ92, SdR87, JZ93, ZHLdR95, ZJ93]

7   Conclusion

Two of the aspect orientation characteristics defined in [FF00, CACM] are illuminating here, the quantification and the implicit invocation. Without these the implementation of a distributed program using CCL is difficult, rich in code tangling and hence not attractive from practical point. Aspect orientation approach separates the CCL concern from the rest of the program. It enables clean integration between distributed program processes composition and the distributed program layers composition. The tyranny of distributed programs processes composition is gently replaced with co-existence of both process composition and layers composition. Code implementing one composition does not tangle with code implementing the other composition.

The roles played by states, assertions and invariants in sequential programming design by contract discipline can be played by virtual global state, virtual assertions and virtual invariants in distributed programming design by contract discipline The effectiveness of this approach increases with the degree of logical cooperation and the degree of communication between the processes. CCL practical implementation relies on an effective handling of crosscutting concerns.

There is still ongoing research to use the WEAVR , a model driven engineering tool developed by Motorola to automate code generation from precise behavioral models to express layers and global assertions. [CE1, CE2, CVE1, CVE2, CVE3]

References

[AOSD] Aspect Oriented Software Development site: http://aosd.net
[CACM] Special section on Aspect Oriented Programming, Guest editors: Elrad, T., Filman, B., Bader, A., Communications of the ACM (October 2001)
[CE1] Cottenier, T., Elrad, T.: Dynamic and Distributed Service Composition with Aspect-Sensitive Services. In: Proceedings of the 1st International Conference on Web Information Systems and Technologies (WEBIST 2005), Miami, USA, pp. 56–63. INSTICC Press (May 2005)

[CE2] Cottenier, T., Elrad, T.: Layers of Collaboration Aspects for Pervasive Computing. In: Proceedings of the 5th Argentine Symposium in Software Engineering (ASSE 2004), Cordoba, Argentina (September 2004)

[CVE1] Cottenier, T., van den Berg, A., Elrad, T.: Motorola WEAVR: Aspect-Orientation and Model-Driven Engineering. Journal of Object Technology 6(7) (2007)

[CVE2] Cottenier, T., van den Berg, A., Elrad, T.: Joinpoint Inference from Behavioral Specification to Implementation. In: Ernst, E. (ed.) ECOOP 2007. LNCS, vol. 4609, pp. 476–500. Springer, Heidelberg (2007)

[CVE3] Cottenier, T., van den Berg, A., Elrad, T.: The Motorola WEAVR: Model Weaving in a Large Industrial Context. In: Proceedings of the Industry Track at the 6th International Conference on Aspect-Oriented Software Development (AOSD 2007) (March 2007)

[EBN96] Elrad, T., Baoling, S., Nastasic, N.: CHESSBOARD: A Synergy of Object Oriented Concurrent programming and Program Layering. In: Jaffar, J., Yap, R.H.C. (eds.) ASIAN 1996. LNCS, vol. 1179, pp. 223–233. Springer, Heidelberg (1996)

[EF82] Elrad, T., Francez, N.: Decomposition of Distributed Programs into Communication-Closed Layers. Science of Computer Programming 2 (1982)

[EFB01] Elrad, T., Filman, B., Bader, A.: Aspect Oriented programming. Guest editors introduction for the Communications of the ACM (2001)

[E84] Elrad, T.: A Practical Software Development for Dynamic Testing of Distributed Programs. In: IEEE Proceedings of the International Conference on Parallel Processing, Bellaire, MI (August 1984)

[EK90] Elrad, T., Kumar, K.: State Space Abstraction of Concurrent Systems: A Means to Computation Progressive Scheduling. In: Proceedings of the 19th International Conference on Parallel Processing (August 1990)

[EK91] Elrad, T., Kumar, K.: The Use of Communication-Closed Layers to Support Imprecise Scheduling for Distributed Real-Time Programs. In: Proceedings of the 10th Annual International Conference on Computers and Communications (March 1991)

[EK93] Elrad, T., Kumar, K.: Scheduling Cooperative Work: Viewing Distributed Systems as Both CSP and SCL. In: Proceedings of the 13th International Conference on Distributed Computing Systems, Pittsburgh (May 1993)

[FF00] Filman, R.E., Friedman, D.P.: Aspect-Oriented Programming is Quantification and Obliviousness. In: Workshop on Advanced Separation of Concerns, OOPSLA 2000, Minneapolis (October 2000)

[FPZ] Fokkinga, M., Poel, M., Zwiers, J.: Modular Completeness for Communication Closed Layers. In: Proceedings Formal Techniques in Real Time and Fault Tolerant Systems. Springer, Heidelberg (1993)

[GS86] Gerth, R.T., Shrira, L.: On Proving Communication Closeness of Distributed Layers. In: Proceedings of the 6th Conference on Foundations of Software Technology and Theoretical Computer Science, New Delhi, India (1986)

[J94] Jessen, W.P.M.: Layered Design of Parallel Systems. In: CIP (1994)

[JZ92] Janssen, W., Zwiers, J.: From Sequential Layers to Distributed Processes, Deriving a Distributed Minimum Weight Spanning Tree Algorithm. In: Proc. 11th ACM Symposium on Principles of Distributed Computing (1992)

[JZ92] Janssen, W., Zwiers, J.: Protocol Design by Layered Decomposition, A Compositional Approach. In: Vytopil, J. (ed.) FTRTFT 1992. LNCS, vol. 571, pp. 307–326. Springer, Heidelberg (1991)

[JZ93] Janssen, W., Zwiers, J.: Specifying and Proving Communication Closeness in Protocols. In: Proceedings 13th IFIP symp. On Protocol Specification, Testing and Verification (1993)

[JPXZ94] Janssen, W., Poel, M., Xu, Q., Zwiers, J.: Layering of Real-Time Distributed Processes. In: Langmaack, H., de Roever, W.-P., Vytopil, J. (eds.) FTRTFT 1994 and ProCoS 1994. LNCS, vol. 863, pp. 393–417. Springer, Heidelberg (1994)

[KHHKPG] Kiczales, et al.: Getting Started with AspectJ. The Communications of the ACM (October 2001)

[M93] Meyer, B.: Systematic Concurrent Object-Oriented Programming. Communication of the ACM (September 1993)

[MP91] Manna, Z., Pnueli, A.: The Temporal Logic of Reactive and Concurrent Systems. Springer, Heidelberg (1991)

[PZ87] Poel, M., Zwiers, J.: Layering Techniques for Development of Parallel Systems. In: Probst, D.K., von Bochmann, G. (eds.) CAV 1992. LNCS, vol. 663. Springer, Heidelberg (1993)

[SdR87] Stomp, F.A., de Roever, W.P.: A Correctness Proof of a Distributed Minimum-Weight Spanning Tree Algorithm. In: Proceedings of the 7th International Conference on Distributed Computer Systems, Berlin, W. Germany (1987)

[ZHLdR95] Zwiers, J., Hannemann, U., Lakhneche, Y., de Roever, W.-P.: Synthesizing different development paradigms: Combining top-down with bottom-up reasoning about distributed systems. In: Thiagarajan, P.S. (ed.) FSTTCS 1995. LNCS, vol. 1026, pp. 80–95. Springer, Heidelberg (1995)

[ZJ93] Zwiers, J., Janssen, W.: Partial order based design of concurrent systems. In: de Bakker, J.W., de Roever, W.-P., Rozenberg, G. (eds.) REX 1993. LNCS, vol. 803, pp. 622–684. Springer, Heidelberg (1994)

No *i*-Sums for Nissim (and Shalom)

Itamar Francez

University of Chicago

Abstract. Lappin and Francez (1994) present a theory of donkey anaphora which, they claim, captures both their existential and their universal readings, while maintaining a uniform representation of donkey pronouns. This paper shows that their analysis does not in fact assign the correct truth conditions to donkey sentences and so does not account correctly for the distribution of readings. An alternative analysis is proposed which retains LF's uniform representation for donkey pronouns, but abandons their analysis in terms of *i*-sums and the corollary derivation of universal readings by means of a maximality constraint. On the proposed analysis, donkey pronouns are uniformly represented with free variables over (Skolemized) choice functions, as in Chierchia's (1992) E-type analysis. The quantification associated with them is inferred quantification over choice functions. Universal readings arise as in Chierchia (1992) when all possible values for the free variable in the representation of a donkey pronoun are salient. For existential readings, a pragmatic account in the spirit of LF's analysis in terms of a cardinality constraint is maintained.

Keywords: Donkey anaphora, E-type pronouns, strong and weak readings, *i*-sums.

1   Introduction

Sentences with intrasentential donkey anaphora are known to give rise to both universal and existential readings. These are exemplified in (1-a) and (1-b) respectively.[1]

(1)   a.   Every farmer who has a credit card hides it in the barn.
            = Every farmer who has a credit card hides **every credit card she has** in the barn.
        b.   Every farmer who has a credit card uses it to buy a donkey.
            = Every farmer who has a credit card uses **some credit card she has** to buy a donkey.

Lappin and Francez (1994) (henceforth, somewhat ironically, LF) present an analysis of donkey anaphora which assigns a uniform representation to donkey

---

[1] Throughout, I restrict myself to donkey sentences with relative clauses. Extensions to other kinds of donkey sentences, e.g. ones with conditional form, are left for another occasion.

O. Grumberg et al. (Eds.): Francez Festschrift, LNCS 5533, pp. 97–106, 2009.

pronouns and on which the distribution of weak and strong readings is pragmatically determined. In this short paper, I show that this analysis does not
in fact account correctly for the distribution of readings, because it assigns the
wrong truth conditions for donkey sentences in which the subject NP involves a
downward increasing determiner, such as (2).

(2)   No farmer who has a credit card uses it to buy a donkey.
    a.   No farmer who has a credit card uses **any** of her credit cards to buy
          a donkey.
    b.   No farmer who has a credit card uses **all** of her credit cards to buy
          a donkey.

The only reading available for (2) is the negated existential reading paraphrased
in (2-a). However, LF's theory does not assign this reading to the sentence
(contrary to their claim), but rather assigns to it the unavailable (or strongly
dispreferred) reading in (3-b).

This is a serious flaw, and calls into question the correctness of the pragmatic approach, which LF advocate over alternatives such as Kanazawa's 1994
semantic theory in terms of monotonicity. I argue however that LF's pragmatic
approach can (and should) be maintained. In particular, I argue that donkey
pronouns should be represented as free variables over Skolemized choice functions as in Chierchia (1992), but that (unlike in Chierchia's theory), they should
be uniformly so represented. It follows that donkey pronouns never have quantificational force. My proposal is that the universal force in universal readings
comes from inferred quantification over choice functions. In this respect, donkey
pronouns are interpreted as free choice items, on a par with free choice indefinites headed by *any*. Existential readings are derived as in LF's account, from
the presence of a world knowledge based "cardinality constraint" associated with
the main predicate in the donkey sentence. In all cases, the quantification is over
choice functions rather than individuals. Thus, the theory proposed is a version of LF's theory in which the representation of donkey pronouns is retained,
but the analysis of quantificational force is modified to incorporate elements of
Chierchia's analysis.

The paper is structured as follows. Section 2 describes LF's i-sums analysis of
donkey sentences. Section 2.1 shows the problem with that analysis. Section 3
describes my alternative proposal. Section 4 summarizes.

2   Lappin and Francez's 1994 Analysis

LF's analysis of donkey anaphora is a version of the E-type analysis found in
Lappin (1989) and Neal (1990), i.e. one in which donkey pronouns are analyzed not as bound variables but as functions from individuals to collections of
individuals. The difference is that LF model collections of individuals as i-sums
(Link (1983), and this allows them to give a formal definition for the notion of a
maximal collection of individuals. An i-sum is a special kind of individual formed
by a summation operation $\vee_i$ on individuals in the domain of quantification E.

Intuitively, an *i*-sum is simply a grouping of one or more atomic individuals. LF introduce the following notations and definitions:

i For any one-place predicate P the denotation of which is a subset of E, LF use the notation $^*P$ for a one-place predicate the denotation of which is the set of *i*-sums in the closure of P under $\vee$.

ii For any binary relation R, $^*R$ is a binary relation between atomic individuals and *i-sums*, such that $\langle a, b \rangle \in {}^*R$ iff a is atomic and b is an *i*-sum of individuals that stand in relation R to a.

iii If $^*R$ is interpreted distributively, then $\langle a, b \rangle \in {}^*R$ iff a bears R to every atomic part of b. If $\neg^*R$ is interpreted distributively, then $\langle a, b \rangle \in \neg^*R$ iff a doesn't bear R to any atomic part of b.

Donkey pronouns are represented as functions from individuals to *i*-sums. For example, a sentence like (1-a), repeated in (3-a), is represented as in (3-b).

(3)   a.   Every farmer who has a credit card hides it in the barn.
       b.   (Farmer $\cap \{ x \; : \; \{ y \; :^* \; \text{has}(x, y) \} \cap \text{credit-card} \neq \emptyset \}) \subseteq \{ z \; :^*$ hides-in-barn$(z, f(z)) \}$

In (3-b), f is a function defined for any farmer who has at least one credit card.[2] To any such farmer, f assigns a member of the set of *i*-sums of credit cards she has (i.e. a member of $\{ y \; :^* \; \text{has}(x, y) \} \cap \text{credit-card} \}$).

To get universal readings, LF assume that the functions f in the representation of donkey pronouns are subject to a default maximality constraint. Since the range of f is always a set of *i*-sums , this set has a supremum, i.e. an element of which all other elements in the set are parts. The maximality constraint requires f to choose the supremum element of its range. The truth conditions in (3-b) then say that the set of credit-card owning farmers is a subset of the set of individuals such that they hide the *i*-sum consisting of all credit cards they own in the barn. This is the universal reading.

LF propose that existential readings of donkey sentences arise when the maximality constraint is suspended. In this case, the function f becomes a choice function. For any individual, it chooses a (possibly non-maximal) *i*-sum from the relevant set of *i*-sums , generating the existential reading.

In this way, a single representation generates both existential and universal readings of donkey sentences. What determines which reading is generated is whether the maximality constraint applies or not. This, in turn, is determined by world knowledge. In some cases, the predicate containing the donkey pronoun describes an eventuality which implies restrictions on the cardinality of the participant(s) denoted by the donkey pronoun. For example, the relevant predicate in (1-b), repeated in (4), is *uses it to buy a donkey*, where *it* is, for any farmer, some *i*-sum of credit cards she has.

(4)   Every farmer who has a credit card uses it to buy a donkey.

---

[2] There is a question as to how to ensure that f is resolved to be the intended function. I ignore this question here.

Events of purchasing something with a credit card are normally associated with a restriction on the number of credit cards used. Normally, only one card is used to make a purchase. To accommodate this cardinality restriction, the maximality constraint is suspended, so that non-maximal i-sums   might be chosen. The predicate in (1-b) on the other hand, *hides it in the barn*, is not associated with any such cardinality restriction – there is no world knowledge limit on how many credit cards one might hide in a barn. Therefore, nothing overrides the default maximality constraint in that case, and the universal reading is preferred.

2.1   The Problem

Where this analysis goes wrong is in cases where the subject determiner is monotone decreasing, such as (2) repeated in (5).

(5)   No farmer who has a credit card uses it to buy a donkey.

As observed by Chierchia (1992), such sentences only receive a negated existential interpretation. Thus, (5) says that there is no credit card owned by a farmer that was used in a donkey purchase. LF claim to capture this fact, and their analysis proceeds as follows. (5) receives the representation in (6), where I use $don for the relation *a used b to by a donkey*

(6)   (Farmer $\cap\{x : \{y :^* \text{has}(x,y)\}\cap\text{credit-card} \neq \emptyset\})\cap\{z :^* \text{\$don}(z, f(z))\} = \emptyset$

This representation is equivalent to the one in (7).

(7)   (Farmer $\cap\{x : \{y :^* \text{has}(x,y)\}\cap\text{credit-card} \neq \emptyset\}) \subseteq \{z : \neg^*\text{\$don}(z, f(z))\}$

Since there is no cardinality restriction associated with the negative predicate in (7) (there is no limit on how many credit cards you can avoid using when purchasing a donkey), the maximality constraint applies, yielding the universal reading.

But this is the wrong universal reading. The required reading is a *negated existential* reading. The reading in (7) is, in effect, a negated universal reading. Both the representation in (7) and the one in (6) are true iff there is no credit card owning farmer such that she used the maximal i-sum   of credit cards she has (i.e. the i-sum   consisting of all of her credit cards) to buy a donkey.

In fact, it would seem LF's analysis cannot generate the correct reading. Consider again the representation in (6). Like their representation of all other donkey sentences, this representation too gives rise to two readings. Strictly speaking, the representation says that no credit card owning farmer used an i-sum   of cards to buy a donkey. If the maximality constraint applies, the relevant i-sum   is the maximal one consisting of, for any farmer a, all of a's credit cards, yielding the undesired negated universal reading. If maximality is suspended, the relevant i-sum   is, for any farmer a, some i-sum   of credit cards owned by a. This is again the wrong reading, requiring only that for every farmer, there is some credit card they did not use to buy a donkey. Thus, LFs analysis produces wrong results regardless of whether maximality applies or not.

What went wrong? The problem is rooted in how universal force is derived. The intuition about donkey sentences with monotone decreasing quantifiers is that they give rise to negated existential readings. A negated existential is (classically) equivalent to a universal ($\neg\exists x[P(x)] \equiv \forall x[\neg P(x)]$). However, the *i*-sums account does not actually involve universal quantification. Instead, it imitates universal force by allowing donkey pronouns to refer to maximal pluralities. Maximal pluralities are scopeless, and this has the effect of "freezing" the scope of their universal force so that they are always interpreted as if a low scope universal quantifier was involved. But the the relevant negated existential readings are equivalent to readings in which a universal has scope *over* negation, not under it.

2.2 Distributivity Is Not Enough

It may seem that this problem is easily overcome within LF's system if all relations between individuals and *i*-sums are interpreted distributively. The definition of distributivity for any such relation $*R$ was given in (ii) above. For example, consider the scope of the quantifier in (7):

(8) $\{z : \neg *\$\mathsf{don}(z, f(z))\}$

Suppose that $f(z)$ is the maximal *i*-sum of credit cards owned by z. If the relation $*\$\mathsf{don}$ is read distributively, the representation in (7) becomes equivalent to (9), which does capture the desired reading. Following LF, I use Π for the atomic part-of relation.

(9)   (Farmer $\cap \{x : \{y :^* \mathsf{has}(x, y)\} \cap \mathrm{credit\text{-}card} \neq \emptyset\}) \subseteq \{z : \forall c(c\Pi f(z))(\neg\$\mathsf{don}(z, c))\}$

Furthermore, it is clear that distributivity is in any case required for other donkey sentences with universal readings, such as (1-a) above or (10), which does not commit the speaker to the existence of any single event in which a farmer sacrifices all of her donkey to Zeus.

(10)   Every farmer who has a donkey sacrifices it to Zeus.

It might even be that singular donkey pronouns *must* be interpreted distributively, given the impossibility of examples like (11):

(11)   #Every farmer who has a pig gathers it in the sty at night.

Unfortunately, allowing or requiring distributive readings of donkey pronouns does not alleviate the problem. This is because the representation in (7), from which (9) is derived, is not a representation of the relevant donkey sentence (2), but merely a logically equivalent representation. The actual representation is the one in (6), repeated in (12).

(12)   (Farmer $\cap \{x : \{y :^* \text{has}(x,y)\} \cap \text{credit-card} \neq \emptyset\}) \cap \{z :^* \$\text{don}(z, f(z))\} = \emptyset$

It is crucial to LF's account that one can move from the representation in (12) to the one in (7) as a matter of pragmatic inference. However, once donkey pronouns are interpreted distributively, the two representations are no longer logically equivalent. This can easily be seen by comparing (9) with a version of (12) in which distributivity is explicitly represented. The two representations are given in (13). (CC stands for credit card.)

(13)   a.   (Farmer $\cap \{x : \{y :^* \text{has}(x,y)\} \cap \text{CC} \neq \emptyset\}) \cap \{z : \forall c(c\Pi f(z))(\$\text{don}(z,c))\} = \emptyset$
        b.   (Farmer $\cap \{x : \{y :^* \text{has}(x,y)\} \cap \text{CC} \neq \emptyset\}) \subseteq \{z : \forall c(c\Pi f(z))(\neg\$\text{don}(z,c))\}$

Generally, $A \cap B = \emptyset$ is equivalent to $A \subseteq C$ iff $C = \overline{B}$. This is clearly not the case in (13). The complement of the set $\{z : \forall c(c\Pi f(z))(\$\text{don}(z,c))\}$ is $\{z : \exists c(c\Pi f(z))(\neg\$\text{don}(z,c))\}$, not $\{z : \forall c(c\Pi f(z))(\neg\$\text{don}(z,c))\}$. Of course, the representation in (13-a) is not the correct one for the relevant sentence. This representation says that no card-owning farmer uses each of her credit cards to buy a donkey. The sentence says no farmer uses *any* of her cards to do so. Thus, even assuming distributivity, LF assign the wrong truth conditions to donkey sentences with a monotone decreasing determiner.

3   An Alternative Analysis

I believe that the key to overcoming the problem with LF's analysis lies in viewing the monotone decreasing cases not as negated existentials but as universals. Recall that what is needed is for universal force to outscope negation. In this section, I suggest that this can be done leaving LFs representation almost unchanged, but that this requires abandoning the idea that donkey pronouns are interpreted as i-sums, as well as the related idea that universal force comes from a maximality constraint. Instead, I adopt Chierchia's (1992) E-type strategy, where the representation of donkey pronouns involves a free variable over Skolemized choice functions (i.e. functions mapping sets to individuals, not to i-sums). Universal readings involve universal quantification over choice functions. Donkey pronouns on a universal reading are therefore interpreted like free choice indefinites. Donkey pronouns with existential readings are interpreted as simple indefinites. Thus, the sentences in (14)-(16) are given the paraphrases in the (a) examples. In particular, (14) is paraphrased with *any* rather than the usual *every*.

(14)   Every farmer who has a credit card hides it in the barn.
        a.   Every farmer who has a credit card hides *any* credit card she has in the barn.

(15)   Every farmer who has a credit card used it to purchase a donkey.
        a.   Every farmer who has a credit card used some credit card she has to purchase a donkey.

(16)   No farmer who has a credit card uses it to purchase a donkey.

    a.   No farmer who has a donkey uses any of the credit cards she has to purchase a donkey.

These sentences then receive the representations in (17)-(19) respectively, where the choice function variable f remains free.

(17) $(\text{Farmer} \cap \{x : \{y : \text{has}(x, y)\} \cap \text{CC} \neq \emptyset\}) \subseteq \{z : \text{hide-in-barn}(z, f(z))\}$

(18) $(\text{Farmer} \cap \{x : \{y : \text{has}(x, y)\} \cap \text{CC} \neq \emptyset\}) \subseteq \{z : \$\text{don}(z, f(z))\}$

(19) $(\text{Farmer} \cap \{x : \{y : \text{has}(x, y)\} \cap \text{CC} \neq \emptyset\}) \cap \{z : \$\text{don}(z, f(z))\} = \emptyset$

Since these representations involve a free variable they cannot be associated with a determinate meaning without the variable being either bound or else provided with a value from context. In the simplest cases, the common ground reduces the range of the choice functions to a singleton set, thus reducing the number of functions that are possible values for f to just one. This would be the case with sentences such as (20), where the only value for f is the function mapping every individual to their nose.

(20)   Every farmer who has a nose uses it to smell the freshly plowed fields after the rain.

However, I follow much of the literature (including Heim, Chierchia, and LF) in assuming that in the cases under consideration, no such uniqueness presupposition is involved. Therefore, in these cases, what context needs to provide is either a particular choice function, or else a binder. Obviously, context does not provide a particular choice function when universal readings are involved, and it seems that usually this is not what happens when existential readings are involved either. For example, the speaker would usually not have a particular way of choosing a credit card in mind when she utters (15). The question is then how universal and existential readings arise, in the absence. An answer should, ideally, also explain why they have the distribution they do.

    My proposal is that the quantification associated with donkey pronouns is over choice functions, not over individuals. The force of quantification is by default universal (as in Chierchia's and LF's theories), with existential readings arising pragmatically through world knowledge based inferences. The proposal thus integrates Chierchia's intuition about universal readings with LF's intuition about existential ones.

3.1   Universal readings

Chierchia (1992:160) makes the point that the standard assumption in logic is that formulae with free variables are true iff they are true under all assignments of values to the variables. The same principle is, according to him, at play when donkey pronouns are interpreted as choice functions. In any donkey sentence, a set of choice functions is made salient, such as the set of functions mapping each farmer to one (or more) of her credit cards. Since all of these functions are equally salient, the sentence is interpreted as true iff it is true relative to all the functions in this set. This is how universal readings arise. Thus, the representation in (17) is true iff it is true for all choice functions from farmers to credit cards they have.

The monotone decreasing cases which formed the problem for LF's account now fall out as a simple case of a universal reading. The representation in (19) is true iff it is true relative to all values for f. These are the correct truth conditions. On this interpretation, the representation becomes equivalent to (21), where F is the set of functions from farmers to credit cards they have. (21) captures the fact that the relevant reading is a negated existential.

(21) $(\text{Farmer} \cap \{x : \{y : \text{has}(x,y)\} \cap CC \neq \emptyset\}) \cap \{z : \exists f \in F[\$\text{don}(z, f(z))]\} = \emptyset$

An attractive feature of this analysis of universal readings is that it automatically captures the fact that donkey sentences with universal readings are interpreted distributively. For example, consider the traditional (22).

(22)   Every farmer who owns a donkey beats it.

A speaker uttering this sentence (say, a medieval monk) is not commited to the existence of any event in which a farmer beats all of her donkeys at the same time. The collectivity of donkeys need not be a participant in the eventuality described. All that the sentence commits the speaker to is the proposition that any relevant donkey is (habitually) beaten by its owner.

3.2   Existential readings

Chierchia's analysis does not extend to existential readings. To generate those, he posits a second strategy of interpretation for donkey sentences involving dynamic binding as in Groenendijk and Stokhoff's (1991) DPL. As LF point out, this is an unintuitive result. How then do existential readings arise? LF explain such readings as arising from the presence of a world knowledge based inference. According to them, the maximality constraint that gives rise to universal readings is cancelled when the main predicate in the donkey sentence is associated with a *cardinality constraint*. A cardinality constraint is a world knowledge based inference about the cardinality of the proto-patient participant of the described eventuality. For example, consider (23).

(23)   Every person who had a dime put it in the meter.

An event of putting dimes in a meter normally involves a limit on how many dimes can be inserted at a time. In other words, in any such event, some dimes will be inserted, others will not, and this information is, usually, in the common ground. In the current analysis, there is no maximality constraint for the cardinality constraint to cancel. However, the same reasoning can be applied to explain why the equal salience of all choice functions as possible values for the free variable in the representation of a donkey pronoun is cancelled. For example, in the case of (23), since the common ground will normally entail that some dimes are used and others are not, it follows that for very person, some choice functions will choose dimes they did *not* put in the meter. Therefore, the default universal quantification over choice functions is overriden and (23) has a preferred existential reading.

Similarly for (15). This sentence describes events of using a credit card to purchase a donkey. While such events are perhaps not very common, we can deduce from other events of credit card purchases that they protoypically involve the buyer choosing a single credit card. Thus, the common ground normally includes the information that, if more than one choice of credit card exists for a farmer, then some of the choices will not be made. Therefore, the context entails that it is not the case that (18) (the representation of (15)) is true relative to all relevant choice functions. The only plausible assertion is then that it is true relative to *some* choice of a credit (though the context need not specify which one).

This contrasts with sentence (14). In order for this sentence to be true, each relevant farmer must participate in an event or a series of events in which she hides credit cards in the barn. But world knowledge does not tell us that there is a limit on how many things one can hide: you can hide whatever you have. Furthermore, if there is reason to hide one credit card, then the same reason is probably good reason to hide all of them. So, when this sentence is uttered, the common ground normally does not entail that there are credit cards that are not hidden, and the sentence is evaluated relative to all relevant functions. Thus, LF's cardinality constraint can be maintained as the explanation for existential readings even if their analysis of universal readings is not.

4   Summary and conclusions

This paper has demonstrated that the theory of donkey anaphora presented in Lappin and Francez (1994) does not assign correct truth conditions to donkey sentences with monotone decreasing quantifiers in the subject. However, I argued that this does not jeopardize the core intuition of their analysis. In particular, donkey pronouns can still be uniformly represented with free variables over functions, and the distribution of readings can still be viewed as arising from pragmatically motivated overriding of defaults. What I suggested should be abandoned is LF's view that donkey pronouns involve functions from individuals to *i*-sums, as well as their derivation of universal readings by means of a maximality constraint. Instead, I suggested following Chierchia in viewing such readings are arising from inferred quantification over Skolemized choice functions. In the default case, all possible values for the free variable are under consideration, and universal readings arise. Thus, the universal force associated with donkey pronouns is similar to the force of free choice indefinites. Existential readings arise when, due to world knowledge about the nature of the eventuality involved, the common ground entails that a universal reading is false. In this case, the interpretation is naturally weakened to an assertion about the existence of at least one value for the free variable.

Much more needs to be said about what kinds of predicates give rise to cardinality constraints and about the distribution of existential and universal readings. In particular, since the publication of LF's paper, several authors have argued the relevance of various other semantic and pragmatic properties of predicates and/or their arguments for determining readings. These include sentence and lexical aspect (Merchant and Giannakidou (1998; Geurts 2002), Yoon's (1996) distinction between total and partial predicates, and Geurt's (2002) intriguing distinction between protoypical and marginal individuals. I find that all of these proposals have convincing aspects. Furthermore, I did not discuss here the semantic approach in Kanazawa (1994) (though LF provide some discussion of his approach, including some objections). Discussion of these works goes beyond the scope of this paper, whose modest goal is only to show how an analysis more or less in line with LF's main ideas can be provided which does not assign wrong truth conditions to donkey sentences with downward increasing determiners.

Acknowledgments. I am grateful to Ashwini Deo, Nick Kroll, Zoltán Gendler Szabó, and an anonymous reviewer for corrections, comments and discussion. This paper is dedicated with much love to my father and mother, one of whom insisted that I sharpen my pencil, and both of whom made me see that it is not the pencil, but the joy of using it, that really matters.

References

Chierchia, G.: Anaphora and dynamic binding. Linguistics and Philosophy 15(2), 111–183 (1992)

Geurts, B.: Donkey Business. Linguistics and Philosophy 25(2), 129–156 (2002)

Groenendijk, J., Stokhof, M.: Dynamic predicate logic. Linguistics and Philosophy 14(1), 39–100 (1991)

Kanazawa, M.: Dynamic generalized quantifiers and monotonicity. In: Kanazawa, M., Piñón, C. (eds.) Dynamics, Polarity, and Quantification. CSLI Publications, Stanford (1994)

Lappin, S.: Donkey pronouns unbound. Theoretical Linguistics 15, 263–286 (1989)

Lappin, S., Francez, N.: E-type pronouns, i-sums, and donkey anaphora. Linguistics and Philosophy 17, 391–428 (1994)

Link, G.: The logical analysis of plural and mass nouns: A lattice-theoretic approach. In: Bäuerle, R., Schwarze, C., von Stechow, A. (eds.) Meaning, Use, and Interpretation of Language, pp. 303–323. de Gruyter, Berlin (1983)

Merchant, J., Giannakidou, A.: An asymmetry in asymmetric donkey anaphora. Studies in Greek Linguistics 18, 141–154 (1998)

Neal, S.: Descriptions. MIT Press, Cambridge, Mass (1990)

Yoon, Y.: Total and partial predicates and the weak and strong interpretations. Natural language semantics 4, 217–236 (1996)

The Power of Non-deterministic Reassignment in Infinite-Alphabet Pushdown Automata

Yulia Dubov[1] and Michael Kaminski[2,*]

[1] Intel Israel
MTM – Advanced Technology Center,
P.O.B. 1659, Haifa 31015, Israel
yulia.dubov@intel.com
[2] Department of Computer Science,
Technion – Israel Institute of Technology,
Haifa 32000, Israel
kaminski@cs.technion.ac.il

Abstract. In this paper we compare two models of pushdown automata over infinite alphabets, one with non-deterministic reassignment and the other with the deterministic one, and show that the former model is stronger than the latter.

1 Introduction

The study of extension of classical models of automata to languages over infinite alphabets that has been started in the earlier 1990s counts numerous works. Being purely theoretical in the beginning, it soon found a number of applications. Naturally, every time a sequence of words is considered – be it messages passing through the network, URLs clicked by the internet surfer, or XML tags - words are treated as atomic symbols, i.e., elements of an appropriate alphabet, and, in absence of a bound on the length of the words, the alphabet becomes infinite.

A number of models of computation over infinite alphabets is known from the literature, e.g., see [1,3,4,5,6,7,8,9,10,11]. A particular example of a model of computation over infinite alphabets tightly related to this paper is context-free grammars over infinite alphabets introduced in [3]. These grammars look very similar to Document Type Definitions (DTDs) used for defining XML documents, see [2], even though they have been invented earlier than the latter.

Actually, this paper deals with two models of pushdown-automata over infinite alphabets, one of which is the semantics counterpart of context-free grammars over infinite alphabets. Both models are analogous to the classical model in that their sets of states are finite. However, the ability to deal with infinite alphabets is achieved by equipping the machine with a finite number of registers, in which the automaton can store any letter from the infinite input alphabet. While reading the topmost stack and the input symbols, the automaton compares them with

---

* A part of this paper was written during the author's sabbatical leave to the Division of Mathematical Sciences of the Nanyang Technological University.

O. Grumberg et al. (Eds.): Francez Festschrift, LNCS 5533, pp. 107–127, 2009.

the content of its registers and proceeds according to the results. It also has a mechanism, called reassignment, for updating the register content.

There are (at least) two possibilities of updating the content of the automaton registers. One possibility is to replace the content of one of the registers with the topmost stack symbol or the currently scanned new input symbol, similarly to the approach in [7], and the other is to replace the content of one of the registers with any new symbol from the infinite input alphabet, independently of the current input symbol or the topmost stack symbol, as was done in [3,8,9]. That is, in the latter case, the automaton does not necessarily have to arrive at the symbol in order to store it in its registers. Such ability will be referred to as a non-deterministic reassignment. Infinite alphabet pushdown automata introduced in [3] are a particular example of the latter (non-deterministic reassignment) type. Thus, it is very natural to ask whether the ability of non-deterministic reassignment is necessary for the equivalence of the automata to context-free grammars over infinite alphabets.[1] That is, whether restricting the computation model by requiring the reassignment to be deterministic would result in the computation model of the same power.

In this paper we show that the answer to this question is negative by presenting a language accepted by infinite alphabet pushdown automata with non-deterministic reassignment (NR-IAPDA), but not accepted by their deterministic counterpart – infinite alphabet pushdown automata with deterministic reassignment (DR-IAPDA).[2]

Our example is based on the following idea. We observe that NR-IAPDA can perform certain tasks, using both its stack and "guessing" (i.e., non-deterministic reassignment) abilities, which are beyond the DR-IAPDA computation power. Namely, we consider a computation involving an unknown future input symbol that appears only once in the input. NR-IAPDA can perform the computation by "guessing" the future input input and using it for comparison while reading the preceding portion of the input word. However, DR-IAPDA seemingly would need to use their pushdown store to perform the same task by storing there the input word, until it reaches the "unknown" input symbols (which it is unable to guess) and then comparing the stack symbols with that symbol by popping them out. Now, if we add a different task that requires the use of the pushdown stack (e.g., comparison of two subwords of the input) and has to be performed simultaneously, both of the tasks together cannot be performed by DR-IAPDA, but can be performed by NR-IAPDA.

This paper is organized as follows. The next section contains the basic notation used throughout the paper. In Sections 3 and 4 we recall the defini-

---

[1] Models of computation are said to be equivalent, if they accept/generate the same class of languages.

[2] One might have the impression that it is quite expected, because, over finite alphabets, nondeterministic pushdown automata are more expressive than deterministic ones. Nevertheless, there are very similar models of finite and pushdown automata over infinite alphabets for which nondeterministic reassignment does not increase the computation power, see [6].

tion (from [3]) of infinite alphabet context-free languages and infinite alphabet pushdown automata (with non-deterministic reassignment), respectively. Infinite-alphabetpushdown automata with deterministic reassignment are defined in Section 5. Finally, in Section 6 we prove that infinite-alphabet pushdown automata with deterministic reassignment are weaker than those with non-deterministic one.

2   Notation

In what follows Σ is a fixed infinite alphabet. An assignment is a word $u_1 u_2 \cdots u_r$ over Σ such that $u_i \neq u_j$ for $i \neq j$. That is, each symbol from Σ occurs in an assignment at most ones. We denote the set of all assignments of length r by $\Sigma^{r\neq}$. Assignments correspond to the content of all registers of an automaton or a grammar.

For a word $\boldsymbol{w} = w_1 w_2 \cdots w_n \in \Sigma^*$, we define the content of $\boldsymbol{w}$, denoted $[\boldsymbol{w}]$, by $[\boldsymbol{w}] = \{w_i : i = 1, 2, \ldots, n\}$. That is, $[\boldsymbol{w}]$ consists of all the symbols of Σ which occur in the word $\boldsymbol{w}$.

Throughout this paper we use the following convention.

- Words are always denoted by boldface letters, possibly indexed or primed.
- Boldface low-case Greek letters denote words over Σ.
- Symbols which occur in a word denoted by a boldface letter are always denoted by the same non-boldface letter with some subscript. That is, symbols which occur in $\boldsymbol{\sigma}$ are denoted by σ_i, symbols which occur in $\boldsymbol{w}$ are denoted by w_i, symbols which occur in $\boldsymbol{X}$ are denoted by X_i, etc.

3   Infinite-Alphabet Context-Free Grammars

In this section we recall the definition of infinite-alphabet context-free grammars which motivate infinite-alphabet pushdown automata (with non-deterministic reassignment) defined in the next section.

Definition 1. ([3, Definition 1]) *An infinite-alphabet context-free grammar is a system $G = \langle V, \boldsymbol{u}, P, S \rangle$ whose components are defined as follows.*

- *V is a finite set of variables disjoint from Σ.*
- *$\boldsymbol{u} = u_1 u_2 \cdots u_r \in \Sigma^{r\neq}$ is the initial assignment.*
- *$P \subseteq (V \times \{1, 2, \ldots, r\}) \times (V \cup \{1, 2, \ldots, r\})^*$ is a finite set of productions. For $A \in V$, $k = 1, 2, \ldots, r$, and $\boldsymbol{a} \in (V \cup \{1, 2, \ldots, r\})^*$, we write the triple $(A, k, \boldsymbol{a})$ as $(A, k) \to \boldsymbol{a}$.*
- *$S \in V$ is the start symbol.*

For $A \in V$, $\boldsymbol{w} = w_1 w_2 \cdots w_r \in \Sigma^{r\neq}$, and $\boldsymbol{X} = X_1 X_2 \cdots X_n \in (\Sigma \cup (V \times \Sigma^{r\neq}))^*$, we write $(A, \boldsymbol{w}) \Rightarrow \boldsymbol{X}$, if there exist a production $(A, k) \to \boldsymbol{a} \in P$, $\boldsymbol{a} = a_1 a_2 \cdots a_n \in (V \cup \{1, 2, \ldots, r\})^*$, and $\sigma \notin [\boldsymbol{w}] \setminus \{w_i\}$ such that the condition below is satisfied.

Let $\boldsymbol{w}' \in \Sigma^{r\neq}$ be obtained from $\boldsymbol{w}$ by replacing w_k with σ. Then all for $i = 1, 2, \ldots, n$ the following holds.

- If $a_i = m \in \{1, 2, \ldots, r\}$, then $X_i = w'_m$.
- If $a_i = A' \in V$, then $X_i = (A', w')$.

For two words $X, Y \in (\Sigma \cup (V \times \Sigma^{r \neq}))^*$, we write $X \Rightarrow Y$, if there exist words $X_1, X_2, X_3 \in (\Sigma \cup (V \times \Sigma^{r \neq}))^*$ and $(A, w) \in V \times \Sigma^{r \neq}$ such that $X = X_1(A, w)X_2$, $Y = X_1X_3X_2$ and $(A, w) \Rightarrow X_3$. As usual, the reflexive and transitive closure of $\Rightarrow$ is denoted by $\Rightarrow^*$ and the language $L(G)$ generated by G is defined by

$$L(G) = \{\sigma \in \Sigma^* : (S, u) \Rightarrow^* \sigma\}.$$

Example 1. Let $G = \langle V, u, P, S \rangle$, where

- $V = \{S, A\}$,
- $u = u_1 u_2 \$$, and
- P consists of
  - $(S, 1) \to S$,
  - $(S, 2) \to A2$, and
  - $(A, 1) \to 3A1 | \epsilon$.

It is not hard to verify that $L(G) = L$, where

$$L = \{\$^{|w|} w \delta : \$ \notin [w] \text{ and } \delta \notin [w] \cup \{\$\}\}.^3 \tag{1}$$

For example, the word $\$^3 \sigma_1 \sigma_2 \sigma_3 \delta$, where $\delta \notin \{\sigma_1, \sigma_2, \sigma_3\}$,[4] is derived as follows.

$$(S, u_1 u_2 \$) \Rightarrow (S, \tau u_2 \$) \Rightarrow (A, \tau \delta \$)\delta \Rightarrow \$(A, \sigma_1 \delta \$)\sigma_1 \delta$$

$$\Rightarrow \$\$(A, \sigma_2 \delta \$)\sigma_1 \sigma_2 \delta \Rightarrow \$\$\$(A, \sigma_3 \delta \$)\sigma_1 \sigma_2 \sigma_3 \delta \Rightarrow \$\$\$\sigma_1 \sigma_2 \sigma_3 \delta,$$

where $\tau \notin \{\delta, \sigma_1, \sigma_2, \sigma_3\}$. That is, by the production $(S, 1) \to S$, the first register is reset to τ, which allows G to store δ in the second register at the following derivation step.

We conclude this section with the following closure property of context-free languages over infinite alphabets that was somehow missed in [3].

Proposition 1. *The class of context-free languages over infinite alphabets is closed under reversing.*

Proof. Let $G = \langle V, u, P, S \rangle$ be an infinite-alphabet context-free grammar. A straightforward induction on the derivation length shows that the reversal of the language $L(G)$ is generated by the infinite-alphabet context-free grammar $G^R = \langle V, u, P^R, S \rangle$, where

$$P^R = \{(A, k) \to a^R : (A, k) \to a \in P\}.$$

---

[3] As we shall see in Section 6, this language is not accepted by infinite alphabet pushdown automata with deterministic reassignment defined in Section 5.

[4] Of course, $\sigma_1, \sigma_2, \sigma_3$ are not necessarily pairwise distinct.

4 Infinite-Alphabet Pushdown Automata with Non-deterministic Reassignment

Here we recall the definition of infinite-alphabet pushdown automata with non-deterministic reassignment which constitute the semantics counterpart of the infinite-alphabet context-free grammars introduced in the previous section.

Definition 2. ([3, Definition 2]) *An infinite-alphabet pushdown automaton with non-deterministic reassignment (NR-IAPDA)[5] is a system $\mathcal{A} = \langle Q, s_0, \boldsymbol{u}, \rho, \mu \rangle$, whose components are defined as follows.*

- Q *is a finite set of states.*
- $s_0 \in Q$ *is the initial state.*
- $\boldsymbol{u} = u_1 u_2 \cdots u_r \in \Sigma^{r\neq}$, *is the initial assignment to the r registers of $\mathcal{A}$.*
- $\rho : Q \rightarrow \{1, 2, \ldots, r\}$ *is a partial function from Q to $\{1, 2, \ldots, r\}$ called the reassignment. Intuitively, if $\mathcal{A}$ is in state q and $\rho(q)$ is defined, then $\mathcal{A}$ may non-deterministically replace the content of the $\rho(q)$th register with a new symbol of Σ not occurring in any other register.*
- μ *is a transition function from $Q \times (\{1, 2, \ldots, r\} \cup \{\epsilon\}) \times \{1, 2, \ldots, r\}$ to finite subsets of $Q \times \{1, 2, \ldots, r\}^*$. Intuitively, if $(p, j_1 j_2 \cdots j_\ell) \in \mu(q, \epsilon, j)$, then $\mathcal{A}$, whenever it is in state q, with content of the jth register at the top of the stack, may replace the top symbol on the stack with the content of j_1th,j_2th,...j_ℓth registers, enter state p (without reading the input symbol), and pass to the next input symbol (possibly ϵ). Similarly, if $(p, j_1 j_2 \cdots j_\ell) \in \mu(q, k, j)$, then $\mathcal{A}$, whenever it is in state q, with content of the jth register at the top of the stack, and the input symbol being equal to the content of kth register, may replace the top symbol of the stack with the content of the j_1th,j_2th,...,j_ℓth registers (in this order, read top-down), enter the state p, and pass to the next input symbol (possibly ϵ).*

An instantaneous description of $\mathcal{A}$ (on a given input word) is a member of $Q \times \Sigma^{r\neq} \times \Sigma^* \times \Sigma^*$. The first component of an instantaneous description is the (current) state of the automaton, the second is the assignment consisting of the content of its registers (in the increasing order of their indexes), the third component is the portion of the input yet to be read, and the last component is the content of the pushdown store read top down.

Next we define the relation $\vdash$ (yielding in one step) between two instantaneous descriptions $(p, v_1 v_2 \cdots v_r, \sigma\boldsymbol{\sigma}, \tau\boldsymbol{\tau})$ and $(q, w_1 w_2 \cdots w_r, \boldsymbol{\sigma}, \boldsymbol{\alpha\tau})$, $\sigma \in \Sigma \cup \{\epsilon\}$ and $\tau \in \Sigma$. We write

$$(p, v_1 v_2 \cdots v_r, \sigma\boldsymbol{\sigma}, \tau\boldsymbol{\tau}) \vdash (q, w_1 w_2 \cdots w_r, \boldsymbol{\sigma}, \boldsymbol{\alpha\tau})$$

if and only if the following holds.

- If $\rho(p)$ is not defined, then $w_k = v_k, k = 1, 2, \ldots, r$. Otherwise, $w_k = v_k$ for $k \neq \rho(p)$, and $w_{\rho(p)} \in \Sigma \setminus \{v_1, \ldots, v_{\rho(p)-1}, v_{\rho(p)+1}, \ldots, v_r\}$.

---

[5] In [3] the model is called just infinite-alphabet pushdown automata.

- If $\sigma = \epsilon$, then for some $j = 1, \ldots, r$, $\tau = w_j$ and there is a transition $(q, j_1 j_2 \cdots j_\ell) \in \mu(p, \epsilon, j)$ such that $\boldsymbol{\alpha} = w_{j_1} w_{j_2} \cdots w_{j_\ell}$.
- If $\sigma \neq \epsilon$, then for some $k, j = 1, \ldots, r$, $\sigma = w_k$, $\tau = w_j$, and there is a transition $(q, j_1 j_2 \cdots j_\ell) \in \mu(p, k, j)$ such that $\boldsymbol{\alpha} = w_{j_1} w_{j_2} \cdots w_{j_n}$.

We denote the reflexive and transitive closure of $\vdash$ by $\vdash^*$ and say that $\mathcal{A}$ accepts a word $\boldsymbol{\sigma} \in \Sigma^*$, if $(s_0, \boldsymbol{u}, \boldsymbol{\sigma}, u_r) \vdash^* (q, \boldsymbol{v}, \epsilon, \epsilon)$, for some $q \in Q$ and some $\boldsymbol{v} \in \Sigma^{r\neq}$.[6] Finally, the language $L(\mathcal{A})$ accepted by $\mathcal{A}$ is defined by

$$L(\mathcal{A}) =$$

$$\{\boldsymbol{\sigma} \in \Sigma^* : \text{for some } q \in Q \text{ and some } \boldsymbol{v} \in \Sigma^{r\neq}, (s_0, \boldsymbol{u}, \boldsymbol{\sigma}, u_r) \vdash^* (q, \boldsymbol{v}, \epsilon, \epsilon)\}.$$

Example 2. Consider the NR-IAPDA $\mathcal{A} = \langle Q, s_0, \boldsymbol{u}, \rho, \mu \rangle$, where

- $Q = \{s_0, s, p, q\}$;
- $\boldsymbol{u} = u_1 u_2 \$$;
- $\rho(s_0) = \rho(q) = 1$ and $\rho(s) = 2$; and
- μ is defined as follows.
  - (i) $\mu(s_0, \epsilon, 3) = \{(s, 3)\}$
  - (ii) $\mu(s, \epsilon, 3) = \{(p, 2)\}$
  - (iii) $\mu(p, 3, 2) = \{(p, 32)\}$
  - (iv) $\mu(p, 3, 3) = \{(p, 33)\}$
  - (v) $\mu(p, \epsilon, 2) = \{(q, 2)\}$
  - (vi) $\mu(p, \epsilon, 3) = \{(q, 3)\}$
  - (vii) $\mu(q, 1, 3) = \{(q, \epsilon)\}$
  - (viii) $\mu(q, 2, 2) = \{(q, \epsilon)\}$

It is not hard to verify that $L(G) = L$, where L is the language from Example 1 defined by (1). The intuition lying behind the accepting run of $\mathcal{A}$ on an input of the form $\$^{|w|} w \delta$, $\delta \notin [w] \cup \{\$\}$, is as follows.

- By the transition of type (i), $\mathcal{A}$ resets the first register to some $\tau \neq \delta$, which allows it to store δ in the second register at the following computation step, cf. Example 1.
- Using the transitions of type (ii), $\mathcal{A}$ "guesses" δ, stores it in the second register, and replaces $\$$ with δ at the bottom of the stack.
- Then, using the transitions of types (iii) and (iv), $\mathcal{A}$ pushes $\$$ into the stack.
- After reading the last $\$$, by the transitions of types (v) or (vi), $\mathcal{A}$ enters the popping state q.
- Then, for each input symbol from $\boldsymbol{w}$ is compared with δ and, if they differ, $\mathcal{A}$ pops $\$$ out of the stack using the transition of type (vii).
- Finally, the transition of type (viii) applies to verify that the last input symbol is indeed δ and that the number of occurrences of $\$$ in the input is indeed $|\boldsymbol{w}|$.

---

[6] Recall that the initial assignment $\boldsymbol{u} = u_1 u_2 \cdots u_r$ is of length r.

For example, $\mathcal{A}$ accepts the word $\$^3 \sigma_1 \sigma_2 \sigma_3 \delta$, where $\delta \notin \{\sigma_1, \sigma_2, \sigma_3\}$, by the following sequence of computation steps.

$$(s_0, u_1 u_2 \$, \$\$\$\sigma_1 \sigma_2 \sigma_3 \delta, \$) \vdash (s, \tau u_2 \$, \$\$\$\sigma_1 \sigma_2 \sigma_3 \delta, \$) \vdash (p, \tau \delta \$, \$\$\$\sigma_1 \sigma_2 \sigma_3 \delta, \delta)$$

$$\vdash (p, \tau \delta \$, \$\$\sigma_1 \sigma_2 \delta_3 \delta, \$\delta) \vdash (p, \tau \delta \$, \$\sigma_1 \sigma_2 \sigma_3 \delta, \$\$\delta) \vdash (p, \tau \delta \$, \sigma_1 \sigma_2 \sigma_3 \delta, \$\$\$\delta)$$

$$\vdash (q, \tau \delta \$, \sigma_1 \sigma_2 \sigma_3 \delta, \$\$\$\delta) \vdash (q, \sigma_1 \delta \$, \sigma_2 \sigma_3 \delta, \$\$\delta) \vdash (q, \sigma_2 \delta \$, \sigma_3 \delta, \$\delta)$$

$$\vdash (q, \sigma_3 \delta \$, \delta, \delta) \vdash (q, \sigma_3 \delta \$, \epsilon, \epsilon),$$

where $\tau \notin \{\delta, \sigma_1, \sigma_2, \sigma_3\}$, cf. Example 1.

Theorem 1. ([3], see also [6, Appendix D].) *A language is generated by an infinite-alphabet context-free grammar if and only if it is accepted by an NR-IAPDA.*

5   Infinite-Alphabet Pushdown Automata with Deterministic Reassignment

In this section we define infinite-alphabet pushdown automata with deterministic reassignment (DR-IAPDA). Instead of the ability to "guess" symbols for reassignment, this model of pushdown automata over infinite alphabets is limited to altering its registers only by replacing their content with the current input symbol (similarly to the finite-memory automata introduced in [7]) or with the symbol at the top of the stack.

Definition 3. *An infinite-alphabet pushdown automaton with deterministic reassignment (DR-IAPDA) is a system $\mathcal{A} = \langle Q, s_0, \boldsymbol{u}, \pi, \rho, \mu \rangle$ whose components are defined as follows.*

- *Q is a finite set of states.*
- *$s_0 \in Q$ is the initial state.*
- *$\boldsymbol{u} = u_1 u_2 \cdots u_r \in \Sigma^r \overline{\mp}$ is the initial assignment to the r registers of $\mathcal{A}$.*
- *$\pi, \rho : Q \to \{1, 2, \ldots, r\}$ are partial functions from Q to $\{1, 2, \ldots, r\}$ called stack-based reassignment and input-based reassignment, respectively.*
- *μ is a transition function from $Q \times (\{1, 2, \ldots, r\} \cup \{\epsilon\}) \times \{1, 2, \ldots, r\}$ to finite subsets of $Q \times \{1, 2, \ldots, r\}^*$.*

Intuitively, a computation step of the automaton $\mathcal{A}$ from state q is composed of the following sequence of "primitive" actions.

- If the top stack symbol occurs in no register, the stack-based reassignment is performed. That is, the content of the register $\pi(q)$ is replaced with this symbol. Otherwise the finite memory of $\mathcal{A}$ remains intact.
- After that, $\mathcal{A}$ may
  - either perform an ϵ-transition, if defined, or
  - proceed to the next input symbol.

In the latter case, if the next input symbol occurs in no register, the input-based reassignment is performed. That is, the content of the register $\rho(q)$ is replaced with this symbol. Otherwise the finite memory of $\mathcal{A}$ remains intact. Then (similarly to its NR-IAPDA counterpart) $\mathcal{A}$ performs a transition involving the next input symbol.

The definition of a DR-IAPDA instantaneous description is exactly like that of NR-IAPDA, and the yielding in one step relation between two DR-IAPDA instantaneous descriptions $(p, v_1 v_2 \cdots v_r, \sigma\boldsymbol{\sigma}, \tau\boldsymbol{\tau})$ and $(q, w_1 w_2 \cdots w_r, \boldsymbol{\sigma}, \boldsymbol{\alpha\tau})$, $\sigma \in \Sigma \cup \{\epsilon\}$ and $\tau \in \Sigma$, is as defined below. We write

$$(p, v_1 v_2 \cdots v_r, \sigma\boldsymbol{\sigma}, \tau\boldsymbol{\tau}) \vdash (q, w_1 w_2 \cdots w_r, \boldsymbol{\sigma}, \boldsymbol{\alpha\tau}),$$

if and only if the following holds.

Let $v'_1, v'_2, \ldots, v'_r$ be defined as follows. If there is a $j = 1, \ldots, r$ such that $v_j = \tau$, then $v'_1 v'_2 \cdots v'_r = v_1 v_2 \cdots v_r$. Otherwise, $v'_{\pi(p)} = \tau$ and $v'_m = v_m$, for each $m \neq \pi(p)$.

- If $\sigma = \epsilon$, then $w_1 w_2 \cdots w_r = v'_1 v'_2 \cdots v'_r$, and for the $j = 1, \ldots, r$ such that $w_j = \tau$, there is a $(q, j_1 j_2 \cdots j_\ell) \in \mu(p, \epsilon, j)$, such that $\boldsymbol{\alpha} = w_{j_1} w_{j_2} \cdots w_{j_\ell}$.
- If $\sigma \in \Sigma$, then

  • either there is a $k = 1, \ldots, r$ such that $u_k = \sigma$ and $w_1 w_2 \cdots w_r = v'_1 v'_2 \cdots v'_r$, or $w_{\rho(p)} = \sigma$ and $w_m = v'_m$, for each $m \neq \rho(p)$; and
  • for the (unique) $k, j = 1, \ldots, r$ such that $w_k = \sigma$ and $w_j = \tau$ there is a $(q, j_1 j_2 \cdots j_\ell) \in \mu(p, k, j)$ such that $\boldsymbol{\alpha} = w_{j_1} w_{j_2} \cdots w_{j_\ell}$.

The language accepted by DR-IAPDA $\mathcal{A}$ is defined by

$$L(\mathcal{A}) = \{\boldsymbol{\sigma} : (s_0, \boldsymbol{u}, \boldsymbol{\sigma}, u_r) \vdash^* (p, \boldsymbol{v}, \epsilon, \epsilon)\},$$

where, as usual, $\vdash^*$ is the reflexive and transitive closure of $\vdash$.

Example 3. In this example we present a DR-IAPDA $\mathcal{A}$ that accepts the language

$$L^R = \{\delta \boldsymbol{w} \$^{|\boldsymbol{w}|} : \$ \notin [\boldsymbol{w}] \text{ and } \delta \notin [\boldsymbol{w}] \cup \{\$\}\}.$$

That is, L^R is the reversal of the language L defined by (1) in Example 1. The idea lying behind the definition of A is quite standard. Namely, the computation of $\mathcal{A}$ on an input of the form $\delta \boldsymbol{w} \$^n$, $\boldsymbol{w} = w_1 \cdots w_n$ and $\delta \notin [\boldsymbol{w}]$, is as follows.

- First $\mathcal{A}$ stores δ in one of its registers.
- Then, for each $i = 1, \ldots, n$, $\mathcal{A}$ compares w_i with δ and, if they are different, pushes $\$$ into the stack.
- Finally, $\mathcal{A}$ compares the stack content with the $\$$-suffix of the input, by popping the $\$$s out of the stack while reading the suffix.

The only little technical problem with the above description of $\mathcal{A}$ is the register in which δ will be stored. Namely, $\mathcal{A}$ cannot store it in register k, if $\delta = u_{k'}$ for some $k' \neq k$. Therefore, $\mathcal{A}$ starts the computation by an ϵ-move, "guessing" an appropriate register for storing δ.

So, let $\mathcal{A} = \langle Q, s_0, \boldsymbol{u}, \rho, \pi, \mu \rangle$, where

- $Q = \{s_0, q\} \cup \{s_1, p_1\} \cup \{s_2, p_2\};$[7]
- $\boldsymbol{u} = u_1 u_2 \$;$
- $\rho(s_k) = 3 - k$ and $\rho(p_k) = k$, $k = 1, 2$;
- π is the empty domain function; and
- μ is defined as follows.
    - (i) $\mu(s_0, \epsilon, 3) = \{(s_k, 3) : k = 1, 2\};$
    - (ii) $\mu(s_k, k, 3) = \{(p_k, 3)\}$, $k = 1, 2$;
    - (iii) $\mu(p_k, k, 3) = \{(p_k, 33)\}$, $k = 1, 2$;
    - (iv) $\mu(p_k, \epsilon, 3) = \{(q, \epsilon)\}$, $k = 1, 2$; and
    - (v) $\mu(q, 3, 3) = \{(q, \epsilon)\}.$

Now, it is not hard to verify that, indeed, $L(\mathcal{A}) = L^R$. The intuition lying behind the accepting run of $\mathcal{A}$ on an input of the form $\delta \boldsymbol{w} \$^n$, $\boldsymbol{w} = w_1 \cdots w_n$ and $\delta \notin [\boldsymbol{w}]$, is as follows.

- By an appropriate transition of type (i), $\mathcal{A}$ enters state s_k such that $\delta \neq u_k$, $k = 1, 2$.
- Then, by the transition of type (ii), $\mathcal{A}$ reads δ, stores it in the register $3 - k$, and enters the state p_i.
- After that, for each $i = 1, \ldots, n$, by the transition of type (iii), $\mathcal{A}$ compares w_i with δ and, if they are different, pushes $\$$ into the stack. Note that, after the entire $\boldsymbol{w}$ has been read, the stack content is $\$^{n+1}$.
- Next, by the transition of type (iv), $\mathcal{A}$ enters the popping state q. Since in this move, the topmost $\$$ has been popped out of the stack, the stack content becomes $\$^n$.
- Finally, using the transition of type (v), $\mathcal{A}$ verifies that the number of $\$$s in the stack is n, by popping them out of the stack while reading the $\$$-suffix of the input.

For example, $\mathcal{A}$ accepts the word $\$^3 \sigma_1 \sigma_2 \sigma_3 \delta$, where $\delta \notin \{\sigma_1, \sigma_2, \sigma_3\}$ and $\delta \neq u_1$, by the following sequence of computation steps, cf. Example 2.

$$(s_0, u_1 u_2 \$, \delta \sigma_1 \sigma_2 \sigma_3 \$\$\$, \$) \vdash (s, u_1 u_2 \$, \delta \sigma_1 \sigma_2 \sigma_3 \$\$\$, \$) \vdash (p_1, u_1 \delta \$, \sigma_1 \sigma_2 \sigma_3 \$\$\$, \$)$$

$$\vdash (p_1, \sigma_1 \delta \$, \sigma_2 \sigma_3 \$\$\$, \$\$) \vdash (p_1, \sigma_2 \delta \$, \sigma_3 \$\$\$, \$\$\$) \vdash (p_1, \sigma_3 \delta \$, \$\$\$, \$\$\$\$)$$

$$\vdash (q, \sigma_3 \delta \$, \$\$\$, \$\$\$) \vdash (q, \sigma_3 \delta \$, \$\$, \$\$) \vdash (q, \sigma_3 \delta \$, \$, \$) \vdash (q, \sigma_3 \delta \$, \epsilon, \epsilon).$$

Theorem 2. *For any DR-IAPDA $\mathcal{A}$, $L(\mathcal{A}) \neq L$.*

The proof of Theorem 2 is presented in the next section.

Remark 1. Since, as we show in this paper, the language L from Example 1 defined by (1) is not a DR-IAPDA languages and L^R is the reversal of L, the class

---

[7] As described above, from the initial state s_0, by an ϵ-move, $\mathcal{A}$ enters the state s_1, stores δ in the second register, and then proceeds to the state p_1, if $\delta \neq u_1$; and enters the state s_2, stores δ in the first register, and then proceeds to the state p_2, if $\delta \neq u_2$.

of the DR-IAPDA languages is not closed under reversing. Note that, by Proposition 1 and Theorem 1, the class of the NR-IAPDA languages is closed under reversing.

Next we show how DR-IAPDA can be simulated by NR-IAPDA.[8] That is, we show how deterministic reassignment can be simulated by non-deterministic one. The idea lying behind simulation of a DR-IAPDA $\mathcal{A}$ by an NR-IAPDA $\mathcal{A}^N$ is rather simple: we just postpone the reassignment of $\mathcal{A}^N$ to the very last moment (in all states), and then force $\mathcal{A}^N$ to move consulting the reassigned register, which would correspond to the move of $\mathcal{A}$ after the reassignment has been made.

Proposition 2. *Every DR-IAPDA language is accepted by an NR-IAPDA.*

Proof. Let $\mathcal{A} = \langle Q, s_0, \boldsymbol{u}, \rho, \pi, \mu \rangle$ be a DR-IAPDA. We shall prove that $L(\mathcal{A}) = L(\mathcal{A}^N)$, where the NR-IAPDA $\mathcal{A}^N = \langle Q^N, s_0^N, \boldsymbol{u}^N, \rho^N, \mu^N \rangle$ is defined as follows.

– $Q^N = Q \cup Q^\pi \cup Q^\rho \cup Q^{\pi\rho}$, where

  - $Q^\pi = \{q^\pi : q \in Q\}$,
  - $Q^\rho = \{q^\rho : q \in Q\}$, and
  - $Q^{\pi\rho} = \{q^{\pi\rho} : q \in Q\}$.[9]

  The intuitive meaning of the sets of states Q, Q^π, Q^ρ, and $Q^{\pi\rho}$ is as follows, see also the definition of ρ^N.

  - If no stack- or input-based reassignment has been made at the next computation step, then $\mathcal{A}^N$ enters the corresponding state from Q.
  - If only an input-based reassignment has been made at the next computation step, then $\mathcal{A}^N$ enters a state from Q^ρ.
  - Otherwise, $\mathcal{A}^N$ enters a state from Q^π, in which it performs the stack-based reassignment of $\mathcal{A}$. If, in addition, $\mathcal{A}$ has made an input-based reassignment, then $\mathcal{A}^N$, by an ϵ-move, proceeds to the corresponding state from $Q^{\pi\rho}$, in which it performs the corresponding input-based reassignment.

– $s_0^N = s_0$.
– $\boldsymbol{u}^N = \boldsymbol{u}$.
– ρ^N is defined by
  - $\rho^N(q^\pi) = \pi(q)$; and
  - $\rho^N(q^\rho) = \rho^N(q^{\pi\rho}) = \rho(q)$.
– For the definition of μ^N, let the functions μ^π and μ^ρ be defined by
  - $\mu^\pi(q, k, j) = \{(p^\pi, j_1 j_2 \cdots j_\ell) : (p, j_1 j_2 \cdots j_\ell) \in \mu(q, k, j)\}$; and
  - $\mu^\rho(q, k, j) = \{(p^\rho, j_1 j_2 \cdots j_\ell) : (p, j_1 j_2 \cdots j_\ell) \in \mu(q, k, j)\}$.

---

[8] Note, that DR-IAPDA are not a special case of NR-IAPDA. This is because nondeterministic reassignment is made *before* seeing the input symbol, whereas the deterministic reassignment is made *after*.

[9] Renaming the elements of Q, if necessary, we may assume that the sets Q, Q^π, Q^ρ, and $Q^{\pi\rho}$ are mutually disjoint.

Then μ^N is defined as follows.

- $\mu^N(q,k,j) = \mu(q,k,j) \cup \mu^\pi(q,k,j) \cup \mu^\rho(q,k,j)$.
- $\mu^N(q^\pi,k,j) = \mu^N(q,k,j)$, if $j = \pi(q)$, and is empty, otherwise. That is, according to the above intuition, after making a stack-based reassignment, $\mathcal{A}^N$ must, immediately, "use" it.
- $\mu^N(q^\pi,\epsilon,j) = \{(q^{\pi\rho},j)\}$.
- $\mu^N(q^\rho,k,j) = \mu^N(q,k,j)$, if $k = \rho(q)$, and is empty, otherwise. Similarly, to a stack-based reassignment, after making an input-based reassignment, $\mathcal{A}^N$ must, immediately, "use" it.
- $\mu^N(q^{\pi\rho},k,j) = \mu^N(q,k,j)$, if $j = \pi(q)$ and $k = \rho(q)$, and is empty, otherwise. Here, $\mathcal{A}^N$ immediately moves according to both stack- and input-based reassignments, simulating in this way the corresponding move of $\mathcal{A}$.

For the proof of the inclusion $L(\mathcal{A}) \subseteq L(\mathcal{A}^N)$, let

$$(q_0, \boldsymbol{u}_0, \boldsymbol{\sigma}_0, \boldsymbol{\tau}_0) \vdash \cdots \vdash$$

$$(q_i, \boldsymbol{u}_i, \boldsymbol{\sigma}_i, \boldsymbol{\tau}_i) \vdash (q_{i+1}, \boldsymbol{u}_{i+1}, \boldsymbol{\sigma}_{i+1}, \boldsymbol{\tau}_{i+1}) \tag{2}$$

$$\vdash \cdots \vdash (q_n, \boldsymbol{u}_n, \boldsymbol{\sigma}_n, \boldsymbol{\tau}_n)$$

be a run of $\mathcal{A}$ on $\boldsymbol{\sigma}$. Introducing a new initial state s with an additional transition $\mu(s, \epsilon, r) = (s_0, r)$, if necessary, we may assume that the first computation step $(q_0, \boldsymbol{u}_0, \boldsymbol{\sigma}_0, \boldsymbol{\tau}_0) \vdash (q_1, \boldsymbol{u}_1, \boldsymbol{\sigma}_1, \boldsymbol{\tau}_1)$ in the above run is by an ϵ-move.

Let q_i^N, $i = 0, \ldots, n$, be defined as follows.

- $q_i^N = q_i$, if no stack- or input-based reassignment has been made at the computation step $(q_i, \boldsymbol{u}_i, \boldsymbol{\sigma}_i, \boldsymbol{\tau}_i) \vdash (q_{i+1}, \boldsymbol{u}_{i+1}, \boldsymbol{\sigma}_{i+1}, \boldsymbol{\tau}_{i+1})$;
- $q_i^N = q_i^\rho$, if a stack-based reassignment, but no input-based reassignment, has been made at the computation step (2); and
- $q_i^N = q_i^\pi$, if an input-based reassignment, but no stack-based reassignment, has been made at the computation step (2).

Then the computation step (2) of $\mathcal{A}$ is simulated by the computation step

$$(q_i^N, \boldsymbol{u}_i, \boldsymbol{\sigma}_i, \boldsymbol{\tau}_i) \vdash (q_{i+1}^N, \boldsymbol{u}_{i+1}, \boldsymbol{\sigma}_{i+1}, \boldsymbol{\tau}_{i+1})$$

of $\mathcal{A}^N$, if at least one of the stack- or input-based reassignments has not been made at (2), and is simulated by two computation steps

$$(q_i^\pi, \boldsymbol{u}_i, \boldsymbol{\sigma}_i, \boldsymbol{\tau}_i) \vdash (q_i^{\pi\rho}, \boldsymbol{u}_i, \boldsymbol{\sigma}_i, \boldsymbol{\tau}_i) \vdash (q_{i+1}^N, \boldsymbol{u}_{i+1}, \boldsymbol{\sigma}_{i+1}, \boldsymbol{\tau}_{i+1})$$

of $\mathcal{A}^N$, otherwise. That is, if both the stack- and the input-based reassignment have been made at (2).

For the proof of the converse inclusion $L(\mathcal{A}^N) \subseteq L(\mathcal{A})$, let

$$(q_0^N, \boldsymbol{u}_0^N, \boldsymbol{\sigma}_0^N, \boldsymbol{\tau}_0^N) \vdash \cdots \vdash$$

$$(q_i^N, \boldsymbol{u}_i^N, \boldsymbol{\sigma}_i^N, \boldsymbol{\tau}_i^N) \vdash (q_{i+1}^N, \boldsymbol{u}_{i+1}^N, \boldsymbol{\sigma}_{i+1}^N, \boldsymbol{\tau}_{i+1}^N) \tag{3}$$

$$\vdash \cdots \vdash (q_n^N, \boldsymbol{u}_n^N, \boldsymbol{\sigma}_n^N, \boldsymbol{\tau}_n^N),$$

$q_i^N = q_i, q_i^\pi, q_i^\rho, q_i^{\pi\rho}$, be a run of $\mathcal{A}^N$ on $\boldsymbol{\sigma}$. We shall translate this run into a run of $\mathcal{A}$ on $\boldsymbol{\sigma}$ as follows.

- If $q_i^N, q_{i+1}^N \notin Q^{\pi\rho}$, then the computation step (3) of $\mathcal{A}^N$ is simulated by the computation step

$$(q_i, \boldsymbol{u}_i, \boldsymbol{\sigma}_i, \boldsymbol{\tau}_i) \vdash (q_{i+1}, \boldsymbol{u}_{i+1}, \boldsymbol{\sigma}_{i+1}, \boldsymbol{\tau}_{i+1})$$

  of $\mathcal{A}$.
- If $q_i^N = q_i^{\pi\rho}$, then the two consecutive computation steps

$$(q_{i-1}^N, \boldsymbol{u}_{i-1}^N, \boldsymbol{\sigma}_{i-1}^N, \boldsymbol{\tau}_{i-1}^N) \vdash (q_i^N, \boldsymbol{u}_i^N, \boldsymbol{\sigma}_i^N, \boldsymbol{\tau}_i^N) \vdash (q_{i+1}^N, \boldsymbol{u}_{i+1}^N, \boldsymbol{\sigma}_{i+1}^N, \boldsymbol{\tau}_{i+1}^N)$$

  of $\mathcal{A}^N$ are simulated by the computation step

$$(q_{i-1}, \boldsymbol{u}_{i-1}, \boldsymbol{\sigma}_{i-1}, \boldsymbol{\tau}_{i-1}) \vdash (q_{i+1}, \boldsymbol{u}_{i+1}, \boldsymbol{\sigma}_{i+1}, \boldsymbol{\tau}_{i+1})$$

  of $\mathcal{A}$.

Correctness of all simulations above immediately follows from the definition of $\mathcal{A}^N$. We omit the proof.

Combining Proposition 2 with Theorem 2 we obtain the corollary below.

Corollary 1. *The class of the DR-IAPDA languages is a proper subclass of the NR-IAPDA languages.*

We conclude this section with two properties of DR-IAPDA needed for the proof of Theorem 2.

To explain the first property we shall need the following notion. For an instantaneous description $\boldsymbol{id} = (q, \boldsymbol{v}, \boldsymbol{x}, \boldsymbol{\alpha})$, the set of symbols $[\boldsymbol{vx\alpha}]$ is called the active alphabet of $\boldsymbol{id}$. Due to the nature of deterministic reassignment, the instantaneous descriptions of DR-IAPDA possess the property of active alphabet monotony, i.e., the active alphabet of a subsequent instantaneous description is always a subset of the active alphabet of the preceding one.

Proposition 3. *Let $\mathcal{A}$ be a DR-IAPDA. If $(p, \boldsymbol{v}, \boldsymbol{x}, \boldsymbol{\alpha}) \vdash_{\mathcal{A}}^* (q, \boldsymbol{w}, \boldsymbol{y}, \boldsymbol{\beta})$, then $[\boldsymbol{wy\beta}] \subseteq [\boldsymbol{vx\alpha}]$.*

The proof of Proposition 3 is by a straightforward induction on the number of computation steps and is omitted.

The following observation will also be used in the sequel.

Proposition 4. *(Cf. [7, Lemma 1].) Let $\iota : \Sigma \to \Sigma$ be a permutation of Σ. Then for any DR-IAPDA $\mathcal{A}$,*

$$(p, \boldsymbol{v}, \boldsymbol{x}, \boldsymbol{\alpha}) \vdash_{\mathcal{A}}^* (q, \boldsymbol{w}, \boldsymbol{y}, \boldsymbol{\beta})$$

if and only if

$$(p, \iota(\boldsymbol{v}), \iota(\boldsymbol{x}), \iota(\boldsymbol{\alpha})) \vdash_{\mathcal{A}}^* (q, \iota(\boldsymbol{w}), \iota(\boldsymbol{y}), \iota(\boldsymbol{\beta})).$$

Again, the proof is by a straightforward induction on the number of computation steps and is omitted.

6 Proof of Theorem 2

The idea lying behind the proof of Theorem 2 is rather simple. Since a DR-IAPDA cannot guess δ at the beginning of the computation, it leaves it with the only option: to store w on the stack until it arrives at δ, and then compare each symbol of w with δ by popping them one by one out of the stack.[10] The problem is, that the stack is also needed to compare the number of $s to the length of w, and apparently, two of these tasks cannot be performed simultaneously. Essentially, this is the same intuition that leads us to the belief that $\{a^i b^i c^i : i = 1, 2 \ldots\}$ is not a context-free language, and that $\{a^i b^i c^j d^j : i, j = 1, 2 \ldots\}$ is not a linear context-free language, say. Nevertheless, usually it requires a significant effort to formalize a clear intuition, because proving a negative result is considerably harder than proving a positive one.

For the proof, we assume to the contrary, that there is an r-register DR-IAPDA $\mathcal{A} = \langle Q, s_0, u, \pi, \rho, \mu \rangle$ that accepts L. In the following section we establish various properties $\mathcal{A}$ would possess, if it existed, which, as we show in Section 6.2, bring us to a contradiction, that will complete the proof.

6.1 Constraints on the Stack Content

In this section we prove a number of properties of $\mathcal{A}$ and introduce notation need for the proof of Theorem 2 in Section 6.2.

Let $z = \$^{|w|} w \delta \in l$ be such that

- $|w| > r$, where r is the number of registers of $\mathcal{A}$,
- none of the elements of Σ occurs w twice or more, and
- none of the symbols of w or δ occurs in the initial assignment u_0 of $\mathcal{A}$.

Let $id_0, id_1, \ldots, id_n$ be an accepting run of $\mathcal{A}$ on z, in which each instantaneous description id_i, $i = 0, \ldots, n$, of $\mathcal{A}$ is of the form $(q_i, u_i, z_i, \alpha_i)$, where

- q_i is the current state of $\mathcal{A}$,
- u_i is the current register assignment,
- z_i is the remaining input (suffix of z), and
- α_i is the content of the stack read top down.

In particular, $id_0 = (s_0, u, z, u_r)$, and $z_n = \alpha_n = \epsilon$.

We shall need the following notation. For each $i = 0, \ldots, n$, such that $z_i \neq \epsilon$, the suffix w_i of $\$^{|w|} w$ is defined by $z_i = w_i \delta$.

Lemma 1. *Let id_i, $i = 0, \ldots, n$, be as above. Then for each i such that $z_i \neq \epsilon$, $[w] \subseteq [u_i z_i \alpha_i]$.*

Proof. Assume to the contrary that for some $i = 0, \ldots, n$, $z_i \neq \epsilon$ and $[w] \not\subseteq [u_i z_i \alpha_i]$, and let $\delta' \in [w]$ be such that

$$\delta' \notin [u_i z_i \alpha_i].$$

---

[10] This intuition is formalized by Lemma 1.

Since $z_i = w_i \delta$, where w_i is a suffix of $\$^{|w|}w$, $id_0 \vdash^* id_i$ implies

$$(s_0, u_0, \$^{|w|}w, u_r) \vdash^* (q_i, u_i, w_i, \alpha_i). \tag{4}$$

Let the permutation ι switch between δ and δ', and fix all other symbols of Σ. Then, applying Proposition 4 to (4), we obtain

$$(s_0, \iota(u_0), \iota(\$^{|w|}w), \iota(u_r)) \vdash^* (q_i, \iota(u_i), \iota(w_i), \iota(\alpha_i)) \tag{5}$$

Since, by the definition of w, $\delta, \delta' \notin [u] \cup \{\$\}$ and $|\iota(w)| = |w|$,

$$(s_0, \iota(u_0), \iota(\$^{|w|}w), \iota(u_r)) = (s_0, u_0, \$^{|\iota(w)|}\iota(w), u_r). \tag{6}$$

Also, by Proposition 3 (and the definitions of w, δ, and δ'),

$$(q_i, \iota(u_i), \iota(w_i), \iota(\alpha_i)) = (q_i, u_i, w_i, \alpha_i). \tag{7}$$

Thus, by (5), (6), and (7),

$$(s_0, u_0, \$^{\iota(|w|)}\iota(w), u_r) \vdash^* (q_i, u_i, w_i, \alpha_i), \tag{8}$$

where $\iota(w)$ is obtained from w by replacing δ' with δ.

Now, appending δ to $\$^{\iota(|w|)}\iota(w)$ and w_i in (8) results in

$$(s_0, u_0, \$^{|\iota(w)|}\iota(w)\delta, u_r) \vdash^* (q_i, u_i, z_i, \alpha_i) = id_i \vdash^* id_n, \tag{9}$$

where $id_i \vdash^* id_n$ is by the definition of $id_0, \ldots, id_n$. The first instantaneous description in (9) is, in fact, the initial instantaneous description of $\mathcal{A}$ on the input $\$^{|\iota(w)|}\iota(w)\delta \notin L$, and since, by definition, id_n is an accepting instantaneous description, $\$^{|\iota(w)|}\iota(w)\delta \in L(\mathcal{A})$, in spite of $\delta \in [\iota(w)]$. This contradicts the equality $L = L(\mathcal{A})$, which completes the proof.

The following series of lemmas formalizes the intuition that somewhere between reading the first and the rth symbols of w, $\mathcal{A}$ enters an instantaneous description whose bottom stack symbol remains intact until the automaton arrives at the end of the input δ.

Let $w = u'u''$, where $|u'| = r$, and let

$$i_0 = \max\{i : z_i = u''\delta\}. \tag{10}$$

That is, at the computation step $id_{i_0} \vdash id_{i_0+1}$, $\mathcal{A}$ arrives at the $(r+1)$st symbol of w.

Remark 2. Note that, by the definition of i_0, w_{i_0} is a suffix of w and $|w| - |w_{i_0}| = r$.

Lemma 2. $[u'] \cap [\alpha_{i_0}] \neq \emptyset$.

Proof. Assume the contrary that $[u'] \cap [\alpha_{i_0}] = \emptyset$. Then, since $[u'] \cap [u''] = \emptyset$, by Propositions 3 and Lemma 1, all symbols occurring in u' are stored in the automaton registers, i.e. $[u_{i_0}] = [u']$. Since $[\alpha_{i_0}] \cap [u'] = \emptyset$, to perform the next computation step $id_{i_0} \vdash id_{i_0+1}$, $\mathcal{A}$ must make the stack-based reassignment. Namely, it has to reassign one of the registers with the first symbol of α_{i_0}. However, the reassignment would cause $\mathcal{A}$ to "loose" a symbol of u' stored in one of its registers, in contradiction with Lemma 1, because $|w| > r$.

Let α be the shortest suffix of α_{i_0} that contains a symbol from $[u']$. In particular, $\alpha \neq \epsilon$.

Remark 3. By definition, only one symbol from $[u']$ occurs in α, and this symbol is the first symbol of the latter.

Lemma 3. $\alpha_{i_0} = \beta_{i_0}\alpha$, where $\beta_{i_0} \neq \epsilon$.

Proof. Assume to the contrary that $\beta_{i_0} = \epsilon$. Then the first symbol of α is the only one from u' occurring in α_{i_0}. Thus, by Lemma 1, the other $r - 1$ symbols of u' must be stored in $r - 1$ registers of $\mathcal{A}$. Consequently, by the same lemma, in the stack-based reassignment before the next computation step, the only register that may be reassigned with the first symbol of α, or already contains it, is the remaining register. Also, after this reassignment, the content of the registers coincides with $[u']$. Since, by the definition of i_0, the computation step $id_{i_0} \vdash id_{i_0+1}$ is not an ϵ-move, one of the registers has to be reassigned with the first symbol of u'' in the input-based reassignment.

This register cannot be the one containing the first symbol of α, because in this case none of the register content would be equal to the top stack symbol, and, therefore, $\mathcal{A}$ would not be able to perform the computation step. Consequently, one of the symbols of $[u']$ is erased from the registers, in contradiction with Lemma 1. That is, $\beta_{i_0} \neq \epsilon$ and the proof is complete.

Lemma 4. *If* $|w| > r + 1$, *then* $\alpha_{i_0+1} = \beta_{i_0+1}\alpha$, *where* $\beta_{i_0+1} \neq \epsilon$.

Proof. By definition, α_{i_0+1} is obtained from α_{i_0} by replacing its first symbol with a word that consists of symbols from $[u_i]$. Since, by Lemma 3, $\alpha_{i_0} = \beta_{i_0}\alpha$, where $\beta_i \neq \epsilon$, α_{i_0+1} is of the form $\beta_{i_0+1}\alpha$, and we only need to show that $\beta_{i_0+1} \neq \epsilon$. If $|\beta_{i_0}| > 1$, this is immediate, because the stack hight may decrease by at most one in one computation step. Otherwise, $\beta_{i_0} = \alpha \in \Sigma$ and we argue that $\alpha \in [u']$.

To show this, assume to the contrary that $\alpha \notin [u']$. Then there is only one register which α may be assigned to without loss of "valuable" symbols. Applying one more time the argument from the proof of Lemma 3, we see that this case is impossible.

So, $\alpha \in [u']$ and we have to show that the word β_{i_0+1} pushed into the stack in the computation step $id_{i_0} \vdash id_{i_0+1}$ is not empty.

For the proof, assume to the contrary that $\beta_{i_0+1} = \epsilon$. Then, after the computation step $id_{i_0} \vdash id_{i_0+1}$ is complete, r of the first $r + 1$ symbols of w are stored in the r registers of $\mathcal{A}$ (and occur only in the registers), and the other (of the $r + 1$ symbols, that is the first symbol of α) is at the top of the stack. Therefore, the stack-based reassignment at the next computation step would erase a symbol stored in the registers, in contradiction with Lemma 1 for id_{i+1}. This is because, by the lemma prerequisite $|w| > r + 1$, the $(r + 2)$nd symbol of w is not δ.

Lemma 5. *If* $|w| > r + 1$ *and* $|w| - |w_i| = r + 1$,[11] *then* $\alpha_i = \beta_i\alpha$, *where* $\beta_i \neq \epsilon$.

---

[11] That is, $id_{i_0+1} \vdash^* id_i$ by ϵ-moves only.

The proof of the lemma is by induction on i. The basis, $i = i_0 + 1$, is Lemma 4, and the induction step is word by word repetition of the proof of Lemma 4 and is left to the reader.

Lemma 6. *If $|w| - |w_i| \geq r + 2$, then $\alpha_i = \beta_i \alpha$, where $\beta_i \neq \epsilon$.*

Proof. The proof is by induction on i. For the basis, let i be such that $|w| - |w_{i-1}| = r + 1$, but $|w| - |w_i| = r + 2$. That is, the number of symbols read from w is exactly $r + 2$. By Lemma 1, each of these symbols occurs either in the stack or in the registers. Namely,

1. r of the symbols are stored in the registers;
2. one symbol occurs in α, because, by Lemma 5, $\beta_{i-1} \neq \epsilon$, which implies that α remains intact at this computation step; and
3. one symbol occurs in β_i.

That is, $\beta_i \neq \epsilon$.

The induction step is treated similarly to the basis. We just replace "$r + 2$" with "at least $r + 2$" and replace "one" in clause 3 with "at least one".

Let

$$i_1 = \max\{i < i_0 : \alpha \text{ is not a proper suffix of } \alpha_{i_1}\} + 1, \tag{11}$$

where i_0 is defined by (10), and let

$$M = \max\{\ell : (p, j_1 j_2 \ldots j_\ell) \in \mu(q; k, j),\ p, q \in Q \text{ and } k, j, j_1, \ldots, j_\ell = 1, \ldots, r\}.$$

That is, id_{i_1-1} is the last instantaneous description before id_{i_0}, whose stack content does not contain α as a proper suffix and M is the maximum length of a word pushed into the stack in one computation step.

Combining Lemmas 3, 4, 5, and 6, we obtain Lemma 7 below.

Lemma 7

1. $|w| - |w_{i_1}| \leq r$.
2. $\alpha_{i_1} = \beta_{i_1} \alpha$, where $0 < |\beta_{i_1}| \leq M$.
3. For each $i \geq i_1$ such that $z_j \neq \epsilon$, $\alpha_i = \beta_i \alpha$, where $\beta_i \neq \epsilon$.

Proof. By definition, $i_1 \leq i_0$, and clause 1 follows from the definition of i_0 according to which $|w| - |w_{i_0}| = r$.

Since α is a proper suffix of α_{i_1}, $\beta_{i_1} \neq \epsilon$. To prove that $|\beta_{i_1}| \leq M$, we proceed as follows. Let $\alpha_{i_1-1} = \alpha \alpha'$. Then, α_{i_1} is obtained from α_{i_1-1} by replacing α with some word α'', i.e.

$$\alpha_{i_1} = \alpha'' \alpha'. \tag{12}$$

By the definition of M,

$$|\alpha''| \leq M, \tag{13}$$

and, since α is not a proper suffix of α_{i_1-1}, α' is a suffix of α. Thus, by (12), β_{i_1} is a proper suffix of α'', and clause 2 of the lemma follows from (13).

Finally, clause 3 follows immediately from Lemmas 3, 4, 5, and 6, and the definition of id_{i_1}, according to which α is a proper suffix of all α_i for $i_1 \leq i < i_0$.

So, by Lemma 7, we have located the instantaneous description id_{i_1}, with α at the bottom of the stack which remains intact until the entire w is read. The following corollary to Lemma 7 shows that, actually, it remains intact until the entire input z is read. Namely, let

$$i_2 = \max\{i : \text{ for all } i', \ i_0 \le i' \le i, \ \alpha \text{ is a suffix of } \alpha_{i'}\}. \tag{14}$$

Corollary 2. *If* $|w| > r + 1$, *then* $z_{i_2} = \epsilon$.

Proof. Let i be the last computation step before reading δ. Then, by clause 3 of Lemma 7, $\alpha_i = \beta_i \alpha$, where $\beta_i \ne \epsilon$. Since at most one symbol may be removed from the stack in one computation step, α is a suffix of α_{i+1}. Therefore, $i_2 \ge i+1$, and the corollary follows.

6.2   The Final Match

Now, to arrive at the desired contradiction and prove Theorem 2, we consider two words, $z' = \$^{|w'|} w' \delta$ and $z'' = \$^{|w''|} w'' \delta$ in L which satisfy the following conditions.

1. $|w'|, |w''| > r + 1$.
2. $|w'| \ne |w''|$.
3. None of the symbols of Σ occur in w' or w'' twice or more, and none of the symbols of w', w'' or δ occurs in the initial assignment of $\mathcal{A}$.
4. The prefixes of w' and w'' of length $r + 1$ are equal to the same word, $\sigma_1 \sigma_2 \cdots \sigma_{r+1}$, say. That is,

$$w' = \sigma_1 \sigma_2 \cdots \sigma_{r+1} v' \tag{15}$$

and

$$w'' = \sigma_1 \sigma_2 \cdots \sigma_{r+1} v'', \tag{16}$$

for the appropriate $v', v'' \in \Sigma^*$.

Let $id'_0, \ldots, id'_{n'}$, $id'_i = (q'_i, u'_i, z'_i, \alpha'_i)$, $i = 0, \ldots, n'$, and $id''_0, \ldots, id''_{n''}$ $id''_i = (q''_i, u''_i, z''_i, \alpha''_i)$, $i = 0, \ldots, n''$, be accepting runs of $\mathcal{A}$ on z' and z'', respectively. Let i'_1, i'_2, i''_1, and i''_2, be the positive integers provided by Lemma 7 and Corollary 2 for w' and w'', respectively. That is, we use the primed and double primed versions of the constants and parameters defined in Section 6.1.

We shall also need the following definitions and notation.

– The alphabet $[u] \cup \{\sigma_1, \sigma_2, \ldots, \sigma_{r+1}\} \cup \{\$\}$ is denoted by Δ.
– Let $\# \notin \Sigma$. We define the function $p : \Sigma \to \Delta \cup \{\#\}$ by

$$p(\sigma) = \begin{cases} \sigma, & \text{if } \sigma \in \Delta \\ \#, & \text{otherwise} \end{cases},$$

and, for a word $v = v_1 \cdots v_m \in \Sigma^*$, we define the word $p(v) \in (\Delta \cup \{\#\})^*$ by

$$p(v) = p(v_1) \cdots p(v_m).$$

That is, $p(v)$ results from v in replacing all its symbols which are not in Δ with $\#$.

– The positive integers r' and r'' are defined by $r' = |w'| - |w'_{i'_1}|$ and $r'' = |w''| - |w''_{i''_1}|$. That is, r' and r'' are the number of symbols read from w' and w'' up to the computation steps $id'_{i'_1} \vdash id'_{i'_1+1}$ and $id''_{i''_1} \vdash id''_{i''_1+1}$, respectively.

Recall that, by clause 1 of Lemma 7, $r', r'' \le r$.

– For a word $v \in \Sigma^*$, $f(v)$ is the first symbol of v, if $v = \neq \epsilon$ and is ϵ, otherwise. That is,

$$f(v) = \begin{cases} v_1, & \text{if } v = v_1 \cdots v_m \neq \epsilon \\ \epsilon, & \text{if } v = \epsilon \end{cases}.$$

– Finally, the tuples

$$t_{z'}, t_{z''} \in Q \times \Delta^r \times \{1, \ldots, r\} \times \Delta^{\le M} \times Q \times (\Delta \cup \{\#\})^r \times \Delta$$

are defined by

$$t_{z'} = (q'_{i'_1}, u'_{i'_1}, r', \beta'_{i'_1}, q'_{i'_2}, p(u'_{i'_2}), f(\alpha'))$$

and

$$t_{z''} = (q''_{i''_1}, u''_{i''_1}, r'', \beta''_{i''_2}, q''_{i''_2}, p(u''_{i''_2}), f(\alpha'')).$$

Since the set

$$Q \times \Delta^r \times \{1, \ldots, r\} \times \Delta^{\le M} \times Q \times (\Delta \cup \{\#\})^r \times \Delta$$

is finite and Σ is infinite, there are words w' and w'' satisfying conditions $1 - 4$ in the beginning of this section such that $t_{z'} = t_{z''}$. That is,

$$(q'_{i'_1}, u'_{i'_1}, r', \beta'_{i'_1}, q'_{i'_2}, p(u'_{i'_2}), f(\alpha')) = (q''_{i''_1}, u''_{i''_1}, r'', \beta''_{i''_2}, q''_{i''_2}, p(u''_{i''_2}), f(\alpha'')). \quad (17)$$

Let $u'_{i'_2} = u'_{i'_2,1} \cdots u'_{i'_2,r}$, $u''_{i''_2} = u''_{i''_2,1} \cdots u''_{i''_2,r}$. and let ι be the permutation of Σ that is defined as follows.

$$\iota(\sigma) = \begin{cases} \sigma, & \text{if } \sigma \notin ([u'_{i'_2}] \cup [u''_{i''_2}]) \setminus \Delta \\ u''_{i''_2,k}, & \text{if } \sigma = u'_{i'_2,k} \in [u'_{i'_2}] \setminus \Delta, \ k = 1, \ldots, r \\ u'_{i'_2,k}, & \text{if } \sigma = u''_{i''_2,k} \in [u''_{i''_2}] \setminus \Delta, \ k = 1, \ldots, r \end{cases}.$$

That is, ι switches $u'_{i'_2,k}$ and $u''_{i''_2,k}$, $k = 1, \ldots, r$, if both are in $([u'_{i'_2}] \cup [u''_{i''_2}]) \setminus \Delta$. Since $u'_{i'_2}, u''_{i''_2} \in \Sigma^{r \neq}$ and $p(u'_{i'_2}) = p(u''_{i''_2})$, ι is well-defined.

Let

$$z = \$^{|w'|} \iota(w'') \delta. \quad (18)$$

Since $|\iota(w'')| = |w''|$ and $|w'| \neq |w''|$, $z \notin L$. However, we shall construct an accepting run of $\mathcal{A}$ on z, in contradiction with our assumption $L(\mathcal{A}) = L$. To construct this run we need one more bit of notation.

Let v_i, $i = 0, \ldots, i'_1$, be defined by

$$z'_i = v_i v' \delta, \quad (19)$$

where v' is defined by (15), and let the instantaneous descriptions $id_i^{(a)}$, $i = 0, \ldots, i'_1$, $id_i^{(b)}$, $i = i''_1, \ldots, i''_2$, and $id_i^{(c)}$, $i = i'_2, \ldots, n'$ be defined as follows.

(a) $id_i^{(a)} = (q_i', u_i', v_i \iota(v'')\delta, \alpha_i')$, where v_is and v'' are defined by (19) and (16), respectively.

(b) $id_i^{(b)} = (q_i'', \iota(u_i''), \iota(z_i''), \iota(\beta_i'')\alpha')$.

(c) $id_i^{(c)} = (q_i', u_i', \epsilon, \alpha_i')$.

Theorem 2 immediately follows from Lemma 8 below.

Lemma 8

$$\underbrace{id_0^{(a)} \vdash \cdots \vdash id_{i_1'}^{(a)}}_{(a)} = \overbrace{id_{i_1''}^{(b)} \vdash id_{i_1''+1}^{(b)} \vdash \cdots \vdash id_{i_2'}^{(b)}}^{(ab)} \underbrace{id_{i_1''}^{(b)} \vdash id_{i_1''+1}^{(b)} \vdash \cdots \vdash id_{i_2'}^{(b)}}_{(b)} = \overbrace{id_{i_2'}^{(c)} \vdash id_{i_2'+1}^{(c)} \vdash \cdots \vdash id_{n'}^{(c)}}^{(bc)} \underbrace{id_{i_2'}^{(c)} \vdash id_{i_2'+1}^{(c)} \vdash \cdots \vdash id_{n'}^{(c)}}_{(c)}$$

$$(20)$$

Indeed, by Lemma 8 (and the definition of the instantaneous descriptions in it, of course), (20) is an accepting run of $\mathcal{A}$ on z. That is, $z \in L$, which contradicts the definition of L, see (1).

Proof of Lemma 8.

Proof of part (a). By the definition of id_i's, (11), and (19),

$$(q_0', u_0', v_0 v'\delta, \alpha_0') \vdash \cdots \vdash (q_{i_1'}', u_{i_1'}', v_{i_1'} v'\delta, \alpha_{i_1'}'),$$

implying

$$(q_0', u_0', v_0, \alpha_0') \vdash \cdots \vdash (q_{i_1'}', u_{i_1'}', v_{i_1'}, \alpha_{i_1'}'),$$

which, in turn, implies

$$(q_0', u_0', v_0 \iota(v'')\delta, \alpha_0') \vdash \cdots \vdash (q_{i_1'}', u_{i_1'}', v_{i_1'} \iota(v'')\delta, \alpha_{i_1'}'),$$

and part (a) of the lemma follows from the definition of $id_i^{(a)}$s.

Proof of part (ab). By definition of $id_{i_1'}^{(a)}$ and $id_{i_1''}^{(b)}$, we have to show that

$$(q_{i_1'}', u_{i_1'}', v_{i_1'} \iota(v'')\delta, \alpha_{i_1'}') = (q_{i_1''}'', \iota(u_{i_1''}''), \iota(z_{i_1''}''), \iota(\beta_{i_1''}'')\alpha').$$

- $q_{i_1'}' = q_{i_1''}''$ immediately follows from (17).
- The proof of $u_{i_1'}' = \iota(u_{i_1''}'')$ is as follows. By Proposition 3, $u_{i_1''}'' \in \Delta^*$. Therefore, by the definition of ι, $\iota(u_{i_1''}'') = u_{i_1''}''$, and again, the equality follows from (17).
- For the proof of

$$v_{i_1'} \iota(v'')\delta = \iota(z_{i_1''}''), \tag{21}$$

we observe that, by (15),

$$z_{i_1'}' = v_{i_1'} v'\delta = \sigma_{r'+1} \cdots \sigma_{r+1} v'\delta,$$

implying $\boldsymbol{v}_{i_1'} = \sigma_{r'+1}\cdots\sigma_{r+1}$. Since

$$z_{i_1''}'' = w_{i_1''}''\delta = \sigma_{r''+1}\cdots\sigma_{r+1}\boldsymbol{v}''\delta$$

and, by (17), $r' = r''$, we have

$$z_{i''}'' = \boldsymbol{v}_{i_1'}\boldsymbol{v}''\delta. \tag{22}$$

Now, applying ι to both sides of (22) we obtain (21).

- The proof of the last equality $\boldsymbol{\alpha}_{i_1'}' = \iota(\boldsymbol{\beta}_{i_1''}'')\boldsymbol{\alpha}'$ is equally easy. Namely,

$$\boldsymbol{\alpha}_{i_1'}' = \boldsymbol{\beta}_{i_1'}'\boldsymbol{\alpha}' = \boldsymbol{\beta}_{i_1''}''\boldsymbol{\alpha}', \tag{23}$$

where the second equality follows from (17). Since, by Proposition 3, $\boldsymbol{\beta}_{i_1''}'' \in \Delta^*$, by the definition of ι, $\iota(\boldsymbol{\beta}_{i_1''}'') = \boldsymbol{\beta}_{i_1''}''$, which, together with (23), completes the proof.

Proof of part (b). By the definition of $\boldsymbol{id}_i''$s,

$$(q_{i_1''}'', \boldsymbol{u}_{i_1''}'', \boldsymbol{z}_{i_1''}'', \boldsymbol{\beta}_{i_1''}''\boldsymbol{\alpha}'') \vdash \cdots \vdash (q_{i_2''}'', \boldsymbol{u}_{i_2''}'', \boldsymbol{z}_{i_2''}'', \boldsymbol{\beta}_{i_2''}''\boldsymbol{\alpha}''),$$

implying

$$(q_{i_1''}'', \boldsymbol{u}_{i_1''}'', \boldsymbol{z}_{i_1''}'', \boldsymbol{\beta}_{i_1''}'') \vdash \cdots \vdash (q_{i_2''}'', \boldsymbol{u}_{i_2''}'', \boldsymbol{z}_{i_2''}'', \boldsymbol{\beta}_{i_2''}''),$$

because, by clause 3 of Lemma 7 and (14), $\boldsymbol{\beta}_{i''}'' \neq \epsilon$, $i = i_1'', \ldots, i_2'' - 1$. Therefore, by Proposition 4,

$$(q_{i_1''}'', \iota(\boldsymbol{u}_{i_1''}''), \iota(\boldsymbol{z}_{i_1''}''), \iota(\boldsymbol{\beta}_{i_1''}'')) \vdash \cdots \vdash (q_{i_2''}'', \iota(\boldsymbol{u}_{i_2''}''), \iota(\boldsymbol{z}_{i_2''}''), \iota(\boldsymbol{\beta}_{i_2''}'')),$$

implying

$$(q_{i_1''}'', \iota(\boldsymbol{u}_{i_1''}''), \iota(\boldsymbol{z}_{i_1''}''), \iota(\boldsymbol{\beta}_{i_1''}'')\boldsymbol{\alpha}') \vdash \cdots \vdash (q_{i_2''}'', \iota(\boldsymbol{u}_{i_2''}''), \iota(\boldsymbol{z}_{i_2''}''), \iota(\boldsymbol{\beta}_{i_2''}'')\boldsymbol{\alpha}'),$$

and part (b) of the lemma follows from the definition of $\boldsymbol{id}_i^{(b)}$s.

Proof of part (bc). By the definition of $\boldsymbol{id}_{i_2''}^{(b)}$ and $\boldsymbol{id}_{i_2'}^{(c)}$, we have to show that

$$(q_{i_2''}'', \iota(\boldsymbol{u}_{i_2''}''), \boldsymbol{z}_{i_2''}'', \iota(\boldsymbol{\beta}_{i_2''}''), \boldsymbol{\alpha}') = (q_{i_2'}', \boldsymbol{u}_{i_2'}', \epsilon, \boldsymbol{\alpha}_{i_2'}').$$

- $q_{i_2'}' = q_{i_2''}''$ immediately follows from (17).
- $\boldsymbol{u}_{i_2'}' = \iota(\boldsymbol{u}_{i_2''}'')$ immediately follows from the definition of ι.
- The equality $\boldsymbol{z}_{i_2''}'' = \epsilon$ is provided by Corollary 2.
- Finally, the equality $\boldsymbol{\alpha}' = \boldsymbol{\alpha}_{i_2'}'$ is by (14).

Proof of part (c). This part of the lemma follows immediately from the definition of $\boldsymbol{id}_i'$s and $\boldsymbol{id}_i^{(c)}$s, because, by Corollary 2, for $i = i_2', \ldots, n'$, $\boldsymbol{z}_i' = \epsilon$.

References

1. Bojanczyk, M., Muscholl, A., Schwentick, T., Segoufin, L., David, C.: Two-variable logic on words with data. In: Proceedings of the 21th IEEE Symposium on Logic in Computer Science (LICS 2006), pp. 7–16. IEEE Computer Society, Los Alamitos (2006)
2. Bray, T., Paoli, J., Sperberg-McQueen, C.M.: Extensible Markup Language (XML) 1.0. W3C Recommendation (1998), http://www.w3.org/TR/REC-xml
3. Cheng, E., Kaminski, M.: Context-free languages over infinite alphabets. Acta Informatica 35, 245–267 (1998)
4. David, C.: Mots et données infinies. Master's thesis, Université Paris 7, LIAFA (2004)
5. Demri, S., Lazic, R.: LTL with the freeze quantifier and register automata. ACM Transactions on Computational logic 10 (2009) (to appear)
6. Dubov, Y.: Infinite alphabet pushdown automata: various approaches and comparison of their consequences. Master's thesis, Department of Computer Science, Technion – Israel Institute of Technology (2008)
7. Kaminski, M., Francez, N.: Finite-memory automata. Theoretical Computer Science 138, 329–363 (1994)
8. Kaminski, M., Tan, T.: Tree automata over infinite alphabets. In: Avron, A., Dershowitz, N., Rabinovich, A. (eds.) Pillars of Computer Science. LNCS, vol. 4800, pp. 386–423. Springer, Heidelberg (2008)
9. Kaminski, M., Zeitlin, D.: Extending finite-memory automata with non-deterministic reassignment. In: Csuhaj-Varjú, E., Ézik, Z. (eds.) Proceedings of the 12th International Conference on Automata and Formal Languages – AFL 2008, Computer and Automation Research Institute, Hungarian Academy of Science, pp. 195–207 (2008)
10. Neven, F., Schwentick, T., Vianu, V.: Finite state machines for strings over infinite alphabets. ACM Transactions on Computational Logic 5, 403–435 (2004)
11. Shemesh, Y., Francez, N.: Finite-state unification automata and relational languages. Information and Computation 114, 192–213 (1994)

Modular Verification of Strongly Invasive Aspects

Emilia Katz and Shmuel Katz

Department of Computer Science
The Technion, Haifa 32000, Israel
{emika,katz}@cs.technion.ac.il

Abstract. An extended specification for aspects, and a new verification method based on model checking are used to establish the correctness of strongly-invasive aspects, independently of any particular base program to which they may be woven. Such aspects can change the underlying base program variables to new states, and after the aspect advice has completed, the base program code continues from states that were previously unreachable. The needed changes in the MAVEN model checker are described, and the soundness of the verification method is proven. An example is shown of its application to aspects that provide various bonus points to student grading programs.

1 Introduction

Aspect-oriented programming is becoming a common approach to extend object systems with modules that cross-cut the usual class hierarchy. Different programming languages, such as the extension of Java, AspectJ [9], are used to define aspect modules for modular treatment of different concerns, such as debugging, security, fault tolerance, and many others. The treatment of such concerns, without aspects, would have been scattered over different classes or methods of the base system, or would have been mixed with code treating other concerns. Every aspect consists of two parts: code for the concern treatment, called *advice*, and a description of when this code should be executed, called a *pointcut*. The pointcut defines the set of all the points in the execution of the program at which the advice is applied. These points are called *join-points*. The process of combining some program with an aspect (or a collection of aspects) is called *weaving*, and we refer to the original program as a *base system*, and to the result as a *woven system*.

Several works have dealt with model checking of aspect systems [4,5,6,8,10,13]. These works either treat a system with aspects woven in, or try to deal with the aspects modularly, relative to a specification. In the later case, the motivation is either to reduce the size of the models, or to allow convenient reuse of aspects in a library. Such an approach requires that the aspect itself have an independent specification that can be shown to hold. In one form or another, the specification of an aspect describes an *assumption* about any base system to which the aspect can be woven, and a *guarantee* about the resultant system after the aspect is

O. Grumberg et al. (Eds.): Francez Festschrift, LNCS 5533, pp. 128–147, 2009.

woven. The aspects are shown correct relative to their specification, and not to interfere with each other [6], and then, for each system to be constructed with the aspects, the base system is shown to satisfy the assumptions of the needed aspects. The construction of a model of the entire concrete woven system (which might be considerably larger than either of those used in the modular verification) and its direct verification do not have to be carried out at all.

So far, when aspects are treated separately from a specific weaving, it has been necessary to add a restriction: that the aspect returns control to the base system in a state that already existed for some computation of the base system without the aspect woven into it. Such aspects are called *weakly invasive* in [7], where the other categories of aspects mentioned in this paper are also defined. The reasoning behind the restriction is easy to understand: the aspect's assumption about the base system only relates to those computation sequences and states (known as *reachable* states) that can occur for some fair execution of the base system without the aspect. When an aspect returns control to the base system code, but in a state of the base variables that does not occur for any computation of the base system that begins from a "normal" initial state, there is no restriction on the behavior of the continuation. Instructions from the base code are executed, but with values that were never expected or tested, and with no restriction on the outcome. Thus the overall behavior of such a system is hard to analyze in a modular manner, separating the reasoning about the base from the reasoning about the aspects to be woven. In such cases, modular reasoning was thought unfeasible.

On the one hand, this restriction still allowed treating most aspects. Several kinds of aspects, including *spectative* ones that merely gather information, and *regulative* ones that merely restrict possible steps, are weakly invasive. Moreover, often the category of such aspects can be identified using dataflow techniques, as described in [7,12,14], and many commonly used aspect examples are weakly invasive. Nevertheless, there are other aspects that definitely are strongly invasive, and that occur in real applications, so that a more complete approach is desirable.

In this paper we show that such a restriction is unnecessary, and that a modular approach can be realized even for so-called *strongly invasive* aspects that do return control to the base system in new states that were unreachable in the base system executing alone. To do this, we take advantage of the usual organization of model checkers for linear time systems, and of the facilities they commonly provide. An extension of the MAVEN aspect verification system is presented, that can treat strongly invasive aspects, and an example of a bonus aspect for student grades is described.

The basic idea of the new approach is to add to the specification an assumption about the base system that restricts the computation segments that may become reachable after a strongly invasive aspect is woven. We then show once-and-for-all that when the aspect is woven into any base system with a reachable part that satisfies the previous type of assumption and an unreachable part that satisfies the added one, the result of the weaving will satisfy the guarantee. For a particular base system, we then have to show that the assumptions are true for both the reachable and unreachable parts (or at least the unreachable part that may become reachable after weaving). These tasks are made feasible due to the fact that many model checkers actually generate a state transition system that includes the unreachable parts of the computation, as a side-effect of the construction, and that marking the reachable states is a built-in operation.

The original MAVEN system [4], over NuSMV [1], builds a single model that can be checked to establish the correctness of a weakly-invasive aspect relative to its assume-guarantee specification, given in Linear Temporal Logic (LTL). (In the examples in this paper we use only the LTL modalities $G\,p$ - for "from now on, p", $F\,p$ - for "eventually, p", and $p\,U\,q$ - for "p is true until q becomes true".) The tableau state machine of the assumption is built using a module of NuSMV, and then the transition system of the aspect advice is woven into it, with pointcuts defining transitions to the beginning of advice state machine fragments, and with transitions back to the states of the base system that match the end states of the advice segments. It is then proven that whenever this particular model satisfies the guarantee assertion, then a woven system with any base satisfying the assumption, and the model corresponding to the aspect woven into it, satisfies the guarantee.

In the following section, precise definitions of the terms involved are presented, the theory behind the verification algorithm is described, and a proof of soundness is given, that extends the one given for the simpler MAVEN system. In Section 3 algorithms are given for computing the last states of the aspect, for determining the category of the aspect, for verifying the aspect, and for checking the base system for the needed assumptions. In Section 4 the specification and verification of an aspect for adding bonus points for student exercises and exams is described, and some concluding remarks are in Section 5.

2   Verification Theory for Strongly Invasive Aspects

Definition 1. *An aspect A is* strongly invasive *relative to a model M if a state of M that was unreachable in M becomes reachable in the woven system M+A and transitions of M are applied to it.*

The last part of the definition is needed to ensure that the aspect advice (sometimes) finishes in a state of M that was previously unreachable, and then the code of M is applied to the new state.

2.1   Refined Aspect Specification

The assumption of a strongly invasive aspect has to contain more information than the assumption of a weakly invasive one: it sometimes needs to define restrictions on the behavior of the unreachable part of the base system into which the aspect can be woven, in order to ensure an appropriate behavior of the woven system from the states that are made reachable by the strongly invasive aspect.

The specification of aspect A is now a triple: (P_A, U_A, R_A), where, as before (in [4]), P_A is the assumption about the reachable part of the base system and R_A is the result assertion guaranteed to hold in the woven base with the aspect. The new U_A statement is an LTL formula defining the restrictions on the unreachable part of the base system which is made reachable by completing an aspect advice fragment. The restriction is posed on computations of the base system that start in the states that might be reached by by completing the aspect advice, which were previously unreachable. We now may define the correctness of an aspect relative to such a specification, relating to a base system $S = S_{reach} \cup S_{unreach}$ where S_{reach} represents the reachable part, and $S_{unreach}$ the unreachable part.

Definition 2. *An aspect A is correct with respect to its refined assume-guarantee specification (P_A, U_A, R_A) if, whenever it is woven (by itself) into a system $S = S_{reach} \cup S_{unreach}$, where S_{reach} satisfies P_A and the part of $S_{unreach}$ that might become reachable after weaving satisfies U_A, the result will satisfy the guarantee, R_A.*

The property of the unreachable part of the system is relevant only for computation segments starting from a state that can be the last state of an advice execution. The reason is that only by an advice execution can a computation of the woven system pass from a state that was reachable in the base system to a state which was unreachable in the base system. Thus in order to check that the unreachable part of the base system satisfies the requirements of the aspect, it is enough to verify a formula of the form $L_A \to U_A$ on it, where L_A is a state formula describing the set of all the possible last states of the advice state machine, projected on the base system variables.

With some abuse of notation, we denote by L_A the set of possible last states of aspect A (identifying the unary predicate with the set it describes). Note that this set consists exactly of all the states in the base system into which a computation can arrive after finishing advice execution.

2.2   Refined Tableau Construction

Given an aspect A and its refined specification, (P_A, U_A, R_A), we need to construct a refined tableau to serve as a representation of all the base systems into which our aspect will possibly be woven. But now in order to build the tableau of the assumption of the aspect, it is not enough to build the tableau of P_A: we need to restrict the unreachable part of the tableau. The tableau needs to represent the systems, the reachable part of which satisfies P_A, and the unreachable part of which satisfies $L_A \to U_A$, where L_A is the predicate defining the set of all the possible return states of the advice. The refined tableau, T, is constructed in three steps:

Step 1: Automatically construct the predicate L_A. The construction is shown in Section 3.1.

Step 2: Use the lt12smv module of the NuSMV model checker to build the tableau T_1 of the LTL formula $(P_A \vee (L_A \wedge U_A))$.

Step 3: Take the tableau T to be the same as T_1 except for the initial states definition. To obtain the $INIT$ predicate of T, restrict the $INIT$ predicate of T_1 to include only states that should be reachable in the base system: $INIT \wedge P_A$.

Note that T_1 is the tableau of $(P_A \vee (L_A \wedge U_A))$ and not of $(P_A \vee (L_A \rightarrow U_A))$, because the only way to reach the part of the base system that does not satisfy P_A is by application of an aspect advice, and this will bring the computation to a state in which L_A must hold. This intuition will be justified during the proof of Theorem 1.

Let us denote the refined tableau constructed as above by $T_{(P_A,(U_A,L_A))}$.

Theorem 1. *Let A be an aspect with the refined assume-guarantee specification (P_A, U_A, R_A), and let L_A be a formula describing the set of all the possible last states of A. Then A is correct with respect to (P_A, U_A, R_A) if the result of weaving A into $T_{(P_A,(U_A,L_A))}$ satisfies R_A.*

We delay the proof of the theorem until after bringing some helpful definitions and lemmas needed for the proof. They appear below, together with the intuition for the proof.

In order to prove the theorem we need to show that if the result of weaving A into $T_{(P_A,(U_A,L_A))}$ satisfies R_A, then for every base system M such that its reachable part satisfies P_A and the unreachable part satisfies $L_A \rightarrow U_A$, the result of weaving A into M satisfies R_A. For this purpose it is enough to show that for every infinite fair path σ in the woven system $M + A$ there exists a corresponding infinite fair path π in the woven tableau, $T_{(P_A,(U_A,L_A))} + A$, such that $label(\sigma) \mid_{AP} = label(\pi) \mid_{AP}$. (Where AP is the set of all the atomic propositions appearing in the specification of A, a label of a state s, $label(s)$, is the set of all the atomic predicates that hold at the state s, and a label of a path τ, $label(\tau)$, is defined to be the sequence of the labels of the states of τ, so that if $\tau = s_0, s_1, s_2, \ldots$, $label(\tau) = (label(s_0), label(s_1), label(s_2), \ldots))$. In that case indeed in order to prove that every path in the woven system satisfies R_A, it is enough to show that every path in the woven tableau satisfies this property.

To simplify the notation, let us denote $T_{(P_A,(U_A,L_A))}$ by T. The task of finding a fair path in $T + A$ that corresponds to the given fair path of $M + A$ will be divided into steps according to prefixes of σ, and at each step a longer prefix will be treated. The following lemma will help to extend the treated prefixes:

Lemma 1. *Let S be a system, and let $s_0, \ldots, s_k$ be states in S such that s_0 and s_k are reachable by a fair path from some initial state of S (the paths and the initial states for s_0 and s_k might be different), and for each $0 \leq j < k$, the transition (s_j, s_{j+1}) exists in S. Then there exists a fair computation in S which contains the sequence of states $s_0, \ldots, s_k$.*

Proof
A computation is *fair* if it visits states from the *Fairness* set of the system model infinitely often. Let π_0 and π_k be fair computations in S in which s_0 and

s_k occur, respectively. Then $\pi_0 = \sigma_0 \cdot s_0 \cdot \ldots$, and
$\pi_k = \ldots \cdot s_k \cdot \sigma_k$ for some σ_0 and σ_k. Let us take
$\pi = \sigma_0 \cdot s_0 \cdot s_1 \cdot \ldots \cdot s_k \cdot \sigma_k$. This is obviously a path in S, and it starts from an initial state, as did σ_0. Moreover, π is a fair computation, because it has the same infinite suffix, σ_k, as the fair computation π_k.   Q. E. D.(Lemma 1)

The following definition will be useful for identifying the "interesting" prefixes of the path σ:

Definition 3. *Any infinite path π in a transition system can be represented as a sequence of* path segments -
$\pi = \pi^0 \cdot \pi^1 \cdot \ldots$, *where each path segment π^i is a sequence of states such that:*

- *If $i = 0$, the first state of π^i is the initial state of π*
- *If $i > 0$, the first state of π^i is either an initial state of an advice or a resumption state of the base system (i.e., a state in the base system into which the computation arrives after an advice execution is finished)*
- *The last state of π^i is either a pointcut state or a last state of an advice (after which the computation returns to the base system), or the last state of the path, if π is finite*
- *There are no pointcut states and no last states of advice inside π^i (i.e., in the states of π that are not the first or the last state)*
- *π is the concatenation of the path segments of π in the order of their indices*

Note that the decomposition of a path to path segments is unique, and that, because of loops, there can be resumption states within a segment. Note also that we could have an infinite (last) segment - in the reachable part of the base, or in the unreachable part, or even in the aspect. In our case all the paths in question are infinite, so the last state of each finite path segment will be either a pointcut or a last state of an advice. A resumption state might be unreachable in the system before weaving - in case of a strongly invasive aspect.

Now if we are given a path of M+A, $\sigma = \sigma^0 \cdot \sigma^1 \cdot \ldots$ where σ^i-s are the path segments of σ, for each finite prefix of σ consisting of a number of path segments we define the set of *corresponding path-segment prefixes* of fair paths in T+A:

$$\Pi_i = \{\pi^0 \cdot \pi^1 \cdot \ldots \cdot \pi^i |$$
$$label(\pi^0 \cdot \ldots \cdot \pi^i) \mid_{AP} = label(\sigma^0 \cdot \ldots \cdot \sigma^i) \mid_{AP},$$
$$\exists \pi \text{ fair path in } T+A \text{ such that } \pi = \pi^0 \cdot \ldots \cdot \pi^i, \ldots\}$$

Each element in Π_i is a prefix of an infinite fair computation of $T + A$ corresponding to the i-th prefix of σ, thus the following lemma will show that for every finite prefix of σ there exists a corresponding prefix of a fair computation in $T + A$:

Lemma 2. *Given a fair computation σ of M+A, and sets of prefixes Π_i-s as defined above, $\forall i \geq 0. \Pi_i \neq \emptyset$.*

Proof
The proof is by induction on i.

Base: $i = 0$ To show that Π_0 is not empty we need to show the existence of π^0 such that $label(\pi^0) \mid_{AP} = label(\sigma^0) \mid_{AP}$ and π^0 is a prefix of some fair path π in T+A. σ^0 is the first path-segment of a fair path in $M+A$, thus there is no advice application before σ^0 or inside it. So σ^0 is also the first path segment of a fair computation in M. According to the assumption on M, $M \models P_A$, thus for every fair path starting from an initial state of M there exists a corresponding fair path in T. In particular, there exists a fair path $\pi = t_0, \ldots, t_k, \ldots$ in T such that $label(\sigma^0) \mid_{AP} = label(t_0, \ldots, t_k) \mid_{AP}$. Then again, as $t_0, \ldots, t_k$ is a beginning of a fair path in T, and there are no pointcuts in it, except maybe for the last state, it is also a beginning of a fair computation in $T+A$. So let us take $\pi^0 = s_0, \ldots, s_k$. We are left to show that π^0 is indeed a path-segment, and then it will follow that $\pi^0 \in \Pi_0$, meaning that Π_0 is not empty.

$label(t_0) = label((\sigma^0)_0)$, thus t_0 is an initial state of $T+A$. There is no pointcut inside σ^0, because it is a path-segment, so the last state of σ^0 cannot be a return state of advice application, which means that it has to be a pointcut state. Due to the agreement on labels, the last state of π^0 will also be marked as a pointcut state. For the same reason, there are no pointcut states among $t_0, \ldots, t_{k-1}$, which, in the same way as for σ^0, implies that there are no advice return states also. Thus both ends of π^0 are legal ends of a path-segment, and there are no pointcut states and no advice return states inside π^0, which makes it, indeed, a legal path-segment.

Induction step. Let us assume that for every $0 \leq i < k$, $\Pi_i \neq \emptyset$. We need to prove that $\Pi_k \neq \emptyset$.

The induction hypothesis holds, in particular, for $i = k - 1$, thus there exists some prefix $\pi^0 \cdot \pi^1 \cdot \ldots \cdot \pi^{k-1}$ of a fair computation of $T+A$, corresponding to the prefix $\sigma^0 \cdot \sigma^1 \cdot \ldots \cdot \sigma^{k-1}$ of $M+A$'s computation, σ. Let us denote by $s\_first(i)$ the first, and by $s\_last(i)$ the last state of $i-th$ path-segment of σ (σ^i), and symmetrically for the states of path segments of $T + A$ - by $t\_first(i)$ the first, and by $t\_last(i)$ the last state of $i-th$ path-segment. There are two possibilities for $s\_last(k-1)$:

1. $s\_last(k-1)$ is a pointcut. Then $t\_last(k-1)$ is also a pointcut, because due to the induction hypothesis
   $label(s\_last(k-1)) \mid_{AP} = label(t\_last(k-1)) \mid_{AP}$. Then in every continuation of the computation both in $M + A$ and in $T + A$ the advice of the aspect will be performed, thus the k-th path-segment will in both cases be the application of the same advice from the same state, and the agreement on the labels of the k-th path-segments will be trivially achieved. Moreover, for the same reason the existence of an infinite fair path with the prefix $\pi^0 \cdot \pi^1 \cdot \ldots \cdot \pi^{k-1}$ implies the existence of an infinite fair path with the prefix $\pi^0 \cdot \pi^1 \cdot \ldots \cdot \pi^k$, because every continuation of the first prefix had to be an advice application. From the above it follows that in this case $\Pi_k \neq \emptyset$.

2. $s\_last(k-1)$ is a last state of the advice. This, in particular, implies that $s\_last(k)$ is a pointcut state, and no advice has been applied between $s\_last(k-1)$ and $s\_last(k)$. Here are again two possibilities:

The first case is that $s\_last(k$-$1)$ is a reachable state in M (more precisely, the state reachable in M is the projection of $s\_last(k$-$1)$ on AP). As no advice is applied between $s\_last(k$-$1)$ and $s\_last(k)$, we have that the whole path-segment σ^k is in the reachable part of M. Moreover, due to Lemma 1, as both $s\_last(k$-$1)$ and $s\_last(k)$ are reachable by some fair paths from some initial states of M, we also have that there exists a fair computation of M containing the sequence $s\_last(k$-$1), s\_first(k), \ldots, s\_last(k)$. All the fair computations of the reachable part of M are represented in the tableau of P_A, which is exactly the reachable part of T. Thus, in particular, the above fair path has a corresponding path in T, and, as there was no pointcut or advice application inside the sequence $s\_last(k$-$1), s\_first(k), \ldots, s\_last(k)$, there are also no pointcuts and advice applications in the corresponding sequence in the computation of T, and thus there exists a corresponding sequence of states in $T + A$, π^k. The first state of π^k, $t\_last(k$-$1)$, is reachable from the initial state of $T + A$ by some fair path, as Π_{k-1} is not empty. Moreover, all the prefixes of such fair pathes appear in Π_{k-1}, thus at least one of them continues to the sequence π^k. So indeed we obtain that there exists a sequence of states π^k corresponding to σ^k in the woven tableau, for which a fair continuation exists. We are left to see that the sequence of states, π^k, is indeed a path segment in the woven tableau computation. But this is true due to the agreement on labels of the states, $label(\pi^k) \mid_{AP} = label(\sigma^k) \mid_{AP}$: the path segment σ^k started from a return state of an advice, ended by a pointcut, and had no advice applications in the internal states, so the same is true for π^k and thus π^k is a path segment.

The last case left is that $s\_last(k$-$1)$ is unreachable in M. Additionally, $s\_last(k$-$1)$ is the last state of the advice, thus $s\_first(k)$ is the return state of the advice, and also is unreachable in M, because according to the weaving algorithm $label(s\_last(k$-$1)) \mid_{AP} = label(s\_first(k)) \mid_{AP}$. From the fact that $s\_first(k)$ is unreachable in M, together with the assumption on the unreachable part of M, we have that $L_A \rightarrow U_A$ holds in the suffix of any path starting from $s\_first(k)$. But from the agreement on labels with $s\_last(k$-$1)$ we also have that $s\_first(k) \models L_A$. Together we obtain that U_A holds in the suffix of any computation in M starting from $s\_first(k)$, and, in particular, for the computation $\sigma\prime$ containing the next path segment of σ, σ^k (because there is no advice application inside σ^k, all its states are states of the original system, M - either in the reachable or the unreachable part). Now let us examine the states of the woven tableau. The tableau of $L_A \wedge U_A$ is included in the refined tableau T, thus every computation satisfying U_A that starts from a state satisfying L_A is represented in T (though its initial state might be unreachable before the aspect is woven into T). Let $\pi\prime$ be a computation that corresponds to the suffix of $\sigma\prime$ that starts from $s\_first(k)$. The first state of $\pi\prime$ agrees on its label with $s\_first(k)$, and thus with $s\_last(k$-$1)$, which, according to the induction hypothesis, implies agreement on labels with $t\_last(k$-$1)$. According to the weaving algorithm, the last state of the advice is connected to all the states in the underlying system with which it agrees on labels. Thus, in particular, $t\_last(k$-$1)$ (which is the last state of the advice, in the same way as $s\_last(k\text{-}1)$), is connected to the first state of $\pi\prime$. So we can take the first state of $\pi\prime$ to be the first state of π^k. Let us then take π^k to be the first path-segment of $\pi\prime$. It is indeed a path segment of a fair computation (due to Lemma 1), it is connected to π^{k-1} and agrees on labels with σ^k, so we found what we needed.

Thus, indeed, the set of possible continuations, Π_i, is never empty.

Q. E. D.(Lemma 2)

Theorem 1 proof: Now let us return to the proof of Theorem 1. Let us be given an infinite fair path σ in the woven system $M + A$. From Lemma 2 it follows that there exists an infinite path π in the woven tableau corresponding to the given path σ - all the prefixes of π appear in the Π_i-s above, and due to the lemma, the Π_i-s are all non-empty. So in order to complete the proof of the theorem we need only to notice that every path constructed from the prefixes in Π_i-s above is fair, for the following reason: There are two possibilities for the infinite suffix of π. It either has infinitely many advice applications, or there exists some infinite suffix in which no aspect state is visited. If there are infinitely many advice applications, some state of the advice must be visited infinitely often, and all the states of the advice are defined as fair. If there is no advice application after some state, then there are only a finite number of path segments of π, and the last path segment is infinite. But, as we know, this path segment belongs to some fair path in $T + A$, so this must be a fair suffix, and so the computation π is indeed fair. This completes the proof of Theorem 1.   Q. E. D.

3   Algorithms

3.1   Computing L_A Automatically

Given a model of the aspect, A, in MAVEN format, we would like to automatically compute the state formula defining the set of all the possible last states of A's advice. The algorithm we propose consists of four steps:

Step 1: Construct a formula φ defining the pointcut of the aspect: take φ to be the disjunction of all the POINTCUT expressions in A.

Step 2: Run MAVEN on a model $A\prime$ which is the same as A except for a change in the specification. The assumption of the aspect is replaced by φ, and the guarantee of the aspect is replaced by *true*. The purpose of this operation is to obtain a system in which all the possible computations of the aspect are represented, and this goal is achieved in the following way:

– At the first step of its work, MAVEN will automatically construct the tableau of the new assumption of the aspect, φ, using the ltl2smv module of NuSMV. Note that in this tableau, T_φ, only the initial states are restricted, and the initial states are exactly all the possible join-points of the aspect.

- At the second step, MAVEN will perform the weaving of the aspect into the constructed tableau. The obtained woven system, $T_\varphi + A$, will contain all the possible computations of the aspect, because the initial states of the tableau are all the possible pointcut states that can occur in either reachable or unreachable parts of the base systems into which A will be woven (as the ranges of all the base variables as defined in the aspect model definition are the maximal possible, and the combinations of variables values are restricted only by the formula φ).

Note that if we added other restrictions on the computations of the tableau T_φ, we may not be able to guarantee that all the possible runs of the advice of A will appear in the woven tableau. For example, if we demand that the computations of the tableau should satisfy P_A, then after the weaving we would not obtain the runs of the aspect from the states that were unreachable in the base system. Since in the unreachable part of the base system which becomes reachable after the weaving there might be join-points of A, we have to model the computations of the advice starting from these states. However, there are cases when additional restrictions might be posed on the computations of the tableau built. For example, there might be some invariant that holds both in the reachable and the unreachable parts of the base system, and then it could be added to φ. Additionally, there might exist an assertion that holds for all the pointcut states, but is not explicitly written as part of the pointcut. Then it would be possible to restrict the initial states of the constructed tableau by this assertion.

Step 3: Take the woven system obtained in Step 2, $T_\varphi + A$, and use the built in functionality of NuSMV to compute the set of all the reachable states of this model, $(T_\varphi + A)_{reachable}$. For each of the states in $(T_\varphi + A)_{reachable}$, check whether it satisfies any of the RETURN conditions of the aspect. If it does, add it to the set L_A.

Step 4: Now L_A is the set of all the possible last states of A. What is left is only to construct the predicate describing this set. This is done by taking the disjunction of all the predicates describing the states in L_A.

Sometimes it might be easy to see a compact description of the possible last states of the aspect. For this case we provide the user a possibility to supply a manually constructed predicate L. But such a predicate should be checked before use, because the intuition of the user might be wrong. Then we use the above algorithm to construct the full L_A predicate, and check that the supplied predicate L is implied by L_A. If indeed $L_A \rightarrow L$ holds, the verification using L will still be sound, because it just might check additional paths, but no relevant path will be left unverified.

3.2   Determining the Aspect Category

Before applying the full verification technique it is very desirable to determine the category of the aspect. If the aspect is of the weakly invasive category (or a simpler category included within the weakly invasive one), then the simpler method described in [4] is applicable to it. Otherwise, the full verification method described in Section 3.3 should be used.

Some ways of determining the category of the aspect using code analysis, dataflow techniques and semantic definitions are described in [7,12,14,3]. If none of them gives a positive answer, the algorithm presented below can help to determine whether the aspect is uniformly strongly invasive, i.e., is always strongly invasive for every possible base system to which it can be woven. But first some definitions and observations are needed:

Remark 2. *From Definition 1 in Section 2 it immediately follows that for any system M in which all the states not reachable from the initial state by some fair path have been removed, if an aspect A is strongly invasive relative to M, there is a deadlock in the system $M + A$: Let s be a last state of advice execution such that there exists no reachable state s/ in M for which $label(s/) = label(s) \mid_{AP}$. Then this state is a deadlock state in the woven system.*

Lemma 3. *Let aspect A have the specification (P_A, U_A, R_A), where AP is the set of all the atomic propositions appearing in the specification and T_P denotes the tableau of P_A. Aspect A is strongly invasive with respect to P_A if when A is woven into T_P, there exists a state s in $T_P + A$ such that:*

- *s is the last state of advice execution, and*
- *there exists no state s/ in T_P such that s/ is reachable by some computation of T_P and $label(s/) = label(s) \mid_{AP}$*

Proof
Immediate from the above remark.

Definition 4. *Given a tableau T of an LTL formula ϕ, the tableau TP obtained from T by removing all the states that are not reachable from the initial state of T by any fair path (and only them) is called the* pruned tableau of ϕ.

Note that the above defined pruned tableau is equivalent to a tableau obtained from T by removing all the states and transitions that only lead to deadlock states.

Lemma 4. *Aspect A with the specification (P_A, R_A) is strongly invasive relative to P_A iff there exists a deadlock in the system $TP_A + A$, where TP_A is the pruned tableau of P_A.*

Proof
The conditions of Remark 2 above hold, in particular, for $M = TP_A$, so there will be a deadlock state in $TP_A + A$.

On the other hand, if there exists a deadlock in the system $TP_A + A$, let s be the deadlock state. Let us denote by s/ the state of TP_A such that $label(s/) = label(s) \mid_{AP}$. There are two possibilities: If s/ is reachable in TP_A, then there exists some infinite computation $\pi = s\prime, s_2, \ldots$ from $s\prime$ in TP_A, because TP_A is a pruned tableau. In particular, there exists a state s_2 in TP_A (the second state of π) to which $s\prime$ is connected. However, in $TP_A + A$ the state s is no longer connected to s_2. According to the construction of $TP_A + A$, the only reason could be that an advice is applied at s. But if an advice was applied at s, s would not be a deadlock state. Thus when we assumed that the projection of s on AP is reachable in TP_A we obtained a contradiction. So we conclude that $s\prime$ is unreachable in TP_A.

But could $s\prime$ still be reachable in T_P? This can only be if $s\prime$ has been removed from T_P during the construction of the pruned tableau. This means that all the paths starting from $s\prime$ led to some deadlock states, and thus $s\prime$ couldn't be reached by any fair computation of T. But according to Lemma 3 this exactly means that the aspect A is strongly invasive relative to its assumption. Q. E. D.

According to Lemma 4, the following algorithm verifies whether the given aspect is strongly invasive relative to its assumption:

1. Construct the pruned tableau TP_A from the tableau of the assumption of A. This is done automatically, by an iterative procedure that we have added to MAVEN. The procedure is as follows:
   - Run NuSMV to detect deadlock states in the tableau.
   - If a deadlock state is detected, construct a predicate describing this state, p
   - Rule out the deadlock state: Add the negation of p to the initial state definition, and to the predicate defining possible next states of the transitions.

   Repeat the procedure until there are no more deadlocks in the tableau.
2. Use MAVEN to weave the aspect into the above constructed tableau.
3. Run NuSMV to check whether there are deadlocks in the woven tableau. If a deadlock is detected, the aspect is strongly invasive relative to its assumption. Otherwise, the aspect A is weakly invasive relative to P_A.

Note that the algorithm presented here gives a positive answer only if the aspect is strongly invasive *relative to the tableau of its assumption*, but not relative to a concrete base system. Thus if the algorithm gives a positive answer, the aspect is strongly invasive relative to all the possible base systems into which it might be woven. But if the algorithm gives a negative answer, there might exist a base system satisfying the assumption of the aspect, with respect to which our aspect is still strongly invasive.

Given a base system S, there is one more way for us to check whether the given aspect, A, is strongly invasive relative to this system. Intuitively, what we would like to do is to look at all the unreachable states of the base system, and check whether there are last states of our aspect among these unreachable states. For that purpose we can check satisfiability of the following formula: $\varphi = S_U \wedge L_A$, where S_U is the formula defining the set of all the unreachable states of S, and L_A is the formula defining the set of all the possible last states of

A. φ can be constructed automatically: the way to construct L_A automatically is shown in Section 3.1, and the way to construct S_U automatically is shown in Section 3.4. Then the satisfiability of φ can be automatically checked using a SAT solver (such as, for example, Chaff [11]). If φ is found unsatisfiable, it means that there are no last states of the aspect A in the unreachable part of S, so A is weakly invasive relative to S, and the simpler model check in [4] can be used. If φ is found satisfiable, it doesn't necessarily imply that A is strongly invasive relative to S, because the predicate L_A is an over-approximation: it contains all the possible last states of the aspect, but maybe some of them will never occur in the computations of the woven system $S + A$, and thus will not bring the computation to states that were unreachable in S. But this over-approximation is a safe one: if we declare some aspect as strongly invasive when it is weakly invasive, we will just have to work harder to prove its correctness than we would if we knew its exact category, but the verification results will be sound.

3.3   Verifying the Aspect

Given an aspect A and its refined assume-guarantee specification, (P_A, U_A, R_A), the verification of correctness of A with respect to (P_A, U_A, R_A) is performed as follows:

1. Construct the refined assumption tableau for A as shown in Section 2.2 - the $T_{(P_A,(U_A,L_A))}$.
2. Use MAVEN to weave A into $T_{(P_A,(U_A,L_A))}$ and to run the NuSMV model checker on the resulting system and check the R_A property on it.

3.4   Base System Correctness Verification

Non-optimized solution. Given a base system S, we need to verify that it satisfies the refined assumption of our aspect, (P_A, U_A):

- Verify that the reachable part of S, S_{reach}, satisfies P_A
- Verify that all the computations starting from the unreachable part of S, $S_{unreach}$, satisfy $L_A \rightarrow U_A$.

The first verification task can be done by usual model-checking of S versus P_A. The meaning of the second task is as follows: we need to examine the model of $S_{unreach}$ and check all the fair computations that start from states satisfying L_A (note that a computation starting from a state in $S_{unreach}$ might return to the reachable part of S at some state). All these computations should satisfy U_A. The verification is performed in three steps:

1. Automatically compute the state formula S_U defining the set of all the unreachable states of S: S_U is the negation of the formula S_R defining all the reachable states of S, and in NuSMV there exists a possibility to compute S_R automatically for a given system S.

2. In the model of the base system, S, automatically replace the initial states definition by the formula $S_U \wedge L_A$
3. Run NuSMV on the obtained model and the formula U_A. If the verification succeeds, it means that the given base system satisfies the restriction on the unreachable part.

Optimization. In some cases, the requirement in the second part of the verification process can be relaxed due to the structure of U_A. For example, in case when U_A is some safety property, i.e., U_A has the form $G\,\varphi$, we do not have to verify that φ holds all along the computations starting from resumption states in the unreachable part of the system. We need to check only the segments between a resumption state and the next join-point or reachable state. So if we denote by *ptc* the predicate defining the pointcut of the aspect, and by *reachable* - the predicate defining the reachable states of the base system, then it is enough to verify the following formula on the unreachable part of the system: $L_A \rightarrow (\varphi \ U \ (reachable \vee (pointcut \wedge \varphi)))$. The reason is that when the computation reaches a join-point, in the woven system the advice will be executed at that point, so the information about the possible continuations of the computation in the base system from that point is useless. And if a computation leaves the unreachable part and arrives to some previously reachable state, its continuation will behave as specified by the assumption of the aspect about the reachable part of the base system, and all these continuations are already checked during the reachable part verification.

As an example of the situation described above, we can take a look at an aspect that is in charge of the scheduling policy of a semaphore-guarded resource. The purpose of the aspect is to implement a possibility of a waiting queue for the semaphore. As a result, the semaphore that could previously have only values 0 or 1 can now have negative values (according to the number of waiting processes). Thus the aspect is indeed strongly invasive. But there is a part of the system invariant that we need to extend to the unreachable part of the base system: regardless of the semaphore value and the concrete scheduling algorithm, we demand that no two processes hold the guarded resource at the same time. So if the formula ψ encodes the fact that two processes hold the resource at the same time, the assumption of the aspect about the unreachable part of the base system should be $U = G\,\neg\psi$. But when verifying the computations starting in the unreachable part of the base system, it is enough to check that after each possible last state of the aspect the computation satisfies $\neg\psi$ until it arrives to a pointcut state or to a reachable state.

4   Example

In this example we discuss an aspect that can be used in any grades-managing system. The aspect B provides a way of giving bonus points for assignments and/or exams (thus making it possible to have assignment/exam grades that are more than 100), but still keeping the final grade within the 0..100 range.

The aspect has two kinds of pointcuts, and two corresponding pieces of advice. The first pointcut of B is the moment when an assignment or exam grade is entered to the system. At this point the original system would accept only grades between 0 and 100, but the aspect offers a possibility of giving a bonus on the grade, and stores the new grade successfully even if it exceeds 100. The second pointcut of B is the moment when the final grade calculation of the base system is performed. Then if the calculation resulted in a grade that exceeds 100, the aspect replaces this grade by 100 (otherwise keeping the grade unchanged).

Aspect B is strongly invasive in the systems into which it can reasonably be woven, because its operation results in states in which some grades are more than 100, which is impossible in the base systems without bonus policies. And this example, though simple, is still of interest to us, because the aspect here exhibits a typical behavior we would like to treat: when it is woven into a system, the calculations there are performed partly in the aspect, and partly in the base system code, but using new inputs, that were impossible before the aspect was woven in.

The specification of B can be formalized as follows:

- The assumption on the reachable part of the base system is that all the grades appearing in the grading system - homework assignment grades (hw_i), exam grades ($exam\_j$), final grade (f) - are between 0 and 100, and after the final grade is ready (f_ready) (i.e., all the assignments and exams that comprise the grade have been checked, and the final grade has been calculated from them according to the base system grading policy), the final grade is published ($f\_published$). The result of the final grade calculation is represented by $calc$.

$$P_B = [\, \mathsf{G}(f\_ready \to ((f = calc) \land \mathsf{F}\, f\_published))$$
$$\mathsf{G}(f\_published \to f = calc) \land$$
$$\mathsf{G}(0 \le f \le 100) \land$$
$$\mathsf{G}(\forall 1 \le i \le 10(0 \le hw\_i \le 100)) \land$$
$$\mathsf{G}(\forall 1 \le j \le 2(0 \le exam\_j \le 100))]$$

Here, for modeling purposes, we have to provide some bounds on the number of assignments and exams, so we assume that there are no more than 10 home assignments and no more than 2 exams in each course. We also show the specification for the grades of a single student (because the grades of different students are independent, and calculations involving them can be viewed as orthogonal). When the model of the aspect is built, the ranges of all the variables - both the aspect variables and the relevant base system ones - are defined. Let us assume, for example, that our aspect gives bonuses in range of 0..20 points, then all the grade variables defined in the model of B are in the range 0..120.

- The assumption on the unreachable part of the base system is in our case a weakening of P_B. We still want the final grades to be published after they are ready, but now the final and the intermediate grades do not have to be bound by 100, but by 120. So we are left with the following property:

$$U_B = [\, \mathsf{G}(\textit{f\_ready} \to ((f = calc) \wedge \mathsf{F}\, \textit{f\_published})) \wedge$$
$$\mathsf{G}(\textit{f\_published} \to f = calc) \wedge$$
$$\mathsf{G}(0 \le f \le 120) \wedge$$
$$\mathsf{G}(\forall 1 \le \ i \le 10(0 \le hw\_i \le 120) \wedge$$
$$\mathsf{G}(\forall 1 \le \ j \le 2(0 \le exam\_j \le 120))]$$

- The guarantee of the aspect now is that regardless of the existence of bonuses on the components of the final grade, the final grade will be the one calculated by the base system function, but rounded down to 100 if needed:

$$R_B = [\, \mathsf{G}(\textit{f\_published} \to f = min(calc, 100))]$$

The guarantee of the aspect might also include a statement about the bonus policy it enforces, saying that the aspect calculates the bonuses as desired. But to simplify the discussion, we omit it here.
- The pointcut of the aspect can be formalized using the following predicates, which define the moments when the grades are entered into the system: $enter\_hw\_i$ for homework grades, and $enter\_exam\_j$ for exam grades.

$$Pointcut_B = [(\bigvee_{i=1}^{10}(enter\_hw\_i)) \vee$$
$$(enter\_exam\_1) \vee (enter\_exam\_2) \vee$$
$$(\textit{f\_ready} \wedge (f > 100))]$$

Let us follow the verification algorithm, applying it to aspect B. The first step is the refined tableau construction. It begins with calculating the predicate L_B, defining all the possible last states of B. In our example, we get

$$L_B = [((\textit{f\_ready} \to ((f = 100) \wedge (calc > 100))) \wedge$$
$$(\neg \textit{f\_published}) \wedge$$
$$\forall 1 \le \ i \le 10(\neg enter\_hw\_i) \wedge$$
$$\forall 1 \le \ j \le 2(\neg enter\_exam\_j) \wedge$$
$$(0 \le f \le 120) \wedge (0 \le calc \le 120) \wedge$$
$$\forall 1 \le \ i \le 10(0 \le hw\_i \le 120) \wedge$$
$$\forall 1 \le \ j \le 2(0 \le exam\_j \le 120)]$$

And here is the explanation: All the combinations of exams and assignments grades values in range 0..120 are possible at the last state of the aspect, because all the grades of assignments and exams are independent. There is a connection between the final grade and the other grades, but only when the final grade is declared to be ready and still is not published. Then the final grade is equal to the minimum between the calculated value (*calc*) and 100. However, as we do not want to restrict the calculation function of the base system, we cannot establish this connection, and at the other states of the computation the value of the final grade is not restricted (except by its range), so effectively we have to enable any combination of the final grade value and the other grades. The values of the other system variables are restricted as follows: The variables *enter_hw_i* and *enter_exam_j* for all *i*-s and *j*-s are *false*, because no grade is entered by the user at the last state of the advice. The variable *f_published* is also *false*, because the aspect does not publish the grades - even if it was called at the moment when the final grade was calculated, it just modifies the calculated grade, but does not publish it. Publishing the grades is done by the base system. The next variable to discuss is *f_ready*. If the aspect was called at the moment of grades entering, the variable *f_ready* is *false* at the join-point. The final grade is not calculated by the aspect in this case, so the variable remains *false* at the last state of the advice. However, if the aspect was called at a join-point when the final grade is calculated, the variable *f_ready* is true there and remains true after the advice finishes its execution. In this case, as we said earlier, we will also have $f = 100$ and $calc > 100$.

Now after the predicate L_B is constructed, the tableau of the $(P_B \vee (L_B \wedge U_B))$ formula is created, its initial states are restricted to those satisfying P_B (that is, the refined tableau $T_{(P_B,(U_B,L_B))}$ is built), and then B is woven into the result. The last part of the verification process is running NuSMV on the woven tableau in order to check the R_B property on it. And for the above described aspect, with the specification given, the verification succeeds, so our algorithm shows that indeed it is correct with respect to its refined assume-guarantee specification. Intuitively, the reason for the success of the verification is that the base system performs only some arithmetic operations on the grades the aspect modifies, and thus we can expect that the result of performing old operations on the new arguments will be as anticipated, if only there is no overflow or type declaration problem. (By a type declaration problem we mean, for example, the case when the type of the grades variables is defined in the base code by some *typedef* to be 0..100, so that larger values cause a fatal type error.) But the assertion U_B ensures that this will not happen, because U_B will not hold for the base system in case such problems arise.

Note that the aspect does not restrict the grade calculation process of the base system, so this aspect is highly reusable, as long as the calculation can handle values greater than 100 (as seen in U_B). Moreover, this aspect can appear in a library of aspects providing different grading policies: different types of bonuses for homework assignments, or factors on the exam grades. All these aspects will have the same requirements from the base system as B does, so when some grading system is checked for applicability of one of the aspects from this library, it is automatically inferred that all the other aspects from the library are also applicable to this base system. Thus the grading policy can be changed as needed at any time, by replacing the applied aspect, without any further checks on the base system.

$$U_B = [\, \mathsf{G}(f\_ready \rightarrow ((f = calc) \land \mathsf{F}\, f\_published)) \land$$
$$\mathsf{G}(f\_published \rightarrow f = calc) \land$$
$$\mathsf{G}(0 \leq f \leq 120) \land$$
$$\mathsf{G}(\forall 1 \leq\ i \leq 10(0 \leq hw\_i \leq 120) \land$$
$$\mathsf{G}(\forall 1 \leq\ j \leq 2(0 \leq exam\_j \leq 120))]$$

- The guarantee of the aspect now is that regardless of the existence of bonuses on the components of the final grade, the final grade will be the one calculated by the base system function, but rounded down to 100 if needed:

$$R_B = [\, \mathsf{G}(f\_published \rightarrow f = min(calc, 100))]$$

The guarantee of the aspect might also include a statement about the bonus policy it enforces, saying that the aspect calculates the bonuses as desired. But to simplify the discussion, we omit it here.
- The pointcut of the aspect can be formalized using the following predicates, which define the moments when the grades are entered into the system: $enter\_hw\_i$ for homework grades, and $enter\_exam\_j$ for exam grades.

$$Pointcut_B = [(\bigvee_{i=1}^{10}(enter\_hw\_i)) \lor$$
$$(enter\_exam\_1) \lor (enter\_exam\_2) \lor$$
$$(f\_ready \land (f > 100))]$$

Let us follow the verification algorithm, applying it to aspect B. The first step is the refined tableau construction. It begins with calculating the predicate L_B, defining all the possible last states of B. In our example, we get

$$L_B = [(f\_ready \rightarrow ((f = 100) \land (calc > 100))) \land$$
$$(\neg f\_published) \land$$
$$\forall 1 \leq\ i \leq 10(\neg enter\_hw\_i) \land$$
$$\forall 1 \leq\ j \leq 2(\neg enter\_exam\_j) \land$$
$$(0 \leq f \leq 120) \land (0 \leq calc \leq 120) \land$$
$$\forall 1 \leq\ i \leq 10(0 \leq hw\_i \leq 120) \land$$
$$\forall 1 \leq\ j \leq 2(0 \leq exam\_j \leq 120)]$$

And here is the explanation: All the combinations of exams and assignments grades values in range 0..120 are possible at the last state of the aspect, because all the grades of assignments and exams are independent. There is a connection between the final grade and the other grades, but only when the final grade is declared to be ready and still is not published. Then the final grade is equal to the minimum between the calculated value (*calc*) and 100. However, as we do not want to restrict the calculation function of the base system, we cannot establish this connection, and at the other states of the computation the value of the final grade is not restricted (except by its range), so effectively we have to enable any combination of the final grade value and the other grades. The values of the other system variables are restricted as follows: The variables $enter\_hw\_i$ and $enter\_exam\_j$ for all i-s and j-s are $false$, because no grade is entered by the user at the last state of the advice. The variable $f\_published$ is also $false$, because the aspect does not publish the grades - even if it was called at the moment when the final grade was calculated, it just modifies the calculated grade, but does not publish it. Publishing the grades is done by the base system. The next variable to discuss is f_ready. If the aspect was called at the moment of grades entering, the variable f_ready is $false$ at the join-point. The final grade is not calculated by the aspect in this case, so the variable remains $false$ at the last state of the advice. However, if the aspect was called at a join-point when the final grade is calculated, the variable f_ready is true there and remains true after the advice finishes its execution. In this case, as we said earlier, we will also have $f = 100$ and $calc > 100$.

Now after the predicate L_B is constructed, the tableau of the $(P_B \vee (L_B \wedge U_B))$ formula is created, its initial states are restricted to those satisfying P_B (that is, the refined tableau $T_{(P_B,(U_B,L_B))}$ is built), and then B is woven into the result. The last part of the verification process is running NuSMV on the woven tableau in order to check the R_B property on it. And for the above described aspect, with the specification given, the verification succeeds, so our algorithm shows that indeed it is correct with respect to its refined assume-guarantee specification. Intuitively, the reason for the success of the verification is that the base system performs only some arithmetic operations on the grades the aspect modifies, and thus we can expect that the result of performing old operations on the new arguments will be as anticipated, if only there is no overflow or type declaration problem. (By a type declaration problem we mean, for example, the case when the type of the grades variables is defined in the base code by some $typedef$ to be 0..100, so that larger values cause a fatal type error.) But the assertion U_B ensures that this will not happen, because U_B will not hold for the base system in case such problems arise.

Note that the aspect does not restrict the grade calculation process of the base system, so this aspect is highly reusable, as long as the calculation can handle values greater than 100 (as seen in U_B). Moreover, this aspect can appear in a library of aspects providing different grading policies: different types of bonuses for homework assignments, or factors on the exam grades. All these aspects will have the same requirements from the base system as B does, so when some grading system is checked for applicability of one of the aspects from this library, it is automatically inferred that all the other aspects from the library are also applicable to this base system. Thus the grading policy can be changed as needed at any time, by replacing the applied aspect, without any further checks on the base system.

5   Conclusions

We have shown that strongly invasive aspects can be specified and shown correct relative to their specification, independently of a particular base system. Moreover, it is reasonable to check the properties needed from the unreachable part of the base system because the possible transitions of the base are considered bottom up, independently of the initial states, thus generating the unreachable part of the base system as a byproduct of model checking. Strongly invasive aspects typically extend the functionality of the base system to situations not originally covered. The examples seen in the paper, of a semaphore with negative values, and of aspects to give bonus points beyond the normal range, are typical. Often some invariants true in the base system alone will no longer hold after weaving such aspects, but other invariants will continue to hold, and are essential to the correctness of the woven system.

The verification method presented here is modular, and thus has an advantage over a straightforward non-modular verification of a woven system: the possibility of reuse without proof. There are two types of such reuse we see, both of which are demonstrated by the aspect described in Section 4. One case is when one and the same aspect is applicable to different base systems. Then the verification of the advice versus the assume-guarantee specification is performed only once, and in order to be able to apply the aspect to a given base system we need only to perform the base system verification described in Section 3.4. Another case is when a library of aspects is given, where all the aspects are built for the same purpose (like defining some action policy) and have a common assumption (P, U) about the base system. Then if we have a base system that satisfies the above assumptions, we can change the policy defined in this system at any time, by applying different aspects from the library - one at a time, of course - without any further checks.

When model-checking is used, the size of the verified system and of the specification is very important, as it strongly affects the verification time, and sometimes, if the model verified is too large, the model-checker can even fail to provide any answer. For the complexity analysis purpose, we denote by m the size of the base system model, by a - the size of the aspect model ($|A|$), by r - the size of all the formulas in A's specification (assuming, without loss of generality, that all the formulas used in verification - P_A, U_A, R_A - are approximately of the same size). When a formula of size k is verified on a model of size m, the space complexity of the model checking is $O(m \cdot 2^k)$ ([2]). Thus the complexity of a straightforward verification of the woven system is $O(2^r \cdot (m \cdot a))$, because a system of size m is verified against a formula of size r (the guarantee of A, in this case). Let us find the complexity of the modular verification method. It is the sum of the following components:

- The verification of the base system. It is of $O(2^r \cdot m)$ for the reachable part, and the same for the unreachable part, so together we obtain $2 \cdot O(2^r \cdot m) = O(2^r \cdot m)$

– Verification of the aspect. Here, first the refined assumption tableau is constructed, $T_{(P_A,(U_A,L_A))}$. The complexity of this step is $O(2^{2r})$ (Note that L_A is always a state formula, and thus does not increase the complexity.) Then the woven tableau is built, and we obtain a system of size $O(2^{2r} \cdot a)$. At the last step, the woven tableau is verified against the guarantee of A, R_A, and this requires complexity of $O(2^r \cdot (2^{2r} \cdot a))$

The total complexity thus is $O(2^r \cdot (2^{2r} \cdot a)) + O(2^r \cdot m)$. But the size of the base system model is usually very large, so $m \geq 2^{2r}$, and thus the complexity of our verification is usually not worse than that of the straightforward woven system check. Even when this is not the case, the possibilities for reuse make the modular approach preferable.

References

1. Cimatti, A., Clarke, E.M., Giunchiglia, F., Roveri, M.: NuSMV: a new Symbolic Model Verifier. In: Halbwachs, N., Peled, D. (eds.) CAV 1999. LNCS, vol. 1633, pp. 495–499. Springer, Heidelberg (1999), http://nusmv.itc.it
2. Clarke Jr., E.M., Grumberg, O., Peled, D.A.: Model Checking. MIT Press, Cambridge (1999)
3. Djoko Djoko, S., Douence, R., Fradet, P.: Aspects preserving properties. In: Proc. of the 2008 ACM SIGPLAN Symposium on Partial Evaluation and Semantic-Based Program Manipulation (PEPM 2008), pp. 135–145. ACM, New York (2008)
4. Goldman, M., Katz, S.: MAVEN: Modular aspect verification. In: Grumberg, O., Huth, M. (eds.) TACAS 2007. LNCS, vol. 4424, pp. 308–322. Springer, Heidelberg (2007)
5. Katz, E., Katz, S.: Verifying scenario-based aspect specifications. In: Fitzgerald, J.S., Hayes, I.J., Tarlecki, A. (eds.) FM 2005. LNCS, vol. 3582, pp. 432–447. Springer, Heidelberg (2005)
6. Katz, E., Katz, S.: Incremental analysis of interference among aspects. In: Proc. of the 7th workshop on Foundations of aspect-oriented languages FOAL 2008, pp. 29–38. ACM, New York (2008)
7. Katz, S.: Aspect categories and classes of temporal properties. In: Rashid, A., Aksit, M. (eds.) Transactions on Aspect-Oriented Software Development I. LNCS, vol. 3880, pp. 106–134. Springer, Heidelberg (2006)
8. Katz, S., Sihman, M.: Aspect validation using model checking. In: Dershowitz, N. (ed.) Verification: Theory and Practice. LNCS, vol. 2772, pp. 373–394. Springer, Heidelberg (2004)
9. Kiczales, G., Hilsdale, E., Hugunin, J., Kersten, M., Palm, J., Griswold, W.G.: An overview of AspectJ. In: Knudsen, J.L. (ed.) ECOOP 2001. LNCS, vol. 2072, pp. 327–353. Springer, Heidelberg (2001), http://aspectj.org
10. Krishnamurthi, S., Fisler, K.: Foundations of incremental aspect model-checking. ACM Transactions on Software Engineering and Methodology (TOSEM) 16(2) (2007)
11. Moskewicz, M.W., Madigan, C.F., Zhao, Y., Zhang, L., Malik, S.: Chaff: Engineering an efficient sat solver. In: Proc. of the 38th Design Automation Conference, DAC 2001, pp. 530–535 (2001)

12. Rinard, M., Salcianu, A., Bugrara, S.: A classification system and analysis for aspect-oriented programs. In: Proc. of International Conference on Foundations of Software Engineering, FSE 2004 (2004)
13. Sipma, H.B.: A formal model for cross-cutting modular transition systems. In: Proc. of Foundations of Aspect Languages Workshop, FOAL 2003 (2003)
14. Weston, N., Taiani, F., Rashid, A.: Interaction analysis for fault-tolerance in aspect-oriented programming. In: Proc. Workshop on Methods, Models, and Tools for Fault Tolerance, MeMoT 2007, pp. 95–102 (2007)

Classes of Service under Perfect Competition and Technological Change
A Model for the Dynamics of the Internet?

Daniel Lehmann[*]

School of Engineering and Center for the Study of Rationality
Hebrew University, Jerusalem 91904, Israel
lehmann@cs.huji.ac.il

Abstract. Certain services may be provided in a continuous, one-dimensional, ordered range of different qualities and a customer requiring a service of quality q can only be offered a quality superior or equal to q. Only a discrete set of different qualities will be offered, and a service provider will provide the same service (of fixed quality b) to all customers requesting qualities of service inferior or equal to b. Assuming all services (of quality b) are priced identically, a monopolist will choose the qualities of service and the prices that maximize profit but, under perfect competition, a service provider will choose the (inferior) quality of service that can be priced at the lowest price. Assuming significant economies of scale, two fundamentally different regimes are possible: either a number of different classes of service are offered (DC regime), or a unique class of service offers an unbounded quality of service (UC regime). The DC regime appears in one of two sub-regimes: one, BDC, in which a finite number of classes is offered, the qualities of service offered are bounded and requests for high-quality services are not met, or UDC in which an infinite number of classes of service are offered and every request is met. The types of the demand curve and of the economies of scale, and not the pace of technological change, determine the regime and the class boundaries. The price structure in the DC regime obeys very general laws.

Keywords: classes of service, Internet.

1 Introduction

1.1 Background

Consider a delivery service. The quality of the service given may be measured by the delay with which the object is delivered. A service guaranteeing same day delivery is of higher quality than one that guarantees only next day delivery, or delivery within three-business days. Assuming that the service can be provided on a continuous range of qualities, it is typically unrealistic to assume that

---

[*] This work was partially supported by the Jean and Helene Alfassa fund for research in Artificial Intelligence.

O. Grumberg et al. (Eds.): Francez Festschrift, LNCS 5533, pp. 148–169, 2009.

a service provider could provide all those different services to the customers requiring them or bill them differentially for the qualities requested. In many situations, the service provider will have to decide on certain discrete qualities, i.e., classes of service, and will provide only services of those specific qualities.

The price paid by the different customers depends only on the quality of the service they receive, not on the (inferior) quality they requested and all customers getting the same service pay the same price. We assume that every customer has sharp requirements concerning the quality of service he requests: he will under no circumstances accept a quality that is inferior to his requirements, even if this is much cheaper, and he will, as long as the price is agreeable, use the service of the least quality that is superior or equal to his request. This assumption is not usual, and not in line with [1] for example. Nevertheless, it is very reasonable in connection with Internet services: there is quite a sharp boundary between delays, jitters and latencies that allow for the streaming of video content and those that do not.

We assume significant economies of scale to the service provider: the cost of providing w services grows less than linearly with w, but we also assume that services of different qualities do not aggregate to generate economies of scale: the cost of providing two classes of service is the sum of the costs of those two different services. This is obviously a severe assumption. In practice, a company offering a few different qualities of service will benefit from economies of scale across those different services: administrative services for example will be shared. But if one considers only the cost of moving packets over the Internet and one assumes, in no way a necessary assumption but a definite possibility (see [2]), that separate sub-networks are affected to the different qualities of service, then the assumption will be satisfied. In the sequel we may therefore assume that a firm provides only a single quality of service: different firms may provide different qualities of service. In deciding what quality or qualities of service to provide, a firm has to avoid two pitfalls: offering a service of poor quality may cater to a share of the market that is too small to be profitable, offering a service of high quality may involve costs that are too high to attract enough customers. A monopoly will set a quality of service and a price that maximize its revenues, but, in a competitive environment, a customer will buy from the provider offering the lowest price for a service whose quality is superior or equal to the quality requested. This will, in general, result in a lower quality of service, a lower price and a higher activity.

Will more than one class of service be proposed? How many? What will the price structure of those classes be? The relation between the traffics in those different classes? Between the revenues gathered in giving those services of different qualities?

Those questions may be asked in many different situations, but they are particularly relevant in connection to the Internet. The Internet delivers packets of information from a source to a destination. The IPv4 protocol, the current protocol for the Internet, reserves three bits in each packet for specifying the quality of service desired, but does not use those bits and treats all packets equally. End users rarely pay per packet and usually pay a flat rate. A number of companies have prepared products for QoS (quality of service), i.e., for controlling an Internet-like network in which packets are treated differentially, but the need for such products is not yet proven. A survey of the different proposals for QoS may be found in [3]. Since the Internet is a loose organization that is the product of cooperation between very diverse bodies, billing for such services would also be a major problem. Many networking experts have therefore claimed that the current *fat dumb pipe* model is best and argued that it will prevail due to the rapid decline in the cost of equipment. A discussion of the different predictions about the evolution of the Internet may be found in [2], where a specific proposal, Paris Metro Pricing (PMP) is advocated. Much interesting information about the economics of the Internet is found at [4]. An important aspect of the economy of the Internet is that the prices of the networking equipment are dropping very rapidly. One wonders about the consequences of this rapid decline on the price structure.

1.2   Main Results

Answers to the questions above are obtained, assuming a one-dimensional ordered continuum of qualities of service, significant economies of scale and a competitive environment. Under some reasonable and quite general assumptions about the demand curve and the cost function, and assuming that a change in price has a similar effect on the demand for all qualities of service, it is shown that one obtains one of two situations.

- (UC regime) If the demand for services of high-quality is strong and the economies of scale are substantive, then a service provider may satisfy requests for service of arbitrary quality: there will be only one class of service catering for everybody's needs; this is the *fat dumb pipe* model.
- (DC regime) In other cases, a number of classes of service will be proposed and priced according to the quality of the service provided. Very high quality services may not be provided at all.

The DC regime just described may appear in one of two sub-regimes:

- (BDC regime) A finite number of classes of service are offered and services of very high quality are not offered at all.
- (UDC regime) An infinite number of classes and all qualities of service are offered.

Notice that, for example, it cannot be the case that a finite number larger than one of classes are offered and that the highest quality of service caters for unbounded qualities. Significant economies of scale and high demand for high-quality services imply a UC regime, independently of the way prices influence demand. When not in a UC regime, high sensitivity of demand to price implies a BDC regime whereas low sensitivity implies an UDC regime. In a BDC regime, when the basic price of the equipment drops, new classes of service will be offered to the high-end customers that could not be profitably catered for previously. In the UC regime, a slowing of the decline in the price of equipment, does not cause the appearance of multiple classes of service. In a DC regime, a change in the price of equipment cannot cause a transition to a UC regime.

A decline in equipment prices, in a DC regime, does not change the boundaries between the different classes of service, but new classes catering to the high-end of the market become available when prices are low enough to make them profitable. The prices drop for all classes, but they drop dramatically for the newly created classes. The ratio of the prices between a class of service and the class just below it decreases and tends to some number typically between 1.5 and 3, depending on the exact shape of the distribution of demand over different qualities and the size of the economies of scale, when prices approach zero. Traffic increases when equipment prices decrease. The direction in which revenues change depends on the circumstances, but, typically, revenues go up and then down when equipment prices decrease. The traffics in neighboring classes of service tend to a fixed ratio when prices approach zero.

2   Previous Work

This work took its initial inspiration in the model proposed by A. Odlyzko in the Appendix to [2]. In [2,5], the authors study the case of two different types of customers each type requesting a specific quality of service. They show that, in certain cases, two classes of service will be proposed. The present work assumes a continuous distribution of types of customers. The model of [1] is more detailed than the one presented here and addresses slightly different questions, but its conclusions are supported by the present work.

The situation described in this paper may seem closely related to the well-studied phenomenon of non-linear pricing [6]. A second look seems to indicate major differences:

- here the pricing is linear: a customer given many services pays for each of those separately and, in the end, pays the sum of the prices of the services received,
- the cost of providing services is assumed to exhibit significant economies of scale, whereas costs are assumed to be additively separable among customers and qualities in most works on non-linear pricing,
- the present work deals with a competitive environment whereas many works on non-linear pricing assume a monopoly situation.

3   Model and Assumptions

The results to be presented are very general, i.e., they hold under very mild hypotheses concerning the form of the cost function and the demand curve, and this is the main reason for our interest in them. Nevertheless, describing exactly the weakest hypotheses needed for the results to hold would lead to a paper loaded with mathematical details distracting the reader from the essentials. Therefore, the conditions under which those general results hold will not be described precisely and those results will be only justified by hand waving. Precise results will deal with two specific parameterized families of functions for cost and demand, described in (2) and (3).

We consider a service that can be given at a continuum of different levels of quality, e.g., the acceptable delay in delivering a packet over the Internet. In this paper, quality is assumed to be one-dimensional (and totally ordered). This is a severe assumption. Quality of service over the Internet is generally considered to be described by a three-dimensional vector: delay, latency and jitter. In this work the quality of service is characterized by a real number $q \geq 1$. A service quality of 1 is the lowest quality available. How does a service of quality q compare with a service of quality 1? The following will serve as a quantitative definition of the quality q of a service:

Quality. The cost of providing a service of quality q is the cost of providing q services of quality 1.

We assume significant economies of scale: the cost of providing w services of quality 1 grows like w^s for some s, $0 < s < 1$: the smaller the value of s, the larger the economies of scale. In [5] one may find a discussion of which s best fits the Internet data: values between $1/2$ and $3/4$ seem reasonable. The cost of providing w services of quality 1 is therefore:

$$cost = c\, w^s$$

where c is some positive value that characterizes the price of the equipment. The number c represents the level of the technology, it is a *technological constant*. Many studies indicate that the price of computer and Internet equipment is decreasing at a rapid pace, probably exponentially. The dynamics of our model is described by a decrease in the value of c.

The cost of providing $f(q)$ services of quality q for every q is:

$$cost = c \left(\int_1^\infty q\, f(q)\, dq \right)^s .$$

As is customary, we include the profits in the costs and will describe the equilibrium by an equation equating costs and revenues. Note that the equation above expresses the fact that economies of scale are obtained only among services of the same quality and do not obtain for performing services of two different qualities.

The second half of our model consists of a demand curve describing the demand for services of quality q at price p. We assume that the demand for such services is described by a density function $d(q, p)$. The demand for services of quality between a and $b > a$ is:

$$demand = \int_a^b d(q, p)\, dq$$

if the price of any such service is p. Notice that we assume that services of different quality are priced identically. It is very difficult to come up with justifiable assumptions about the form of the demand function $d(q, p)$. One may assume that, for any fixed q, the value of $d(q, p)$ is decreasing in p, and even approaches zero when p tends to infinity. The rate of this decrease is much less clear. For a fixed p, how should $d(q, p)$ vary with q? Equivalently, what is the distribution of the demand over different qualities of service? This is not clear at all. Both cases of $d(q, p)$ increasing and decreasing in q will be considered. Fortunately, our results are very general and need not assume much about those questions. There are two assumptions we must make, though. They will be described and discussed now. The first one is that the effect of price on the demand is similar at all quality levels, or that the distribution of demand over q is the same at all prices. We assume that the function d is the product of two functions, one that depends only on q and one that depends only on p. In the absence of information (difficult to obtain) on the exact form of the demand curve, this sounds like a very reasonable first order approximation.

$$\textbf{Decoupling} \quad d(q, p) = f(q)\, h(p) \tag{1}$$

The function f describes the distribution of the demand over different qualities of service. The (decreasing) function h describes the effect of price on the demand. Our results depend on this Decoupling Assumption but should be stable under small deviations from this approximation. The problem of studying systems in which this assumption does not hold is left for future work. Our second assumption is that the function f is constant in time, i.e., does not depend on the technological constant c. Technological progress modifies demand only through price. Our results about the dynamics of the model rely on this assumption.

To be able to prove mathematically precise results, we shall, when needed, assume that:

$$f(q) = q^\alpha \tag{2}$$

for some α: the larger the value of α, the larger the relative size of the demand for high-quality services. Is it clear whether α should be positive or negative? Suppose e-mail packets and streaming-video packets can be sent at the same price, would the demand for streaming-video packets be larger than that for e-mail? Probably. It seems that a positive value for α is more realistic, but we do *not* make any assumption on the sign of α. It may be the case that a more realistic function f should, first, for q's close to one, increase, and then decrease for q's above the quality required for the most demanding applications at hand. We shall not try to discuss this case here, but the conclusions of this paper offer many qualitative answers even for such a case.

For h, we shall always assume that $h(p) \geq 0$ for every $p \geq 0$, that h is decreasing when p increases and that, when p approaches $+\infty$, $h(p)$ approaches zero. Since $h(0)$ characterizes the demand at price 0, we shall also assume, for proper scaling, that $h(0) = 1$. The following are examples of possible forms for h:

$$h(p) = \frac{1}{1 + (a\, p)^\beta}, \text{ for some } \beta > 0,\ a > 0 \tag{3}$$

In (3), β describes the asymptotic rate of decrease of the demand when price increases, but a describes the transient behavior of the demand: the larger β, the more sensitive the demand is to price. For β close to zero, the demand is little influenced by price, i.e., the demand is incompressible. One may also consider the following:

$$h(p) = e^{-p}, \tag{4}$$

$$h(p) = e^{-p^2}. \tag{5}$$

As explained at the start of this Section, precise mathematical results will assume that the functions f and h are defined by (2) and (3), but the qualitative results hold for a much larger family of functions.

4   Beginnings

4.1   Equilibrium Equation

As explained in Section 1, we assume that the service provider cannot provide differentiated services: it has to provide the same service and to charge the same price for every service it accepts to perform. Let us, first, assume that there is no service available and that a service provider considers whether or not to enter the market. In a sense, the provider is a monopolist for now. But we shall show below that, because of possible competition, it cannot maximize its profit. It is clear that a provider will choose a quality of service b and a price p, and offer services of quality b at price p to any potential customer. The potential customers are all those customers who request services of quality $q \le b$. Customers requesting quality above b will not be served. If there are no competitors, one can hope to attract the totality of the demand for services of quality $q \le b$ at price p. The size of the demand for service of quality q ($q \le b$) is $f(q)h(p)$. Since we provide a service of uniform quality b, the equivalent number of services of quality 1 that we shall perform is:

$$\int_1^b b\, f(q)\, h(p)\, dq = b\, h(p)\, A(1, b), \tag{6}$$

where we define:

$$A(a, a') = \int_a^{a'} f(q)\, dq.$$

Note that $A(q, q')$ is the demand for services of quality between q and q' if the price for those services is zero. The cost of providing the services of (6) is:

$$cost = c\, b^s\, h^s(p)\, A^s(1, b). \tag{7}$$

The total demand for services is:

$$demand = \int_1^b f(q)\, h(p)\, dq = h(p)\, A(1, b). \tag{8}$$

Since every service pays the same price p, the revenue is:

$$revenue = p\,h(p)\,A(1,b). \tag{9}$$

Equilibrium is characterized by:

$$p\,h(p)\,A(1,b) = c\,b^s\,h^s(p)\,A^s(1,b),$$

or equivalently by:

$$p\,h^{1-s}(p) = c\,b^s\,A^{-(1-s)}(1,b). \tag{10}$$

Equation (10) characterizes the price p as a function of the quality of service b. Most of the remainder of this paper is devoted to the study of this equation.

4.2   Existence of an Equilibrium

Given b and c, we want to solve (10) in the variable p. The solutions of (10) are obtained as the intersections of the curve representing the left hand side and a horizontal line representing the right hand side. The left hand side is equal to zero for $p = 0$ and is always non-negative. It is reasonable to assume that, considered as a function of p, it has one of the two following forms.

- Sensitive case: $ph^{1-s}(p)$ increases, has a maximum and then decreases. This seems to be the most typical case: the price has a strong influence on the demand and therefore h is such that $h^{1-s}(p)$ decreases more rapidly than $1/p$ when p is large. This is the case, for any s, when h has the form described in (4) or (5). If h follows (3), it is the case iff $\beta > \frac{1}{1-s}$. Fig 1 presents a typical sensitive case: $s = 2/3$, $\beta = 6$ and $a = 1$, i.e., $h(p) = \frac{1}{1+p^6}$.
- Insensitive case: $ph^{1-s}(p)$ increases always and tends to $+\infty$ when p tends to $+\infty$. This is the case if the demand is relatively insensitive to the price. It happens if h follows (3) and $\beta < \frac{1}{1-s}$. The case in which $ph^{1-s}(p)$ increases and approaches a finite value when p tends to $+\infty$ is possible ($\beta = \frac{1}{1-s}$) but not generic. Fig 2 presents a typical insensitive case: $s = 2/3$, $\beta = 2$ and $a = 1$, i.e., $h(p) = \frac{1}{1+p^2}$.

Equation (10) behaves quite differently in those two cases. Whereas in the insensitive case, (10) has a unique solution whatever the values of c and b, in the sensitive case we must distinguish between two situations.

- If the horizontal line is high, e.g., if the technological constant c is large, the horizontal line does not intersect the curve and there is no price p to solve (10). In this case providing a service of quality b is unprofitable, even in the absence of competition and no such service will be provided.
- If the horizontal line representing the right hand side intersects the curve, for each intersection there is a price p solving (10). In this case, the provider may consider providing a service of quality b at any such price p, but before any more sophisticated analysis, it is clear that the price at which service of quality b will be proposed (if proposed) is the smallest of the p's solving (10). If a provider proposes another solution of the equation, a competitor will come in and propose a service of the same quality at a lower price.

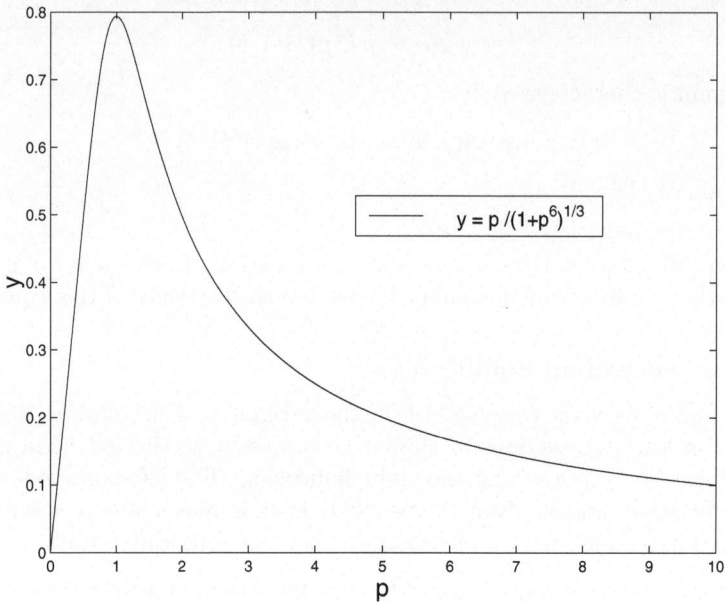

Fig. 1. Left hand side of (10), $s = 2/3$, $\beta = 6$

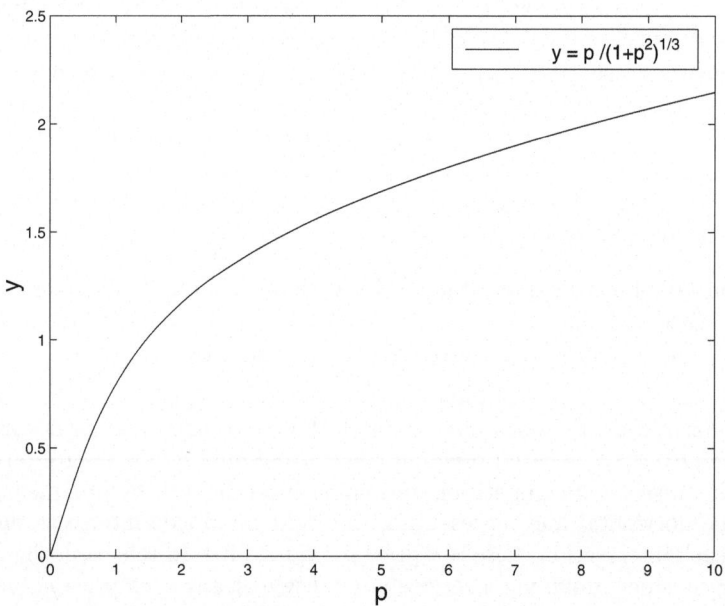

Fig. 2. Left hand side of (10), $s = 2/3$, $\beta = 2$

In other terms: if (10) has no solution, no service of quality b can be profitably provided, and if (10) has at least a solution then the service may be provided at a price that is the smallest solution of (10).

4.3 Quality of the Service Provided in a Competitive Environment

Suppose a service provider finds it is in a situation in which (10) has solutions for some values of b. It may consider providing a service of quality q for any one of those values. The price it will charge, the traffic and the revenue obtained depend on the quality q chosen. To understand the choice that a provider has to make, we need to study (10) in some more depth. Assume that c is given. One would like to know which qualities of service can be profitably provided, i.e., one wants to consider the set of b's for which the Equation has a solution. The left hand side of the Equation has been studied in Section 4.2: let us study the right hand side. Under very general circumstances, for $b \to 1^+$, $A^{-(1-s)}(1, b)$ approaches $+\infty$. Figure 3 describes the behavior of the right hand side of (10) as a function of b for different choices of f (i.e. α) and s. The number c is chosen to be 1 in all cases. In the leftmost graph $\alpha = 1$ and in the rightmost graph $\alpha = -1$.

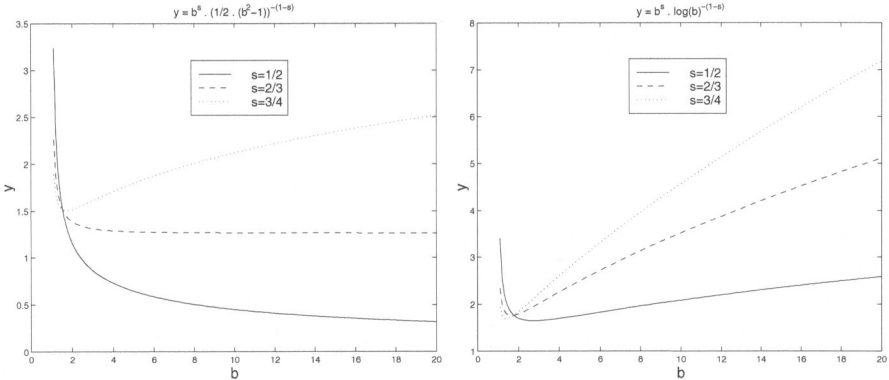

Fig. 3. Right hand side of (10)

The function $b^s A^{-(1-s)}(1, b)$ also approaches $+\infty$ for b approaching 1 and, essentially, it may present any one of two possible behaviors: either it decreases always, or it decreases until it gets to a minimum and then increases without bounds. The case it decreases to a minimum and then increases and approaches a finite value when b tends to $+\infty$ is possible but not generic. These two possible behaviors for the function $b^s A^{-(1-s)}(1, b)$ determine two very different regimes.

Definition 1. *In the first regime, the Universal Class regime (UC regime), the right hand side of (10) is a decreasing function of b. In this regime the set of b's for which (10) has a solution is an interval of the form $[q_1, +\infty[$, for some*

$q_1 > 1$. *Any service of quality at least q_1 may be profitably provided. In the second regime, the Differentiated Classes regime (DC regime), the right hand side of (10) decreases, reaches a minimum and increases. One must then distinguish between the sensitive and the insensitive cases. In the sensitive case, the set of b's for which (10) has a solution is either empty or an interval of the form $[q_1, q_2]$, for some $q_2 \geq q_1 > 1$. This regime (DC, sensitive case) will be called the BDC (bounded DC) sub-regime. In the insensitive case, (10) has a solution for any b in the interval $[1, +\infty[$. This regime (DC, insensitive case) will be called the UDC (unbounded DC) sub-regime.*

Suppose a service provider finds it is in a situation in which (10) has solutions for some interval of qualities. We know it should choose a quality q in this interval, offer a service of quality q and propose this service at the price p that is the smallest solution to (10). Which q should our provider choose? To each choice for q corresponds a price p_q. The number of services performed is given by

$$demand = h(p_q)\, A(1, q).$$

Since every service performed is a service of quality q, and every such service is equivalent to q services of quality 1, the weighted traffic is:

$$traffic = q\, h(p_q)\, A(1, q).$$

The revenue is:

$$revenue = p_q\, h(p_q)\, A(1, q).$$

Consider, first, a case typical of the DC regime. In Figure 4 the price p_q, the traffic and the revenue are described as functions of the quality q of the service chosen by the provider. The service can be provided only for a bounded interval of qualities. Notice that, in this interval, the price decreases rapidly and then increases slowly, that traffic and revenue increase and then decrease, and that the minimum price is obtained for a quality inferior to the quality that maximizes the revenue.

A case typical of the UC regime is described in Figure 5. By our discussion of (10), it is clear that p_q is a decreasing function of q. Notice that traffic and revenue increase with q.

What is the quality of service most profitable? Since the revenue includes the profits, one may assume, at first sight, that the first provider to propose such a service will choose the quality q that maximizes its revenue. This analysis is clearly incorrect in a competitive situation. If a provider decides to offer the service that maximizes its revenue, a competitor provider may come in and offer a service of lower quality at a lower price and steal the demand for the lower quality services. Therefore, the price at which the service will be offered is the minimum price possible: the minimum of the function p_q on its interval of definition.

Claim. The quality q of the service provided is the q for which p_q is minimal.

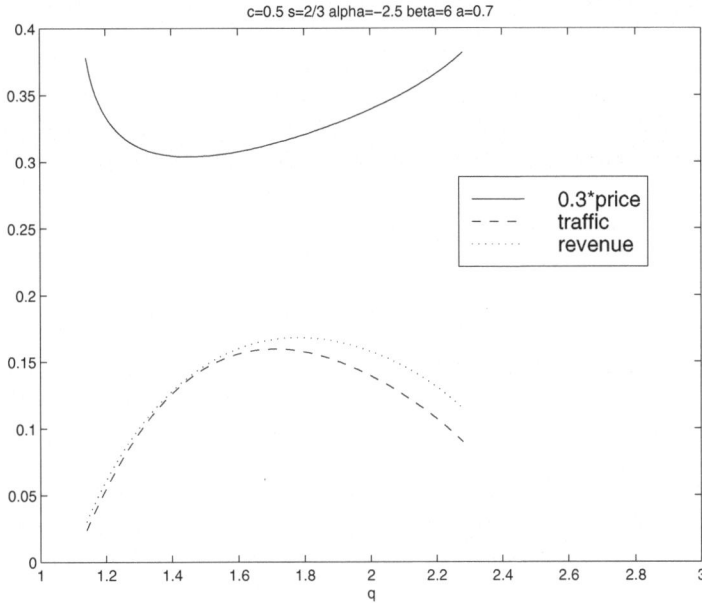

Fig. 4. Price, Traffic, Revenue as functions of q in the DC regime. Service can be provided only for q in a bounded interval.

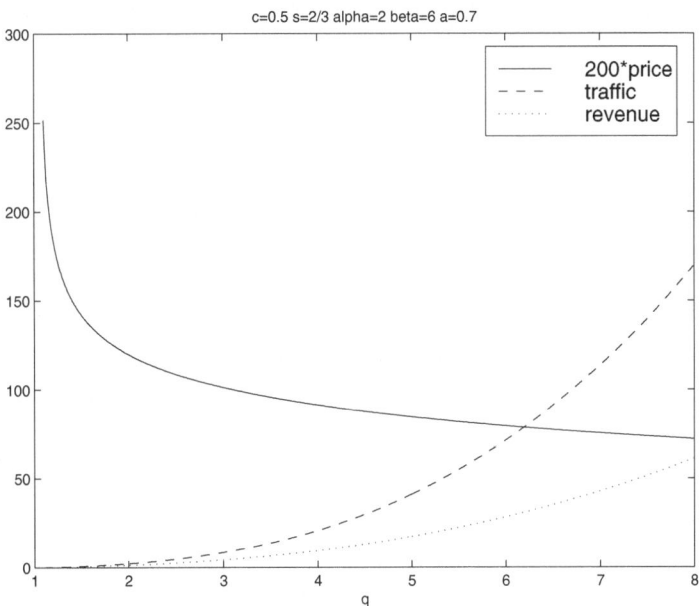

Fig. 5. Price, Traffic, Revenue as functions of q in the UC regime

This remark turns out to be of fundamental importance because this quality q, contrary to the quality that maximizes revenue, does not depend on the technological constant c or on the function h that describes the dependence of the demand on the price charged. It depends only on the function f that describes the distribution of the demand over the different qualities of service and on the size of the economies of scale described by the number s. Indeed, this quality q is the value of b that minimizes $b^s A^{-(1-s)}(1, b)$ in the DC regime and it is $+\infty$ in the UC regime. There is, therefore, a sharp distinction between the two regimes: in the DC regime a service of finite quality is provided, whereas, in the UC regime, the quality of service provided is high enough to please every customer, i.e., unbounded.

Law 1 (First Law). *Whether or not some service is provided does not depend on the function h that describes the way demand depends on price. It depends on the technological constant c, the size of the economies of scale s, and on the function f that describes the distribution of demand over different qualities of service. If some service is provided its quality depends only on s and f, it does not depend on c or h. In the UC regime, a single class serves requests for services of arbitrary quality and it is given for free. In the DC regime, the service of lowest quality has a finite quality and does not cater for high-end customers. The prevailing regime depends on the form of the economies of scale and on the form of the distribution of the demand over different qualities, it does not depend on the price of the equipment or on the way prices influence demand.*

In Section 5, we shall, first, mark the boundary between these two regimes. Then, both regimes: UC and DC will be studied in turn.

5   Between the Two Regimes

In this Section, we shall assume that f has the form described in (2), for some α, and study the exact boundaries of the UC and DC regimes. The regime is determined by the behavior of the function:

$$w(b) = b^s \left(\int_1^b \frac{dq}{q^\alpha} \right)^{-(1-s)} ,$$

for $b \geq 1$. If $w(b)$ has a minimum, then the regime is DC, if it always decreases, the regime is UC. Three cases must be distinguished.

$$\text{If } \alpha < -1, \quad w(b) = \frac{b^s}{(-\alpha - 1)^{-(1-s)}} \left(1 - b^{\alpha+1} \right)^{-(1-s)}.$$

$$\text{If } \alpha = -1, \quad w(b) = b^s \ln^{-(1-s)} b,$$

and

$$\text{if } \alpha > -1, \quad w(b) = \frac{b^s}{(\alpha + 1)^{-(1-s)}} \left(b^{\alpha+1} - 1 \right)^{-(1-s)}.$$

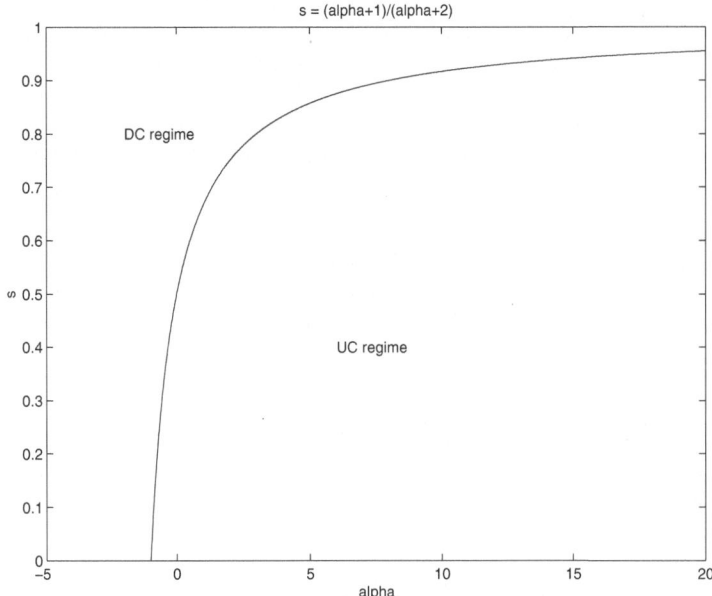

Fig. 6. Boundary between the UC and DC regimes

For $\alpha < -1$, we have:

$$w'(b) = \frac{b^{-(1-s)} \left(1 - b^{\alpha+1}\right)^{-(2-s)}}{(-\alpha - 1)^{-(1-s)}} \left[s(1 - b^{\alpha+1}) - (1 - s)b(-\alpha - 1)b^{\alpha} \right],$$

$$w'(b) = \frac{b^{-(1-s)} \left(1 - b^{\alpha+1}\right)^{-(2-s)}}{(-\alpha - 1)^{-(1-s)}} \left[(-(2 + \alpha)s + \alpha + 1) b^{\alpha+1} + s \right],$$

which approaches s and is therefore positive for large b's. We conclude that for any $\alpha < -1$ the regime is DC.

For $\alpha = -1$,

$$w'(b) = b^{s-1} \ln^{-(2-s)} b \left[s \ln b - (1 - s) \right],$$

which is also positive for large b's. For $\alpha = -1$ the regime is DC.

For $\alpha > -1$, we have:

$$w'(b) = \frac{b^{-(1-s)} \left(b^{\alpha+1} - 1\right)^{-(2-s)}}{(\alpha + 1)^{-(1-s)}} \left[s(b^{\alpha+1} - 1) - (1 - s)b(\alpha + 1)b^{\alpha} \right]$$

$$w'(b) = \frac{b^{-(1-s)} \left(b^{\alpha+1} - 1\right)^{-(2-s)}}{(\alpha + 1)^{-(1-s)}} \left[((2 + \alpha)s - \alpha - 1) b^{\alpha+1} - s \right]$$

For large b's, $w'(b)$ has the sign of $(2 + \alpha)s - \alpha - 1$. We may now conclude our study.

Claim

$$\text{the regime is DC iff } \alpha \leq -1 \text{ or } s > \frac{\alpha+1}{\alpha+2},$$

$$\text{the regime is UC iff } \alpha > -1 \text{ and } s \leq \frac{\alpha+1}{\alpha+2},$$

The boundary between the two regimes can be seen in Figure 6. Informally, if the economies of scale are very substantial (small s) and if high quality services are in big demand (f grows rapidly), the UC regime prevails. Otherwise, there is an upper bound to the quality of profitable services. There is no profitable way to provide for a service of arbitrary quality because its cost would deter too many potential customers.

6   The Universal Class Regime

6.1   A paradoxical Regime

Let us study the UC regime. As discussed in Section 4.3, Figure 5 and in our first Law, in this regime, the right hand side of (10) is as close to zero as one wishes if one takes b to be large enough. For any value of the technological constant c, even a very high one, it is profitable to provide a service of infinite quality at zero price, and therefore, for any value of the technological constant c, a unique class of service of unbounded quality is offered for free. This seems quite a paradoxical situation. The demand (8) and the cost of providing the service (7) are both infinite. The consideration of the revenue is interesting. The revenue (see (9) and (10)) is infinite, but depends linearly on the technological constant c. The revenue drops linearly with the price of the equipment.

This regime seems to fit the fat dumb pipe model of the Internet: there are enough resources to give every request a treatment of the highest quality, there is no need to give differentiated services and the price is very low since the demand is very high and the economies of scale very substantive.

The dynamics of the UC regime under technological progress, i.e., a drop in the constant c, is quite paradoxical. Such a drop does not significantly affect the, already negligible, price of the service, but it affects adversely the (infinite) revenues of the provider.

The First Law implies that the boundary between the UC and DC regimes does not depend on the technological constant c, therefore the answer to the question whether the fat dumb pipe model prevails or whether a number of classes of service will develop depends only on the relation between the economies of scale and the distribution of demand over different qualities, it does not depend on the price of the equipment. Both regimes are stable under technological progress.

6.2   A More Realistic Model

The paradoxical aspects of the UC regime as described above stem from the consideration of a demand curve that assumes significant demand for services

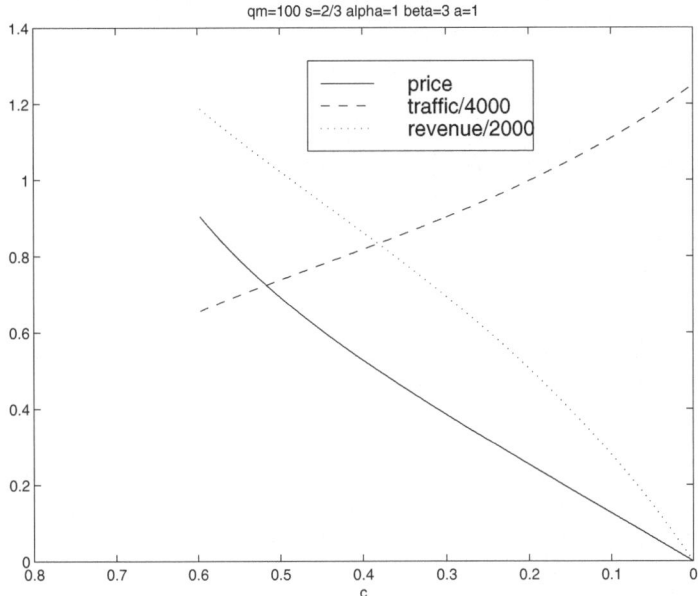

Fig. 7. Evolution of price, traffic and revenue in the UC regime

of unbounded quality. It seems more realistic to assume an upper bound to the quality requested from services. A specific example will be given now, to explain what the UC regime may look like in practice. Assume that there is an upper bound q_m to the quality of service requested and that the distribution of demand f over different qualities is described by:

$$f(q) = \begin{cases} q \text{ if } 1 \leq q \leq q_m \\ 0 \text{ if } q_m \leq q \end{cases} \tag{11}$$

Assume also that $s = \frac{2}{3}$ and that $h(p) = \frac{1}{1+p^3}$. It follows from our discussion above that the service provided, if provided, will be of quality q_m. Therefore the equilibrium equation is:

$$\frac{p}{(1+p^3)^{1/3}} = 2^{1/3} c\, q_m{}^{2/3} \left(q_m{}^2 - 1\right)^{-1/3}. \tag{12}$$

The left hand side of (12) has a maximum value of 1. For any c such that

$$c \geq 2^{-1/3} q_m{}^{-2/3} \left(q_m{}^2 - 1\right)^{1/3}$$

there is no profitable service. For any c smaller than this value, a provider will choose to provide a service of quality q_m. The price, traffic and revenue are functions of c and described in Figure 7.

As soon as some service is provided, it caters for all needs. A unique class of service will still prevail in the presence of a drop in the price of equipment, i.e., a decrease in c. Notice that the revenue decreases when the technology constant drops.

7   The Differentiated Classes Regime

Let us study the DC regime now. The best approach is to take a dynamic perspective and consider the evolution of the system under a decrease in the technological constant c.

7.1   The BDC Sub-regime

In the first part of this discussion we study the BDC sub-regime, i.e., the *sensitive* case described in Section 4.2 in which the left hand side of (10) increases to a maximum and then decreases. We have seen that for high c's no service is provided. When c falls below a certain value, c_0, a service of some finite quality q_0 is provided. From the First Law we know that, even in the face of further declines in c, the quality q_0 of the service provided will not change. Two main questions arise:

– how do the price, the traffic and the revenue of the provider of a service of quality q_0 evolve under a further decrease in c?
– will services of higher quality be offered? which? when? at what price?

The first question is relatively easily disposed of. Since the right hand side of (10) is a linear function of c, the assumption that the left hand side increases to a maximum implies that the price drops rapidly at first (for c close to c_0), and then at a slower pace. For the functions h considered, the slope of the left hand side is one at $p = 0$ and the price p drops essentially linearly with c and approaches zero when c approaches zero. The demand (or the traffic) varies as $h(p)$ and therefore increases very rapidly at first and then approaches some finite value. The revenue varies as $ph(p)$ or $ch(p)^s$. Typically it may exhibit, first, an increase, for values of c at which a drop in the price of the service is over-compensated by an increase in demand and then a decrease. For small c's, the revenue drops linearly with c. Figure 8 describes the evolution of price, traffic and revenue for the lowest class of service, the first to be offered on the market.

The second question is deeper. Since a service of quality q_0 will be provided at a price we have just studied, it is clear that the only customers who could be interested in another service are those requesting a service of quality superior to q_0. Will such a service, of quality $b \geq q_0$ be provided? The equilibrium equation now becomes:

$$p\, h(p)^{1-s} = c\, b^s\, A(q_0, b)^{-(1-s)}. \tag{13}$$

Notice that (13) is very similar to (10). A service of quality b may be profitably offered iff (13) has a solution. See Section 4.2 for a discussion: the situation is

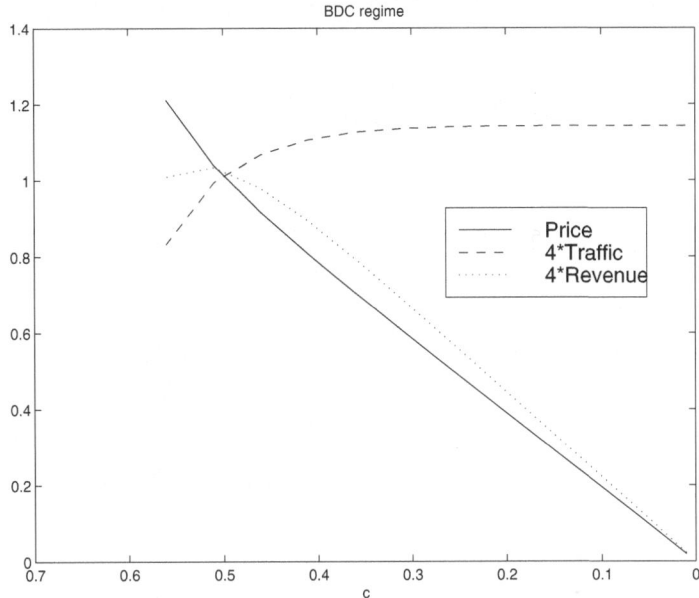

Fig. 8. Evolution of price, traffic and revenue in the BDC regime, lowest class of service for $\alpha = -2.5, s = 2/3, \beta = 6$ and $a = 0.7$

the same. There is a value c_1 above which there is no solution and below which there is a solution. Since we are in the sensitive case, the left hand side has a maximum and the price of the service, of quality b, if provided, will be the smallest p satisfying (13).

The discussion of which service will be provided is similar to the discussion found in Section 4.3. The behavior of the right hand side of (13) as a function of b is the same as that of (10). We have assumed we are in the DC regime, and therefore there is a finite value q_1 for b that minimizes $b^s A(q_0, b)^{-(1-s)}$. The service provided will be of a quality q_1.

In the case f has the form described in (2), the computation of q_1 is simplified by a change of variable $q = q_0 q'$.

$$A(q_0, b) = \int_{q_0}^{b} f(q) \, dq = \int_{1}^{b/q_0} f(q_0 q') \, q_0 \, dq' = q_0^{\alpha+1} A(1, b/q_0)$$

and the equation becomes:

$$p \, h(p)^{1-s} = c' \, (b/q_0)^s \, A(1, b/q_0)^{-(1-s)}, \quad \text{for } c' = c \, q_0^{-(\alpha+1)(1-s)+s}.$$

Notice that

$$-(\alpha + 1)(1 - s) + s = (\alpha + 2)s - (\alpha + 1) > 0$$

by Claim 5 since we are in the DC regime and therefore $c' > c$. Equation (13) is the same as (10), for a larger technological constant, after our change of variable.

This implies that the service provided will be of quality q_1 with

$$q_1/q_0 = q_0, \text{ i.e., } q_1 = q_0^2.$$

It will appear when

$$c' = c_0, \text{ i.e., } c_1 = c_0 \, q_0^{(\alpha+1)(1-s)-s}.$$

The general dynamic picture is now clear. Different classes of service appear as the technological constant c decreases. The n'th class $(n \geq 0)$ appears when $c = c_0 \, q_0^{n(-s+(\alpha+1)(1-s))}$ and it is of quality $q_0{}^n$. For any $c > 0$, only a finite number of qualities of service are proposed and no service is provided for request of high quality service. Since, from Section 5, we may also easily compute q_0 and see that:

$$\text{if } \alpha < -1, \quad q_0 = \left(\frac{s}{(2+\alpha)s - \alpha - 1} \right)^{\frac{1}{\alpha+1}},$$

$$\text{if } \alpha = -1, \quad q_0 = e^{(1-s)/s},$$

$$\text{if } \alpha > -1, \quad q_0 = \left(\frac{s}{(2+\alpha)s - \alpha - 1} \right)^{\frac{1}{\alpha+1}},$$

the picture is complete. Figure 9 describes the price, traffic and revenue in the second class of service as functions of the technological constant c, for the same constants as in Figure 8. Compare with Figure 8 and see that the qualitative evolution is similar in both classes. The ratio of prices in two neighboring classes is interesting to study. The ratio between the prices of service in a class and the class just below, as a function of c, has the same behavior and the same values (for different c's) for all classes. Let the classes of services are described by qualities: $q_0 < q_1 < \ldots < q_i < \ldots$ and appearing for the values of c: $c_0 > c_1 > \ldots > c_i > \ldots$. Let $p_0(c) < p_1(c) < \ldots < p_i(c) < \ldots$ be the prices of the corresponding services. The behavior of $p_1(c)/p_0(c)$ for c between c_0 and c_1 is the same as the behavior of $p_2(c)/p_1(c)$ for c between c_1 and c_2. The price of a newly created class (of high quality) drops (with c) more rapidly than that of an older class of lower quality and therefore the ratio $p_{i+1}(c)/p_i(c)$, larger than 1, decreases with c. Under the assumptions above, it approaches

$$\left(\frac{s}{(2+\alpha)\,s - \alpha - 1} \right)^{\frac{(2+\alpha)\,s-\alpha-1}{\alpha+1}}$$

when c approaches zero. For $\alpha = -2$ and $s = 1/2$, this gives a value of 2, the ratio used by the Paris Metro system when it offered two classes of service, and quite typical of other transportation systems.

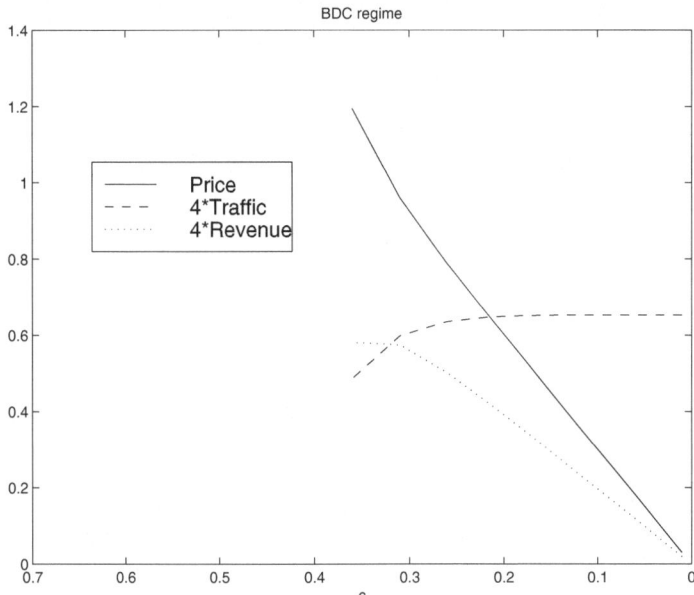

Fig. 9. Evolution of price, traffic and revenue in the BDC regime, next to lowest class of service for $\alpha = -2.5, s = 2/3, \beta = 6$ and $a = 0.7$

7.2   The UDC Sub-regime

We may now understand what happens in the insensitive case: the left hand side of (10) grows without bound. In this case, for any c, an infinite number of qualities: $q_0, q_1, \ldots, q_i, \ldots$ of service is offered. The qualities offered are constant, they do not depend on c. Figures 10 and 11 show the first two classes for a generic case.

7.3   The General Picture

Law 2 (Second Law). *Under a decrease in the technological constant c, i.e., under technological progress and a drop in the price of equipment:*

- *In the UC regime, at any time, i.e., for any value of c, a unique class serves requests for arbitrary quality.*
- *In the DC regime,*
  - *in the BDC sub-regime, no service is provided at first, then a service of finite quality is provided, later on another service of higher quality will be proposed, and so on. For any quality q a service of quality q will be proposed at some time (and on), but at any time there are qualities of services that are not provided.*
  - *in the UDC sub-regime, at any time, an infinite number of classes of service are proposed and requests for arbitrary quality are taken care of.*

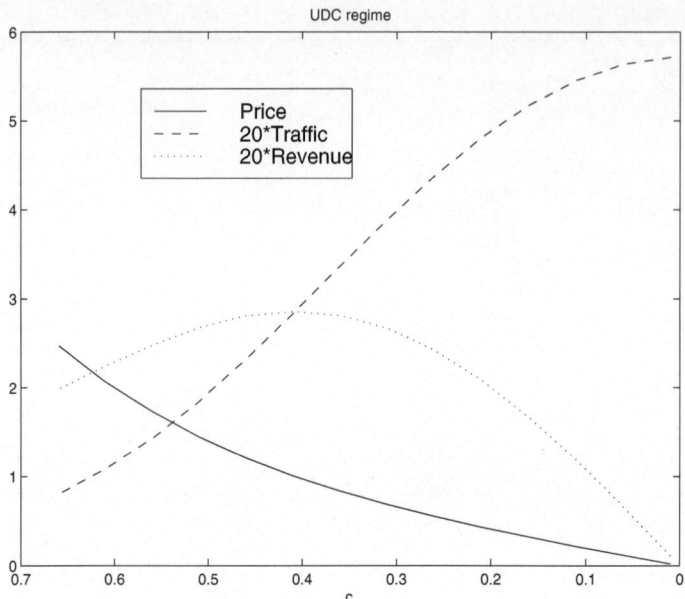

Fig. 10. Evolution of price, traffic and revenue in the UDC regime, lowest class of service for $\alpha = -2.5, s = 2/3, \beta = 2$ and $a = 1$

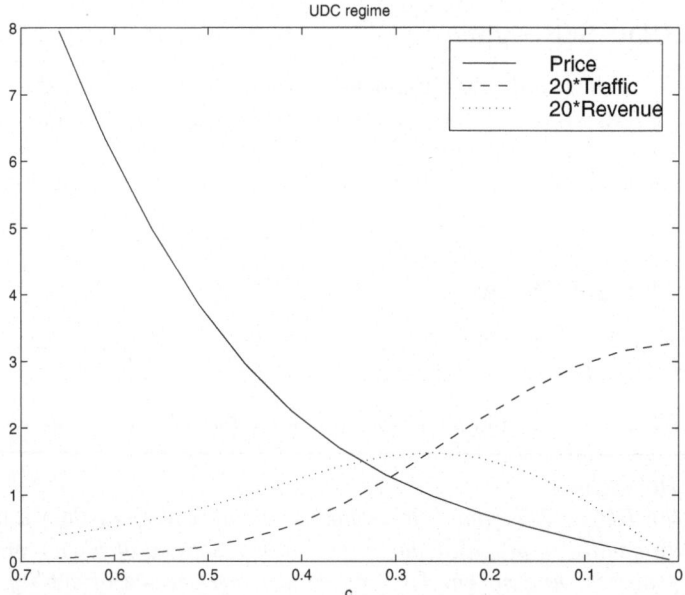

Fig. 11. Evolution of price, traffic and revenue in the UDC regime, next to lowest class of service for $\alpha = -2.5, s = 2/3, \beta = 2$ and $a = 1$

This Second Law contradicts the folk wisdom that says that the Fat Dumb Pipe (FDP) model for the Internet will collapse when prices become low enough to enable very high quality services. It says that if the FDP model is the one that prevails when c is high, it will continue to prevail after a decrease in c. A drop in the price of equipment cannot cause the breakdown of a system based on a universal class of service.

8   Further Work and Conclusions

The most intriguing challenges are a theoretical one and an experimental one. The first one concerns the study of the case in which the Decoupling of Equation 1 does not hold, or holds only approximately. The second one is the study of real systems, in particular the Internet, and the evaluation of the parameters of interest for them.

The main conclusion of this work is that the parameters that determine the type of the prevailing market are the size of the economies of scale and the distribution of demand over the range of qualities. The sensitivity of demand to price and the changes in the price of equipment bear a lesser influence.

This paper made two severe assumptions. It assumes perfect competition, and it assumes that economies of scale do not aggregate over different qualities of service. Further work should relax those assumptions.

References

1. Gibbens, R., Mason, R., Steinberg, R.: Internet service classes under competition. Technical report, University of Southampton (September 1999),
   http://www.soton.ac.uk/~ram2/
2. Odlyzko, A.M.: Paris metro pricing for the internet. In: Proceedings of the ACM Conference on Electronic Commerce EC 1999, Denver, CO, pp. 140–147. ACM Press, New York (1999)
3. Ferguson, P., Huston, G.: Quality of Service: Delivering QoS on the Internet and in Corporate Networks. Wiley, Chichester (1998)
4. Varian, H.R.: The economics of the Internet, information goods, intellectual property and related issues. Reference Web pages with links:
   http://www.sims.berkeley.edu/resources/infoecon/
5. Fishburn, P.C., Odlyzko, A.M.: Dynamic behavior of differentiated pricing and quality of service options for the internet. In: Proceedings of the First International Conference on Information and Computation Economics ICE 1998, pp. 128–139. ACM Press, New York (1998), http://www.research.att.com/~amo
6. Wilson, R.B.: Nonlinear Pricing. Oxford Press (1993)

On the Ontological Nature of Syntactic Categories in Categorial Grammar

Rani Nelken

Harvard University

Abstract. We address the ontological nature of syntactic categories in categorial grammar. Should we interpret categories as being a model of some actual grammatical reality, or are they merely part of an empirical model, one that attempts to capture the range of data, but without making any ontological commitments. This distinction, which is not often made, is important for determining the goals of formal grammar research. We evaluate this question within the context of a particular grammatical phenomenon, the modeling of premodifier ordering. We compare the categorial grammar treatment of this phenomenon with empirical statistical approaches to the same problem. We show that the whereas both models are equally empirically adequate, the statistical model is more generalizable and learnable, and thus advantageous as a realistic model.

1 Introduction

In this paper we address the question of the ontological nature of syntactic categories. Theories couched in the categorial grammar (CG) framework posit a certain set of syntactic categories, consisting of basic types plus a recursive mechanism for constructing more complex types from the primitive ones. How should we interpret these categories from an ontological standpoint, are they real, or are they just conventional devices?

At first thought, doubting the independent existence of categorial grammar categories seems counterintuitive to say the least. After all, the syntactic categories in the purer forms of categorial grammar–such as the Lambek Calculus–are not only very natural, but also truly minimalistic. Thus, it would seem that the nature of categories is uncontroversial. Nevertheless, in this paper we propose to raise some doubts about this seeming certainty. Let us call the somewhat extreme view under which CG categories are supposed to reflect actual reality, the "Realist View". Under this view, the categories are inherently self-existent, e.g., there really is an **n** category in grammar.

We can define the realist view as claiming that there is one distinguished model of grammar. Other models, which might be empirically just as adequate, would be dispreferred as long as they are essentially different, and not just notational variants. For a realist, the preferred model is one that would eventually map to the actual psychological mechanism that implements grammar, though one can be a realist without committing to how such a mapping would work. For instance, one can claim that grammar really uses nouns without saying anything about how nouns are represented in the brain or actually "implemented" within the language faculty. Similarly, the division into categories does not have to be based on external, objective criteria, but might be entirely based on cognitive principles, as would be argued by [1].

O. Grumberg et al. (Eds.): Francez Festschrift, LNCS 5533, pp. 170–176, 2009.

Admitting a type for sentences, **s**, and nouns, **n**, appears innocuous enough. CG is minimalist in positing just two very natural basic types and a recursive mechanism for constructing new types. The fact that languages can talk about things (type **n**), and make statements (type **s**) is uncontroversial. Moreover, a capacity for recursive construction in natural language is one of the core principles of the Chomskyan revolution [2]. The exact nature of this recursion, its propensity in all natural languages, and its restriction to natural languages within the animal kingdom is not without debate [3], but the existence of some combinatory mechanism for constructing complex syntactic categories from more basic ones is a prerequisite for any grammatical theory.

What about the more complex categories posited by CG, such as categories decorated with multiple modal operators [4]? It is harder to argue for such categories being self-existent from first principles, but they become necessary for modeling (and perhaps explaining) some of the more complex phenomena encountered in languages. Under the realist view, there truly exist such complex categories, and it is the goal of the grammar researcher to discover the correct set of the grammatical categories.

While this question is applicable to any grammar formalism, it is arguably particularly pertinent for CG and related systems, which impose a very close connection between syntactic and semantic categories through the Curry-Howard isomorphism. Thus, if we assign ontological significance to syntactic categories, this would seem to immediately require us to assign a similar reality to semantic categories, projecting from one to the other, and vice versa.

A more pragmatic view of grammatical categories would relax this realist view to claim that syntactic categories might not be independently self-existent. Instead, their role is to provide a model for natural language phenomena that does not rule out alternative models as equally viable. Thus, there is no external distinguished reality a pragmatist would want to match. The only interest is in a theory that best explains the empirical phenomena.

Note that we are focusing here solely on the question of whether the grammar of a specific language should be interpreted as realistic. It is a separate question whether all languages would have the same model—at least at some level of abstraction.

2 Desiderata for a Realistic Theory of Categories

As language researchers, the question of whether we are seeking a realistic model of language or just a pragmatic model that can effectively cover the empirical data is philosophical question of the researchers' convictions. We can nevertheless define some minimal desiderata that a realistic model should satisfy.

Clearly, **empirical adequacy** is a necessary condition for any theory, whether pragmatic or realistic. Empirical adequacy is of course not a sufficient condition. Given a CG grammar, we could always "flip" categories around, yielding a grammar that is weakly equivalent (accepting the same language), but not strongly equivalent (assigning different derivations). For instance, standard CG models impose an asymmetry between nouns which are basic categories of **n**, and verbs which are complex categories of type **s/n**. We could theoretically flip this asymmetry, by viewing verbs as basic, and defining nouns as **s/v**. Nouns under this reconstruction would be equivalent to type-raised nouns, i.e., functions accepting verbs as arguments. It is unclear whether we can distinguish between these two models based solely on empirical adequacy.

More radically, Partee has recently raised some doubts about whether the distinction between Montague's two basic semantic types **e** and **t** is really necessary, and not simply traditional [5]. Could we build a grammar based on a single type? Partee offers a preliminary sketch of how such a system might work, based on one single type, **p**, the type of "properties of situations," where a situation is somewhat loosely described as covering both entities and eventualities. Under this approach, both noun phrases and sentences denote such properties. Although Partee's sketch is far from constituting a worked-out grammar, it does lend credence to the pragmatic view that alternative formulations of grammar in the CG tradition can be equally plausible.

Note that empirical adequacy requires scalability. Much work in the formal grammar tradition is based on introspection, and is thus limited to a "human-scale" range of examples. For a theory to be truly empirically adequate we must verify that the same principles can readily apply to a much broader set of examples.

The second important criterion is **theoretical parsimony**. While parsimony is important as good methodological practice for any theory, it is doubly important for a realistic theory. When we start considering—even in the most informal sense (cf. [6] for a review of more formal notions of learnability)—how a grammar might be cognitively processed or learned, it is clear that a theory with a smaller set of parameters is to be preferred. To examine these questions in more detail, we now turn to a concrete example, that of premodifier ordering.

3   Premodifier Ordering: A Case Study

Several well-known observations point to the fact that certain adjective orderings and adverb orderings are significantly preferable to others [7]. For instance, consider the NP "my big fat Greek wedding," which is acceptable, but other permutations of the same adjective list, e.g., "*my Greek fat big wedding" would not be. A common hypothesis is that adjectives belong to semantic classes having a characteristic order. For instance, a size adjective must precede a shape adjective. In fact, as [8] shows, multiple factors at different levels of linguistic analysis—from phonology to pragmatics—affect adjective order. Moreover, there is evidence that there are cross-language regularities in these orderings [9]. It would thus seem natural to impose a division of adjectives and adverbs into sub-categories that reflect these classes, e.g., a category for "size adjectives," which is distinct from the category of "shape adjectives".

Let us examine one such proposal in more detail. Recently, Nilsen [10] has suggested a multimodal type-logical grammar [4] account of adverb ordering in Norwegian. On Nilsen's account, we can use facts about the ordering tendencies of adverbs to determine a partial order on them, though not necessarily a linear order. To give an English example [11], "paradoxically" should precede "completely" when they co-occur, a fact which should be reflected in the categories we assign to these adverbs. According to Ernst, "completely" modifies events, while "paradoxically" modifies facts, and thus they are essentially a different type of adverb. Nilsen proposes a method that induces a division of adverbs into such sub-classes from empirical data about their typical orderings. The model works by collapsing two adverbs into the same class—for reasons of parsimony—as long as this does not violate any ordering constraints. Nilsen proceeds to provide an implementation of this ordering within multi-modal type-logical grammar, using the mechanism of modal operators to capture such dependencies.

For the sake of simplicity, we will not repeat the details of the multimodal account here. They are quite typical of multimodal CG, which allows one to still view all these different adverbs as being flavors of adverbs, but their grammatical interaction is limited by using a set of modal operators which act as a sort of "lock-and-key" mechanism. Using these operators, Nilsen imposes the partial order constraints as part of the grammar. While Nilsen's solution clearly works for the range of adverbs he considers, it relies on an implicit assumption that the adverbs can indeed be neatly organized into a partial order. Nilsen illustrates this assumption for a handful of Norwegian adverbs, which he orders based on introspection.

Should we interpret this division into finer adverbial categories as merely a pragmatic solution to the problem, or as an attempt to model a more fundamental realistic model of grammar? For instance, while a difference in semantic interpretation yields different classes of adverbs, does the converse hold? Thus, if we examine the set of adverb classes that we would get from Nilsen's approach, would each distinct syntactic sub-class have a distinct semantic interpretation?

Interestingly, empirical natural language processing has also approached the problem of premodifier ordering from a different perspective. Several authors have attempted to predict the right order of a list of premodifiers. Under the supervised learning paradigm, we can collect a large set of adjective pairs, train some machine learning method to predict their order, and then apply it to a held-out set of sequences. In particular, Malouf [12] presents some extremely simple—yet surprisingly effective—methods of predicting the order of adjective pairs. One such simple method is the following "direct evidence" method. Given a pair of adjectives, a and b, in the test data, look up all examples in the training data consisting of both a and b in any order. If a precedes b more often than vice versa, then it would be reasonable to predict that a should precede b for the test case as well. Conversely, if b precedes a more often, we would predict the order accordingly. Of course, this leaves the cases where we have not encountered the exact pair a, b in the dataset, or we have encountered both orderings equally frequently. For these, we just flip a coin.

Malouf tests this method on a set of more than a quarter of a million adjective pairs collected from the BNC corpus,splitting the set of pairs into a training and a held-out test set (of sizes 90%, 10%, respectively). This simple method achieves almost 80% accuracy on the test set.

The remaining errors stem from the sparseness of the data; a large proportion of the adjective pairs occur only once in the dataset, so if we encounter them in the test set, we would not have seen them previously in the training set, in which case we do no better than chance. Thus, the chief challenge becomes to generalize from the training data to previously unseen adjective pairs. Malouf suggests the following even simpler "positional probabilities" method. For each adjective in the training data, a, we record the frequency of a appearing first in a pair, regardless of what the second adjective is. Similarly, we find what the frequency of a appearing second is. For a newly encountered pair, a, b, we find what is more probable, the combination of a appearing first and b

appearing second, or of *b* appearing first and *a* appearing second. Surprisingly, this simple method performs even better than the direct evidence method, achieving almost 90% accuracy. This method is also much more efficient, since it only requires one to keep information about the individual adjectives and not the full set of adjective pairs. Needless to say, Malouf's system is evaluated solely on its prediction accuracy. Malouf describes the role of grammar as split between a formal "competence" grammar, and an essentially independent statistical "performance" system.

4   Comparing the Two Theories

Let us compare the two approaches according to the desiderata for a realist theory discussed above, exploring the implications to the ontological nature of grammatical categories. First, in terms of empirical adequacy, we must verify that both models explain the data, and indeed, assuming fully observed data, Nilsen's model readily extends to the full range of Malouf's set of adjective pairs. Recall that the model relies on the assumption that for each pair of adjectives, *a*, *b*, we either have *a* mostly preceding *b*, *b* mostly preceding *a*, or no preference. We can obtain these preferences directly from the data by simply counting which ordering is more frequent. This method coincides nicely with Malouf's "direct evidence" baseline, and thus both models are equally empirically adequate.

It is only when we consider theoretical parsimony for a large set of data that we see a marked difference. While Nilsen's model looks plausible for a handful of adverb categories, when we consider the full range of data, it becomes clear that we would have to divide the categories into thousands or even tens of thousands of adjective categories, which becomes considerably less appealing.

Moreover, what of learnability? Given access to the full set of data we can easily compile the categories, but what if we only get access to partial data? The question of partial data becomes important when we consider how a language learner might acquire the categories. Unless we are somehow born with the full classification of adjective sub-categories pre-equipped with this set of sub-categories (even for adjectives that were not invented yet), how would a language learner generalize from partially observed data to the full range of data? Note that Nilsen's system is quite rigid in the sense that it does not allow gradual refinement as more adjectives are added. Thus if a learner has compiled some set of sub-categories based on previous linguistic experience, the addition of a new adjective might require a drastic reshuffling of the categories.

How realistic is the statistical approach as a model of adjective ordering? By our adequacy criteria, this model is advantageous not only pragmatically, but also has when taking into account realistic constraints of generlizability and learnability. To be sure, a simplistic model such as positional probabilities is by no means an adequate explanatory theory of adjective ordering. In particular, it does nothing to explain the semantic and other regularities in adjective ordering. Judging by Wulff's results [8], it would seem that an explanatory model would have to take into account the combination of multiple such factors, using some combinatory rule that would appropriately weight the relative importance of the different factors. Of course, such a model strays even further from the discrete CG tradition.

Could we create a CG model that would have the capacity for such generalization? At the very least, we require the capability of expressing ordering tendencies such as *a* "tends to appear first" to some quantifiable extent. This requirement seems to be beyond the expressive power of CG even with multimodal operators. Furthermore, even if such an encoding would be technically possible, it would likely be quite awkward, especially compared to a simple single positional probability. We thus require the incorporation of a stochastic element within grammar, for instance as in [13]. Thus, we are not just advocating adding a "performance" aspect on top of a discrete grammar, as Malouf suggests but are suggesting viewing the stochastic component as fundamental to grammar as argued previously by [14].

5   Conclusion

In conclusion, we believe that as language researchers we should ask ourselves what problem we are trying to solve. Are we setting out to find a pragmatically viable model that best explains the data, or are we seeking a realistic model of how grammar might actually work. We find the example of premodifier ordering to be instructive in this context as it highlights the tension between finding a theory that would account for the discrete semantic generalizations as well as explain a large set of data in a compact, generalizable, and learnable way. The latter criteria apparently require the ability to express gradient constraints which go beyond the expressive power of CG.

Acknowledgments

This paper is part of a Festschrift for Nissim Francez on the occasion of his 65'th birthday. We thank an anonymous reviewer for his comments on an earlier version of the paper. This work was partially supported by a Google research award: "Mining Wikipedia's Revision History".

References

1. Lakoff, G.: Woman, Fire, and Dangerous Things: What Categories Reveal about the Mind. University of Chicago Press, Chicago (1987)
2. Hauser, M.D., Chomsky, N., Fitch, W.T.: The faculty of language: What is it, who has it, and how did it evolve? Science 298, 1569–1579 (2002)
3. Jackendoff, R., Pinker, S.: The nature of the language faculty and its implications for evolution of language (reply to Fitch, Hauser, and Chomsky). Cognition (September 2005)
4. Moortgat, M.: Categorial type logics. In: Handbook of Logic and Language, pp. 93–177. Elsevier, Amsterdam (1997)
5. Partee, B.H.: Do we need two basic types? In: Gaertner, H.-M., Regine Eckardt, R.M., Stiebels, B. (eds.) Puzzles for Manfred Krifka, Berlin, pp. 40–60 (2006)
6. Lappin, S., Shieber, S.M.: Machine learning theory and practice as a source of insight into universal grammar. Journal of Linguistics 43(2), 393–427 (2007)
7. Cinque, G.: On the evidence for partial n-movement in the romance DP. In: Cinque, G., Koster, J., Pollock, J.Y., Rizzi, L., Zanuttini, R. (eds.) Paths Towards Universal Grammar. Studies in Honor of Kayne, R.S. Georgetown University Press (1994)

8. Wulff, S.: A multifactorial corpus analysis of adjective order in English. International Journal of Corpus Linguistics 8(2), 245–282 (2003)
9. Sproat, R., Shih, C.: The cross-linguistic distribution of adjective ordering restrictions. In: Georgopoulos, C., Ishihara, R. (eds.) Interdisciplinary Approaches to Language. Essays in Honor of Kuroda, S.-Y., pp. 565–593. Kluwer Academic Publishers, Dordrecht (1991)
10. Nilsen, O.: Adverb order in type logical grammar. In: van Rooy, R., Stokhof, M. (eds.) Proceedings of the Amsterdam Colloquium 2001 (December 17-19, 2001)
11. Ernst, T.: The scopal basis of adverb licensing. ms. Amherst (2001)
12. Malouf, R.: The order of prenominal adjectives in natural language generation. In: ACL (2000)
13. Osborne, M., Briscoe, T.: Learning stochastic categorial grammars. In: CoNLL 1997, pp. 80–87. ACL (1997)
14. Abney, S.: Statistical methods and linguistics. In: Klavans, J., Resnik, P. (eds.) The Balancing Act: Combining Symbolic and Statistical Approaches to Language. The MIT Press, Cambridge (1996)

Masking Gateway for Enterprises

Sara Porat, Boaz Carmeli, Tamar Domany, Tal Drory, Ksenya Kveler,
Alex Melament, and Haim Nelken

IBM Haifa Research Lab,
Haifa University, Haifa 31905, Israel
(porat,boazc,tamar,tald,ksenya,melament,nelken}@il.ibm.com

Abstract. Today's business world revolves around the need to share data. On
the other hand, the leakage of sensitive data is becoming one of our main security threats. Organizations are becoming more aware of the need to control the
information that flows out of their boundaries and must more strictly monitor
this flow in order to comply with government regulations. This paper presents
an SOA-based solution called Masking Gateway for Enterprises (MAGEN),
which allows the sharing of data while safeguarding sensitive business data.
The major novelty lies in architecting a single system that handles a wide range
of scenarios in a centralized and unified manner.

Keywords: Data Leakage Prevention (DLP), data masking, de-identification,
anonymization, Service Oriented Architecture (SOA), Optical Character Recognition (OCR).

1 Introduction

In the modern business world, organizations often need to share and expose data. Outsourcing, off-shoring, servicing, partnerships, data mining, and statistical analysis—to
name a few—are all cases where data must be shared to advance business operations.
Moreover, in many situations, data sharing is mandatory to create a competitive edge.
However, data cannot simply flow as-is since it may include sensitive business data,
ranging from personal information to intellectual property and trade secrets. The issue
of data leakage—the intentional or accidental exposure of sensitive information—is
becoming a major security issue. A recent IDC survey [1] claims that data leakage is
the number one threat, ranked higher than viruses, Trojan horses, and worms.

Governments and industry-coalitions are developing sector-specific regulations to
govern information sharing and protect sensitive business data. HIPAA in the US,
FIPPA in Canada, the Sarbanes-Oxley Act, and the Gramm-Leach-Bliley Act are just
a few examples. Organizations found disobeying these regulations can be fined and
forced to establish compliance solutions.

Organizations are therefore becoming more aware of the need to control the information that flows out of their boundaries. This has yielded a wide range of approaches
and solutions to address this evolving demand. A Gartner report [2] describes the
market of Data Leakage Prevention (DLP) through tools that use content inspection
techniques over network communications. Other reports ([3], [4]) refer to techniques

O. Grumberg et al. (Eds.): Francez Festschrift, LNCS 5533, pp. 177–191, 2009.

that prevent data leakage in non-production environments. These usually apply to solutions that hide sensitive data in databases by changing their contents, yet keep the modified data testable. In addition to these two approaches, there are quite a few techniques that are applied to cope with removing sensitive data from text documents. It's interesting to note that all the existing solutions in the market address a limited range of scenarios. As far as we know, there is no centralized holistic solution that addresses all possible scenarios that an organization may need to tackle.

Often the sensitive data cannot be simply removed since the party receiving the data needs to use it. In these cases, the sensitive data has to be changed, while still maintaining specific characteristics. For example, let's assume a company needs to share data for testing purposes, but it must avoid sharing the social security numbers of its customers. If these numbers are used as foreign keys, the numbers may have to be changed to fake numbers while still maintaining referential integrity. Another example is when sharing numbers about salaries for statistical analysis purposes. The exact amounts should not be exposed, but the distribution of possible values should be kept. In some cases, the process of changing the data assumes that it is going to be utilized by diverse entities; thus, different changes may need to be employed per possible utilizations.

As Feiman notes in the Gartner report [4], there are many names to describe the changes used to hide sensitive entities such as data masking, data obfuscation, and data scrambling. We prefer the term "data masking" and use the terms "data masking" or "de-identification" interchangeably. The latter indicates changes to hide identifiable attributes.

Data masking solutions should be designed to cope with evolving and changing regulations. This is usually done through rule-based mechanisms, where the rules indicate what to hide, when and how. The system is then configured to apply transformations according to these pre-defined rules.

In this paper, we present a solution called Masking Gateway for Enterprises (MAGEN), which falls under the network-based Gartner [2] definition for DLP, but goes far beyond its scope to cover a wider range of scenarios where data is shared and sensitive data has to be protected. MAGEN's novelty lies in its architecture, which provides a single system that incorporates the capabilities of network-based DLP solutions, as well as the capabilities of traditional data masking solutions such as IBM Optim [5], Oracle data masking [6], Camouflage [7], etc. We describe the architecture and the benefits of having a single, enterprise-wide, centralized solution, instead of several disparate solutions.

MAGEN is an evolution of the IBM De-Identification Framework [8], also referred to as the Universal De-identification Platform (UDiP). While UDiP was designed to enable the sharing of health and medical information (similar to HIDE [9]), MAGEN is a general solution upon which vertical solutions for various industries can be easily implemented.

The paper is organized as follows. In Section 2 we describe the landscape of data masking techniques and introduce MAGEN as a solution that combines all techniques under a single architecture. This section also provides a high-level overview of MAGEN through its major use cases. Section 3 describes the MAGEN SOA-based architecture and presents the underlying management platform upon which MAGEN is built. Section 4 presents a library of transformations that can be used to hide sensitive and identifiable data and still keep the disclosed data useful for the receiving parties. Section 5 takes a deep dive into a special incarnation of MAGEN for masking screens using OCR technologies. We conclude in Section 6 with some observations and future directions for our work.

2 A Single Centralized Data Masking Solution

There are many existing techniques to control cases where data is shared internally or externally. In this chapter, we describe the landscape of possible techniques, each of which enables the sharing of data while safeguarding sensitive information. At first glance, the various techniques look very different, either because the mode of operation differs from one to another or because each handles unique data types. We divide the landscape of data masking techniques into three categories:

1. **Batch data masking techniques**. These methods process the data and generate secured copies without the sensitive entities. Solutions in this space are typically applied in non-production environments so that copies being made (to support test and development) do not expose sensitive information. These methods also apply in scenarios where companies share their data with research institutes for statistical or other analyses. Since these techniques do not involve real-time aspects, we refer to them as methods for securing data at rest.

2. **On-line (or on-the-fly) data masking techniques**. These techniques focus on preventing leakage at the point of use, as opposed to avoiding leakage at the point of access. The idea is to process the data when it is used, for example, when copying confidential data into a mail or when sending an HTML web page over the wire. Such techniques are usually part of transactional, messaging, mailing, or streaming systems. Other scenarios where these techniques are applicable happen when a company outsources its operations and external agents run the company's applications from their remote workstations. The masking process is then done to avoid exposure of sensitive data to unauthorized external agents. This kind of masking does not involve copying the original store, since the transformation is done as the data flows from one point to another.

3. **Data redaction techniques**. These techniques target unstructured data such as text files and scanned documents. They usually apply natural language processing (NLP), optical character recognition (OCR), and image processing to recognize sensitive data in text and images. These are generally part of content and information management systems. A relatively new arena that involves such techniques is that of e-discovery, where digital documents and images are shared as relevant evidence in law suits.

Although these categories seem to be very different from each other, they share some fundamental features. Obviously, they all apply methods to define rules (usually through GUI mechanisms) and thereby configure the systems to comply with certain regulations. Moreover, a mandatory process in all data masking solutions is that of managing, monitoring, auditing, and reporting on violations of rules and policies. Figure 1 illustrates the scope of data masking techniques and outlines their commonalities.

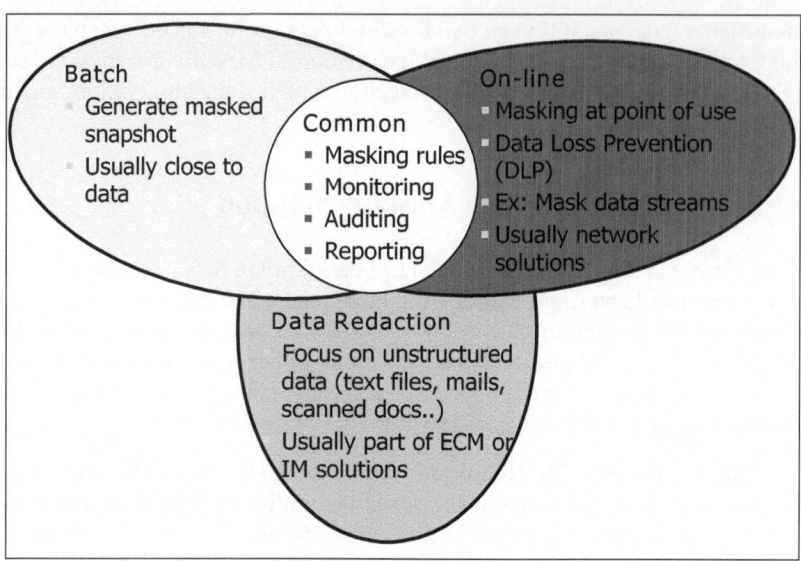

Fig. 1. Data masking landscape

We believe that a company should strive to have one centralized management system to control all the scenarios where data is exposed and shared, instead of having several different systems, each of which addresses a subset of all possible scenarios. Suppose the company has a solution to enable the sharing of databases for testing purposes, and then it decides to outsource some operations to an external call center. It makes no sense to have separate and disparate systems, each with its own GUI for defining masking rules. Moreover, each system would have to be re-configured every once in a while to cope with evolving regulations. The only way to enforce and control policies in a unified and accurate manner is through a centralized system. The challenge is to architect a single, unified system that can turn the company's policies into enforcement logic across all possible scenarios.

MAGEN is an innovative solution that provides a central, policy-based, enterprise-wide system to address all three categories described above. The system enables enterprises to avoid leakage of sensitive data, regardless of where the data resides, in what shape it comes, and where in the data flow the masking should occur. MAGEN can be activated in batch and on-line modes. The input data can be extracted from various sources, whether from stationary storages such as databases and file systems or from streams of data in motion, such as messages and bitmaps of application displays. The input data can be structured (e.g., relational tables), semi-structured (e.g., XML files, complex healthcare objects such as DICOM objects or HL7 messages), or unstructured data (e.g., screen images and scanned documents).

Next, we explain the constituents of an end-to-end data masking solution. We distinguish between three phases in which users interact with the system: the design phase, the runtime phase, and the monitoring and reporting phase.

The **design** phase aims to discover what and where the sensitive data is, decide what should be done with that data, and configure the system to enforce those decisions during runtime. The design phase should be carried out by experts who have an in-depth knowledge of the data, of the business processes that need to use the data, and of the organization's privacy and security regulations. The first challenge is to understand what exactly is going to be shared; where the data is stored (e.g., centralized, distrib-uted); how is it stored (e.g., database, files); what is its structure (e.g., tabular, XML); how it is going to be shared; and who can access it, how and when. The next step is to figure out which data elements (e.g., columns in tables, attributes in XML files) are sensitive and which can be safely shared. Another decision taken during this phase is how to hide the sensitive information. Can the data be simply removed or should the system generate fictional values? Can such replacement values be random or should they maintain specific characteristics such as format, referential integrity, correlation, or aggregated characteristics? And should the process be reversible or repeatable? To make these decisions, one must understand how the masked/de-identified data will be used. Finally, the most challenging step is to translate policies and regulations into a list of role-based masking rules that describe which roles in the organization are au-thorized to see the data, what should be done with each sensitive entity, and in which situations. These rules are enforced during the run-time phase.

The *run-time* phase enforces the policies and role-based masking rules. This en-forcement has to be accurate and coherent to avoid any data corruption and data leak-age. This phase involves acquiring the data, analyzing it to locate sensitive elements, and changing the contents according to the pre-defined rules. This phase can be car-ried out either as a batch process or as a dynamic on-the-fly process, where the mask-ing is applied while the data flows from one point to the other. In the latter case, there are usually end-users that operate on some system and use or view the data; therefore, the challenge in this case is to minimize the latency caused by the extra discovery and masking process.

The *monitoring and reporting* phase includes mechanisms to watch user activities and data flows, and produce informative reports on the execution of the system. This helps the management team understand and track which data elements were accessed and by whom, and what changed in the data during the process. Full-fledged solutions incorporate monitoring and alerting components to protect against the violation of policies, and methods to help understand the level of vulnerability and verify that policies are correctly enforced.

MAGEN uses a central GUI for configuring, controlling, monitoring, and auditing the end-to-end data masking processes. This enables a unified modeling of policies across all cases, and supports easy adaption to evolving and changing regulations.

3   SOA-Based Architecture for End-to-End Data Masking

The MAGEN design follows the principles of Service Oriented Architecture (SOA). It provides a comprehensive portfolio of services integrated into a single powerful solution, which supports the variety of tasks performed during the masking life cycle. Services interoperate by exchanging messages via the Service Integration Bus. MA-GEN executes workflows, each of which defines the path that the message takes as the Service Integration Bus moves it between services to address specific tasks. New flows can be defined and existing flows can be modified to better address evolving business needs.

In addition to following the general principles of SOA, MAGEN leverages a unique **management platform**, which has already been incorporated into a variety of IBM content management offerings. This management platform enables users to interact with the MAGEN system to carry out system configuration tasks like deploying new services or defining new workflows. It also enables users to continuously monitor the status of the system and its health. Through a comprehensive notification system, the management platform captures important events like service start time and how long a specific operation takes. It also allows users to detect failures and investigate their origin, and to track performance issues. The subject interaction with the MAGEN system, namely the system configuration and system monitoring, is done by administrators with a deep knowledge of the MAGEN internals.

One of the significant strengths of this underlying management platform is that it exposes a single and central point for managing all tasks—both the system management tasks as well as the masking-related tasks. The same mechanism is applied to configure the system and to configure the masking process during the design phase. All services are configured through the same mechanism, using a single web-based user interface, which hides the actual system topology details. Similarly, the same implementation is used to monitor the system health as well as the masking activities (e.g., by logging which rule was applied by whom and when).

Figure 2 illustrates the architecture and several of the MAGEN services. It shows sample services that communicate through the Service Integration Bus. Some of the services communicate with GUI components. These GUI elements are illustrated on the right side of the figure with the squared background.

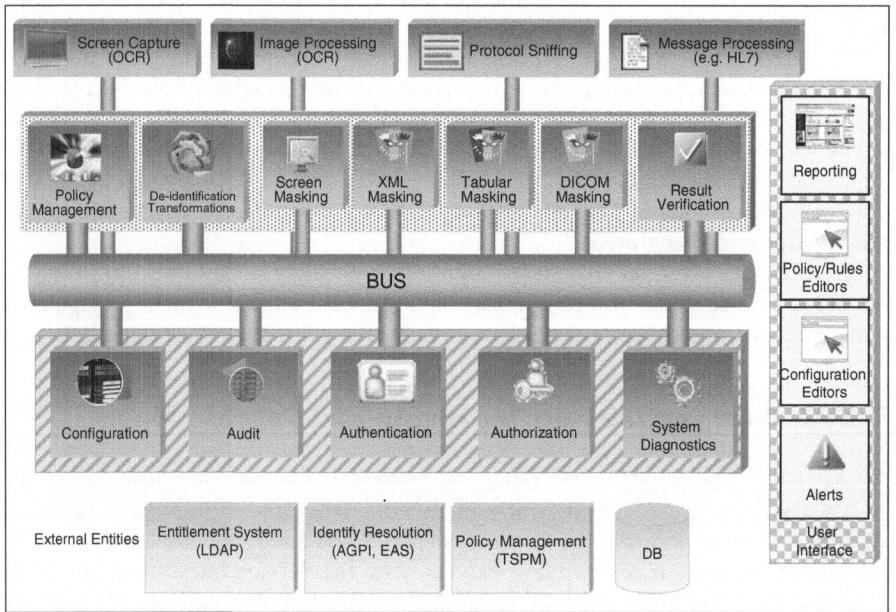

Fig. 2. MAGEN SOA-based architecture

Some of the services are part of the underlying management platform and are illustrated in Figure 2 through the set of quadrates below the BUS (with the striped background). The paragraphs below describe some of these services.

The *Configuration Service* encapsulates the logic for configuring MAGEN services. This service is used to configure the system (e.g., deploy new services) and to generate masking rules during the design phase. It communicates with various GUI components to deliver user input. The interaction is performed through the *Configuration Editors* to configure services and workflows, or through the *Policy/Rule Editors* to create masking rules. Services can register to listen for changes in their own configuration or for changes in the configuration of other services. Once the users save their changes (e.g., adjusting certain service configurations or creating new masking rules), the new data is sent to the Configuration Service, which propagates it to all services that are listening for these changes. An important and valuable feature of this service, named *hot configuration*, ensures that the new settings take effect immediately.

The *Audit Service* and the *System Diagnostics Service* are used to capture service notifications, analyze them, and provide detailed reports. The services are used to monitor the system execution, examine its health, and report on its performance. The same services are used in the monitoring and reporting phase to produce reports on data masking activities (such as which rules were modified and when, statistics about rule usage, etc.) and to raise alerts about violations of masking-related policies.

The *Authentication Service* and the *Authorization Service* authenticate and authorize users that interact with the system. The Authentication Service interacts with the preferred entitlement system for authenticating the user, and when security and privacy levels need to be associated with the authenticated users. The Authorization Service is responsible for assigning the correct privileges based on the user's privacy level and the predefined policies. This mechanism allows control over the use of masking rules according to roles in the organization. To fully enable cross-enterprise information sharing, MAGEN implements the Cross Enterprise User Assertion (XUA) standard. The Authentication Service creates a mechanism whereby a request from outside the enterprise boundaries is authenticated by an entity trusted by both enterprises.

Using SOA together with the unique underlying management platform described above provides a number of advantages:

1. *Extensive flexibility*. The same system can be used to accommodate a wide range of masking needs, while easily configured to use only those services that are needed. Evolving business needs can be answered by deploying new services and creating new workflows.
2. *Scalability, high-availability, and load balancing*. The system can be scaled according to specific environment needs. If the bottleneck is the service computations, then several instances of each service can be initiated to allow parallel processing. If the load is too high for a single machine, the services can be deployed on different machines using IBM WebSphere Network Deployment to balance the load and to achieve high-availability.
3. *Integration with existing systems*. The flexibility of the MAGEN architecture eliminates situations where you are locked into working with a single vendor when interacting with existing third-party products. For example, for authentication purposes, MAGEN can interact with any existing entitlement system,

such as LDAP systems [10], IBM RACF [11], or CA ACF2 [12]. Another important integration point is to identity resolution engines, also called Anonymous Global Personal Identification (AGPI) providers, such as IBM Entity Analytics [13] or MitemView [14]. These maintain the correlation between records across multiple data silos. (See more details on AGPI below.)

4. *Hot configuration.* Hot configuration means that any changes take effect immediately without the need to restart the services or the system. This capability is especially important when editing masking rules after the system is already in use. While end-users continue working with the system, the design phase can be re-initiated to define new rules or update existing ones. Once completed, the modified set of rules immediately comes into effect.

The next two sections detail major concepts that are implemented in MAGEN: the comprehensive set of data transformation mechanisms for safeguarding sensitive information and privacy, and the use of MAGEN for dynamically masking application displays using OCR technologies.

4 Data Transformation and De-identification

At the heart of the masking concept lies the conflict of sharing versus hiding data. Sharing is aimed at bringing value to both the data owners and the data receivers. Data owners need to comply with regulations and are thus interested in hiding as much as possible. On the other hand, the data must remain useful for the receiving parties so they can properly carry out their tasks (e.g., using the data for testing purposes, carrying out outsourcing services, or performing statistical analysis).

From the data owner's point of view, the de-identification process must ensure that no sensitive business data is leaked and that personal data is anonymized so that it cannot be associated with a specific individual. In contrast, the requirements from those that receive the data can vary and depend on the use of the data. For example, if the data is used for testing, one requirement is to keep the data looking similar to the original. This may involve keeping names with capital letters at the beginning, replacing credit card numbers with legal numbers, and so forth. If medical data is used for research, then it is mandatory to blur personal details but still keep the data truthful; for example, the system should avoid changing the age of a person from 35 to 80, but make it more general by indicating a range from 30 to 40. If the data is used for statistical analysis, it is important to maintain the statistical characteristics of the aggregated data [15]. In many cases, it is compulsory to maintain the correlation between records that belong to an individual and preserve certain relationships between entities. Say, for example, the data contains dates that indicate when a customer opens a bank account, when she closes it, and when she makes deposits or withdrawals. If the masking process changes the dates, it should ensure that there will be no withdrawals or deposits from an account that does not exist. Additional requirements that should be supported are reversibility, which is the ability to reveal sensitive data or re-identify the person if the need arises, and repeatability, where the same result is generated for the same input.

The de-identification process attempts to fulfill the requirements of the data owners as well as those of the data receivers. The two sets of requirements may be contradictory, and the challenge is to find the right balance and suggest the proper compromise. MAGEN introduces a comprehensive set of powerful transformations that makes it possible to de-identify personal information and prevent leakage of sensitive or confidential data, but still keep certain characteristics as per the requirements set by those that receive the data. The transformations are grouped into a library of functions. Some functions can be applied on single data elements, whereas others operate on a group of elements or on complete data stores. The library includes dozens of functions that fall under several families of transformations. The following list describes some of these families.

- Random scrambling: Replaces the input values with random values while preserving the original format, such as uppercase, lower-case, and digits.
- Repeatable scrambling: Replaces the input values with different values that look similar to the original values, and guarantees to generate the same output for identical input.
- Generalization: Replaces input values with more generalized values, e.g., replaces date with year, concrete age with range, and truncates prefix/suffix.
- Swapping: Performs swapping between group of values, e.g., swaps values of a specific column in a table.
- K-anonymization: Changes the values in a way that each record is indistinguishable from at least k-1 other records, with respect to certain "identifying" attributes ([16], [17], [18]).
- Special functions for credit card, social security, and international identification numbers. These functions can be configured to guarantee that the generated numbers remain valid numbers.

Figure 3 illustrates the use of MAGEN transformation routines to anonymize a table that contains private information about customers.

Cust. id	first name	last name	g	address	city	phone num	email	insert date
735127	James	Bolton	M	183 Park Ave.	Newton	617-7529340	jbolton@abc.us	1995-03-16
570034	Adam	Harvey	M	1892 Highland Ave.	Newton	617-4779204	adamh@def.us	1997-05-07
992404	Gabriela	Rose	F	632 Broadway St.	Waltham	781-6692188	grose@ghi.us	1980-01-22
205236	Todd	Flint	M	82 Spruce Rd.	Brookline	617-4308196	todflint@xyz.us	1984-04-27

| Repeatable Scramble | Randomly Replace with Real Names (Lookup) | | Scramble Preserving Format | | Scramble Preserving Area Code | Mask Local Part Only | Random Replace |

Cust. id	first name	last name	g	address	city	phone num	email	insert date
240126	Martin	Poindexter	M	370 Yooh Ave.	Newton	617-3085840	*******@abc.us	2002-12-27
461537	Nellie	Ngo	M	1877 Pcbmkbzt Ave	Newton	617-5111385	*****@def.us	1998-10-11
981908	Monique	Belcher	F	939 Ynkiecbv St.	Waltham	781-4116286	*****@ghi.us	1995-07-21
910214	Eli	Crain	M	46 Xkedol Rd.	Brookline	617-3025500	*******@xyz.us	1991-01-19

Fig. 3. Anonymization example: customer table

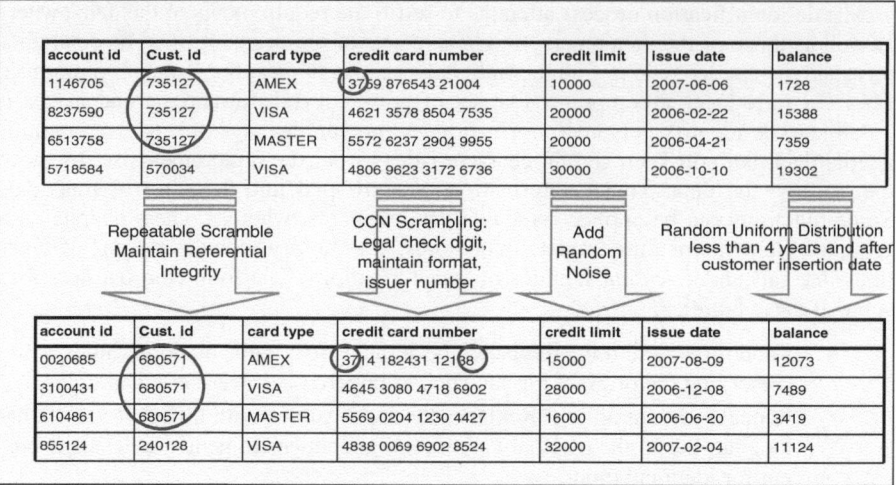

Fig. 4. Anonymization example: accounts table

Figure 4 illustrates the use of MAGEN transformation routines to remove sensitive data from a table with data on bank accounts.

MAGEN can address scenarios that involve cross-enterprise information sharing. For example, MAGEN can be used in systems that correlate biomedical information coming from several clinical and medical repositories. To maintain a correlation between records that belong to the same individual, MAGEN integrates with a trusted Anonymous Global Personal Identification (AGPI) server. The AGPI server associates a unique ID with each individual. It resolves problems such as misspelled or inconsistent names, missing data, or incompatible fields. For example, in the healthcare space, different health standards require different identification fields.

MAGEN has a pluggable mechanism to integrate with any AGPI provider, subject to fulfilling the following criteria:

1. The provider always gives the same ID to the same individual, regardless of the source of the record, the spelling of the individual's name, etc.
2. The provider never gives the same ID to more than one individual.

5   Masking Application Displays Using OCR Technology

This section deals with real-time masking of application displays, i.e., interfering with the flow of data before it gets displayed, and applying masking rules (whenever appropriate) to locate and transform sensitive entities. Existing technologies address the challenge of masking data in motion by intercepting streams of data, scanning and parsing them, and ultimately reconstructing secured streams that can be shared. In contrast, MAGEN provides a unique solution for on-the-fly masking of application displays, by analyzing and modifying the visual representation of the data—rather than the data itself. In other words, MAGEN has the capability of modifying the rendered pixels that form the images to be displayed and does not need to make copies of, or transform, the original data. MAGEN is therefore independent of the protocols used to transmit the data.

While protocol-based streams follow well-defined schemas that describe the structure of the transmitted displays (e.g., an indication that a certain text phrase is a label or that a group of text phrases constitute a table), the displays themselves lack this kind of metadata. To be able to apply masking rules that designate sensitive entities through GUI-related concepts (such as masking all fields labeled with "SSN"), MAGEN has to reveal the structure of the transmitted screens from the unstructured bitmaps.

MAGEN sits between the application that generates the data to be displayed and the end-user machine where the image is displayed. The MAGEN run-time phase ensures that each screen is masked according to the rules that are defined during the design phase. The operation on each new screen starts with OCR processing that captures the image before it is displayed, analyzes it, and recognizes the text patterns that appear on it. This is followed by a unique analysis process that generates a structured representation of this screen, including information about its layout and contents. The masking rules instruct the system how to find the sensitive entities in the resulting structured screen representation and which transformations to apply (as described in the previous Section). Finally, the modified screen representation is translated into a new image that is sent to display.

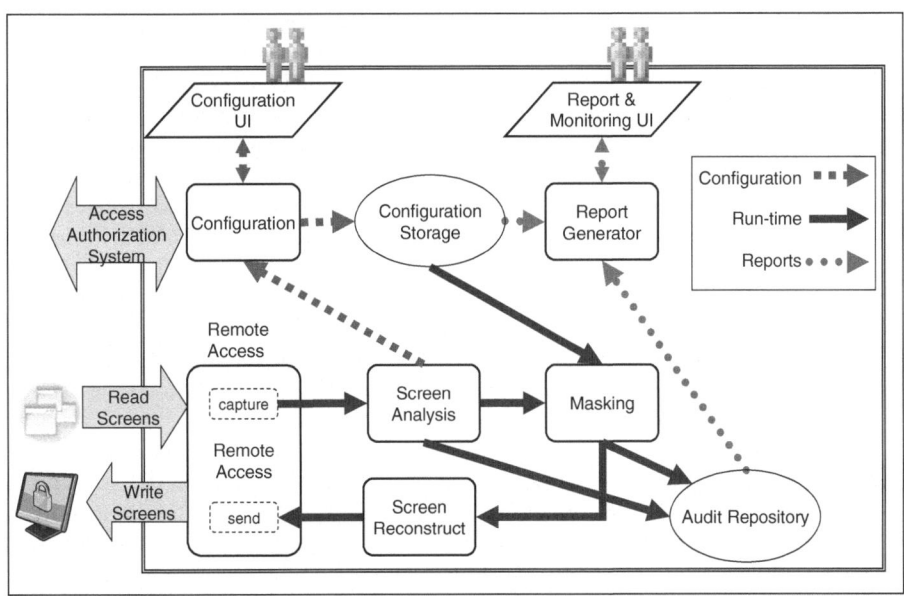

Fig. 5. System components diagram

Figure 5 illustrates the components that take part in the screen masking process. The squares represent the operational components in the system, the ovals represent the persistent storage, and the diamond shapes represent the user interfaces. The arrows indicate when the component is used and the other components with which it interfaces.

Although using OCR technology for masking sensitive data offers many advantages, it introduces some new technical challenges. The first challenge is to ensure that the system is optimized and to reduce extra latency. Classical OCR techniques are not suitable due to their nature of probabilistic results [19]. Therefore, MAGEN uses a newly developed OCR engine that leverages the fact that screen text is noise free. The OCR engine is optimized in several ways to reduce latency to a minimum while guaranteeing a very high recognition rate:

1. **Optimized character classification** when dividing text pixels into individual characters. Since characters rendered on a particular platform are always the same, the optimization is achieved by looking for the exact sequences of pixels that compose these characters in the training set.
2. **Incremental recognition** ensures that text recognition is executed only on the modified rectangles of the screen. The results are used to incrementally build the entire contents of the screen.

The next challenge is to translate each screen into a structured representation that includes enough metadata to enforce MAGEN masking rules. The outcome of the OCR engine may not be enough, as this recognition by itself reveals only the text phrases that appear on the screen, their coordinates (location and size), and colors.

In order to understand the complexity of the screen analysis beyond the basic OCR analysis, we need to understand the MAGEN rule language. We distinguish between several types of screen masking rules. The simplest way to designate sensitive entities is by providing their exact locations or by referring to their special color or font size (if relevant). We refer to such rules as location- or font-based rules. Context-based rules designate sensitive elements through their appearance in GUI constructs and by referring to constant parts of these GUI constructs (e.g., names of columns). Content-based rules specify sensitive entities by characterizing the set of possible values (e.g., specifying legal credit card numbers). The formal definition of the MAGEN rule language is beyond the scope of this paper. In what follows, we briefly outline the major principles of context-based rules, as we later focus on the analysis needed to apply such rules.

Each context-based rule is associated with one GUI element and refers to a list of screen templates to which it may be applied. It includes a string that will be used in the search process, and indicates where to perform the actual transformation. The search refers to a constant part of the GUI construct, such as the label of a labeled field or a column name within a table, while the transformation is performed on the variable part, such as the text area of a labeled field or the set of values in a column. In addition, the rule specifies how to search for strings, whether to look for exact string matching or for sub-strings, whether to use a regular expression, etc. A rule may also include conditions on the surrounding GUI elements. For example, a rule can say 'mask the contents of the column "Telephone" in any table that also contains a column "Address"' or 'mask any labeled fields with the label "Name" that appears after the title "Client details"'.

Applying location- or font-based rules is fairly straightforward, since the corresponding entities can be located easily by scanning the results of the OCR engine. This is by no means sufficient for applying context- or content-based rules. Here, we need further analysis to reveal the structure of the screen and the semantics of its entities. In this paper, we focus on the analysis needed to apply context-based rules. It includes two complex tasks:

1. **Recognizing the set of GUI constructs** that appear on the screen and their relative positioning. MAGEN incorporates several unique analysis techniques to achieve this task. The analysis searches for sets of text phrases whose relative positioning determines a specific GUI construct. For example, the analysis recognizes tables by searching for several lines that conform to the same partition into columns. These techniques can be further aided by platform- or application-specific configurations (e.g., information on size of fonts, separators, and color schemes). Moreover, the relative positioning of all the GUI constructs has to be revealed. This is induced by a containment relationship between constructs, or by their relative order from left to right or top to bottom.

2. **Recognizing screens with identical structures**, yet possibly with different contents (e.g., screens that present a structured employee form, while each appearance may contain different employee details). This is used to define rules for one screen and apply them to all screens that adhere to the same structure. Since two screens that have the same structure may have different content, a simple comparison does not work and the comparison is done only on the constant parts of the screens. MAGEN again applies unique analysis techniques to identify the text phrases that appear in the constant parts of the GUI constructs such as table column headers and field labels. This summary information is used to calculate an *ID* for each processed screen. All screens that have the same structure are mapped to a single ID.

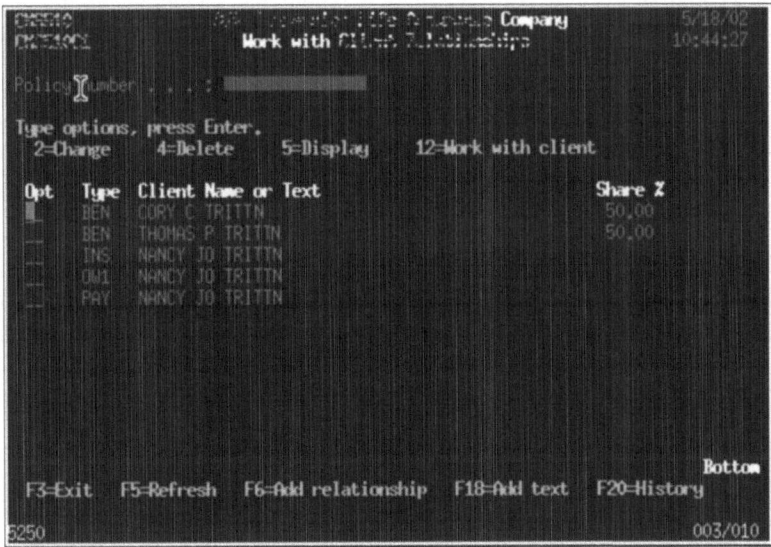

Fig. 6. Masked green screen

One can define a context-based rule for screens, such as the one shown in Figure 6, to specify that the number that appears right after the label "Policy number" has to be masked. Suppose that MAGEN captures the screen shown in Figure 6. It computes the screen ID and thereby understands that the subject rule should be applied. It also recognizes all the GUI constructs that appear in this screen and, in particular, the labeled field with the label "Policy number". Figure 6 illustrates the result of applying the rule and substituting the pixels of the policy number with red pixels.

MAGEN utilizes the OCR engine and the subsequent screen analysis during the design phase to simplify and automate the process of generating masking rules. Instead of manually editing rules, MAGEN incorporates a process that runs through a set of possible application displays, and allows the users (with privileges to generate rules) to select certain entities on screens. These selections are automatically translated into masking rules.

A detailed description for using the MAGEN system to mask application displays appears in [20].

6 Summary and Future Work

Data masking is a required element of information management and data governance, enabling enterprises to effectively and efficiently reduce the risks of unauthorized disclosure. MAGEN suggests a flexible SOA-based platform that can handle a wide range of scenarios where information is shared, while keeping the configuration and management capabilities centralized, thus allowing for the uniform design of rules and the monitoring of policy violations.

The MAGEN set of transformation routines is designed as a stand-alone library that can be integrated into and enrich existing data masking solutions. Indeed, it is shipped as part of several IBM products or solutions, including Optim, which is IBM's primary privacy-protecting product [5], and DataStage, which is IBM's leading ETL solution [21].

For dynamically masking application displays, MAGEN suggests a unique approach that uses OCR technology as the basis for content acquisition and discovery. This has the tremendous advantage of allowing MAGEN to be platform- and protocol-agnostic.

MAGEN uses a generic rule language to indicate what, when, and how to mask sensitive entities. The rule language is designed to make it as easy as possible for users to create rules. MAGEN follows the concept of transforming an unstructured document into an XML format. The latter reveals semantics and metadata of the original form, and the rules are translated into a discovery and transformation process over that XML representation. The idea is thus to apply a unified method to mask all unstructured and semi-structured data formats. In the future, we plan to examine integration into standard policy languages such as Platform for Privacy Preferences (P3P) [22] and eXtensible Access Control Markup Language (XACML) [23].

Much research is still needed to develop verification techniques that help prove compliance with regulations. The challenge is to ensure that policies about protecting sensitive data are properly translated into practical rules that specify which data is sensitive and how to transform it. We plan to further develop techniques to verify that the result of applying such rules satisfies the regulations.

References

1. Burke, B.E.: Information Protection and Control Survey: Data Loss Prevention and Encryption Trends. IDC Special Study, Doc. # 211109 (2008)
2. Proctor, P.E., Mogull, R., Ouellet, E.: Magic Quadrant for Content Monitoring and Filtering and Data Loss Prevention. Gartner report, ID Number: G00147610 (2007)
3. Yuhanna, N.: Protecting Private Data with Data Masking. Forrester report (2006),
   http://www.forrester.com/Research/Document/Excerpt/
   0,7211,39116,00.html
4. Feiman, J.: Data Obfuscation (Masking, Privacy, Scrambling): Many Names for the Same Technique. Gartner report, ID Number: G00157661 (2008)
5. IBM Optim., http://www.optimsolution.com/
6. Oracle,
   http://www.oracle.com/enterprise_manager/data-masking.html
7. Camouflage, http://www.datamasking.com/camouflage.html
8. Nelken, H., Vortman, P., Cohen, S., Melament, A.: De-Identification Framework. IBM White Paper (2004)
9. Gardner, J.L.: HIDE: An Integrated System for Health Information DE-identification. In: 21st IEEE International Symposium on Computer-Based Medical Systems, CBMS 2008, pp. 254–259. IEEE Press, Los Alamitos (2008)
10. Carter, G.: LDAP System Administration. O' Reilly, Sebastopol (2003)
11. IBM RACF,
    http://www-03.ibm.com/servers/eserver/zseries/zos/racf/
12. CA ACF2, http://www.ca.com/us/products/product.aspx?id=111
13. IBM EAS,
    http://www-01.ibm.com/software/data/ips/products/
    masterdata/eas/
14. MitemView, http://www.mitem.com/products/mitemview.asp
15. Agrawal, R., Srikant, R.: Privacy Preserving Data Mining. ACM's Special Interest Group on Management Of Data (SIGMOD) 29(2), 439–450 (2000)
16. Sweeney, L.: k-anonymity: A Model for Protecting Privacy. International Journal of Uncertainty, Fuzziness and Knowledge-Based Systems 10(5), 557–570 (2002)
17. Bayardo, R.J., Agrawal, R.: Data Privacy through Optimal k-anonymization. In: Proceedings of the 21st IEEE International Conference on Data Engineering. ICDE 2005, pp. 217–228 (2005)
18. Shand, B., Rashbass, J.: Protecting privacy in tabular healthcare data: explicit uncertainty for disclosure control. In: Proceedings of the 2005 ACM workshop on Privacy in the Electronic Society (WPES 2005), pp. 20–26 (2005)
19. Goshtasby, A., Ehrich, R.W.: Contextual word recognition using probabilistic relaxation labeling. Pattern Recognition 21(5), 455–462 (1988)
20. Porat, S., Carmeli, B., Domany, T., Drory, T., Geva, A., Tarem, A.: Dynamic Masking of Application Displays Using OCR Technologies. IBM Systems Journal, special issue on Global Service Delivery Technologies (2008) (submitted)
21. IBM DataStage,
    http://www-01.ibm.com/software/data/infosphere/
    datastage/
22. Cranor, L.F.: Web Privacy with P3P. O'Reilly, Sebastopol (2002)
23. OASIS eXtensible Access Control Markup Language (XACML),
    http://lists.oasis-open.org/archives/wsia/200205/
    pdf00001.pdf

No Syllogisms for the Numerical Syllogistic

Ian Pratt-Hartmann

School of Computer Science,
University of Manchester,
Manchester M13 9PL, U.K.
ipratt@cs.man.ac.uk

Abstract. The *numerical syllogistic* is the extension of the traditional syllogistic with numerical quantifiers of the forms at least C and at most C. It is known that, for the traditional syllogistic, a finite collection of rules, similar in spirit to the classical syllogisms, constitutes a sound and complete proof-system. The question arises as to whether such a proof system exists for the numerical syllogistic. This paper answers that question in the negative: no finite collection of syllogism-like rules, broadly conceived, is sound and complete for the numerical syllogistic.

1   Introduction

The *numerical syllogistic* is the set of English sentences of the forms

At least C p are q	At least C p are not q
At most C p are q	At most C p are not q,

where p and q are common (count) nouns, and C is a (decimal) digit string representing a natural number in the usual way. We here ignore, and henceforth silently correct, details of English number-agreement and plural morphology, since these matters have no bearing on the ensuing discussion. The argument

$$
\begin{array}{l}
\text{At least 13 artists are beekeepers} \\
\text{At most 3 beekeepers are not carpenters} \\
\underline{\text{At most 1 carpenter is not a dentist}} \\
\text{At least 9 artists are dentists,}
\end{array} \tag{1}
$$

whose premises and conclusion all belong to the numerical syllogistic, is evidently valid: any circumstance in which all the premises are true is one in which the conclusion is true. For suppose the premises are true. Take any collection of thirteen artists who are beekeepers; since at most three of these are not carpenters, the remaining ten are; and since, of these ten, at most one is not a dentist, the remaining nine are.

The numerical syllogistic generalizes the *traditional syllogistic*, which, for our purposes, we may take to be the set of English sentences of the forms

Some p are q	Some p are not q
All p are q	No p are q.

O. Grumberg et al. (Eds.): Francez Festschrift, LNCS 5533, pp. 192–203, 2009.

To see this, note that the sentence Some p are q may be equivalently written At least 1 p is a q; likewise, All p are q may be equivalently (if somewhat unidiomatically) written At most 0 p are not q; and so on. Since the standard system of syllogisms presented in Aristotle's *Prior Analytics* can be shown—with a few relatively minor adjustments—to license exactly the valid arguments in the traditional syllogistic [1,2,3,4], it is natural to ask whether a similar situation holds for the numerical syllogistic.

To understand what this question means more concretely, consider the rule

$$\frac{\text{At most } C\ p \text{ are not } q \qquad \text{At least } D\ o \text{ are } p}{\text{At least } E\ o \text{ are } q}\ (0 \leq E \leq D - C), \qquad (2)$$

which we interpret as licensing an inference from any instances of the sentence-schemata above the line to the corresponding instance of the sentence schema below the line, subject to the side-condition $0 \leq E \leq D - C$. Clearly, this rule is valid: it never leads from true premises to a false conclusion. For suppose the premises are true. Take any collection of $D\ o$ which are p; since at most C of them are not q, the remaining $D - C$ are. In fact, by chaining two instances of Rule (2) together, we can formally demonstrate the validity of Argument (1), thus:

$$\frac{\dfrac{\text{At most 3 beekeepers are not carpenters} \qquad \text{At least 13 artists are beekeepers}}{\text{At most 1 carpenter is not a dentist} \qquad \text{At least 10 artists are carpenters}}{\text{At least 9 artists are dentists.}}$$

Rule (2) might reasonably be regarded as a "numerical syllogism". Indeed, the traditional syllogism *Darii* is simply the special case obtained by putting $C = 0$ and $D = E = 1$:

$$\frac{\text{All } p \text{ are } q \qquad \text{Some } o \text{ are } p}{\text{Some } o \text{ are } q}.$$

Thus, we are led to ask whether there exists a finite collection of such numerical syllogisms—broadly conceived—that licenses all (and only) the valid arguments in the numerical syllogistic? We show in the sequel that there is not.

Despite its obviousness as a generalization of the traditional syllogistic, the numerical syllogistic seems not to have attracted the attention of logicians before the Nineteenth Century. The first systematic investigation known to the author is that of de Morgan [5] (Ch. VIII), though this work was closely followed by treatments in Boole [6] (reprinted as [7], Sec. IV) and Jevons [8], (reprinted as [9], Part I, Sec. IV). For a historical overview of this episode in logic, see [10]. De Morgan presented a list of what he took to be the valid numerical syllogisms. Latterly, various other proof-systems have been proposed, also based on numerical generalizations of the traditional syllogisms. Good examples are those of Murphree [11,12], and of Hacker and Parry [13]. The negative results presented below apply to all of these systems. These same results constitute a strengthening (and simplification) of earlier observations made by the author in [14], Sec. 5.

At the same time, the present paper can be seen as a contribution—though perhaps something of a negative one—to an established tradition of attempts to provide logical calculi more or less closely modelled on aspects of natural languages. Examples of work in this tradition include Fitch's use of combinatory logic [15], Suppes' use of relation algebra [16], Purdy's 'natural logic' [17,18,19], and Fyodorov *et al.*'s inference calculus based on monotonicity features [20].

The plan of the paper is as follows. Section 2 presents the syntax and semantics of a formal language, $\mathcal{N}$, which faithfully reconstructs the numerical syllogistic, together with a natural extension of $\mathcal{N}$, which we denote $\mathcal{N}^\dagger$. Section 3 reconstructs—as liberally as possible—the notion of a numerical syllogism, and states the main result of this paper: that neither $\mathcal{N}$ nor $\mathcal{N}^\dagger$ admit a finite system of numerical syllogisms that licenses exactly the valid inferences. In stating this result, we pay particular attention to indirect proof and the rule of *reductio ad absurdum*. Section 4 proves the result.

2 Syntax and Semantics of $\mathcal{N}^\dagger$ and $\mathcal{N}$

Fix a countably infinite set $\mathbf{P}$. We may assume $\mathbf{P}$ to contain all English common count-nouns such as man, animal *etc.* An *atom* is an element of $\mathbf{P}$; a *literal* is an expression of either of the forms p or $\bar{p}$, where p is an atom. A literal which is an atom is said to be *positive*; all other literals are said to be *negative*. If $l = \bar{p}$ is a negative literal, then we take $\bar{l}$ to denote the positive literal p. An $\mathcal{N}^\dagger$-*formula* is an expression of either of the forms $(\leq C)[l, m]$ or $(> C)[l, m]$, where C is a decimal string representing a non-negative integer, and l, m are literals. To avoid cumbersome circumlocutions, we henceforth ignore the distinction between natural numbers and the decimal strings representing them. An $\mathcal{N}$-*formula* is an $\mathcal{N}^\dagger$-formula at least one of whose literals is positive. We denote the set of $\mathcal{N}^\dagger$-formulas by $\mathcal{N}^\dagger$; and similarly for $\mathcal{N}$. A subset $P \subseteq \mathbf{P}$ is a *signature*. If P is a signature, we denote by $\mathcal{N}^\dagger(P)$ the set of $\mathcal{N}^\dagger$-formulas involving no atoms other than those in P; and similarly for $\mathcal{N}(P)$.

We provide formal semantics for the language $\mathcal{N}^\dagger$—and hence for $\mathcal{N}$—as follows. A *structure* $\mathfrak{A}$ is a pair $\langle A, \{p^{\mathfrak{A}}\}_{p \in \mathbf{P}} \rangle$, where A is a non-empty set, and $p^{\mathfrak{A}} \subseteq A$, for every $p \in \mathbf{P}$. The set A is called the *domain* of $\mathfrak{A}$. Given a structure $\mathfrak{A}$, we extend the map $p \mapsto p^{\mathfrak{A}}$ to all literals by setting $\bar{p}^{\mathfrak{A}} = A \setminus p^{\mathfrak{A}}$. We define the truth-relation $\models$ between structures and $\mathcal{N}^\dagger$-formulas by declaring

$$\mathfrak{A} \models (\leq C)[l, m] \text{ iff } |l^{\mathfrak{A}} \cap m^{\mathfrak{A}}| \leq C$$

$$\mathfrak{A} \models (> C)[l, m] \text{ iff } |l^{\mathfrak{A}} \cap m^{\mathfrak{A}}| > C.$$

Note that these truth-conditions are symmetric in the literals l and m. Accordingly, we henceforth identify formulas differing only with respect to the order of their literals, silently performing any transpositions required.

These semantics justify the following English glosses for $\mathcal{N}$-formulas, where p and q are English count nouns (and hence also elements of $\mathbf{P}$):

$(\leq C)[p, q]$	At most C p are q	$(\leq C)[p, \bar{q}]$	At most C p are not q
$(> C)[p, q]$	At least $C + 1$ p are q	$(> C)[p, \bar{q}]$	At least $C + 1$ p are not q.

We also provide pseudo-English glosses for $\mathcal{N}^\dagger$-formulas in a similar way, except that 'negated' subjects are required when both literals are negative:

$$(\leq C)[\bar{p}, \bar{q}] \quad \text{At most } C \text{ non-}p \text{ are not } q$$
$$(> C)[\bar{p}, \bar{q}] \quad \text{At least } C + 1 \text{ non-}p \text{ are not } q.$$

The use of $\leq$ and $>$ (rather than the $\leq$ and $\geq$ employed in Section 1) simplifies various technical details in the ensuing presentation. Nothing of substance hinges on this decision, however; the results obtained below would not be materially altered by expanding our languages to include formulas of the form $(\geq 0)[l, m]$.

If Θ is a set of formulas, we write $\mathfrak{A} \models \Theta$ if, for all $\theta \in \Theta$, $\mathfrak{A} \models \theta$. A formula θ is *satisfiable* if there exists $\mathfrak{A}$ such that $\mathfrak{A} \models \theta$; a set of formulas Θ is *satisfiable* if there exists $\mathfrak{A}$ such that $\mathfrak{A} \models \Theta$. If, for all structures $\mathfrak{A}$, $\mathfrak{A} \models \Theta$ implies $\mathfrak{A} \models \theta$, we say that Θ *entails* θ, and write $\Theta \models \theta$. We take it as uncontroversial that $\Theta \models \theta$ constitutes a rational reconstruction of the pre-theoretic judgment that a conclusion θ may be validly inferred from premises Θ. For example, the valid argument (1) corresponds to the entailment

$$\{(> 12)[\text{artst}, \text{bkpr}] \,, \, (\leq 3)[\text{bkpr}, \overline{\text{crpntr}}]$$
$$(\leq 1)[\text{crpntr}, \overline{\text{dntst}}]\} \models (> 8)[\text{artst}, \text{dntst}]. \quad (3)$$

No formula of the form $(> C)[l, \bar{l}]$ is satisfiable: that is, for all $\mathfrak{A}$, $\mathfrak{A} \not\models (> C)[l, \bar{l}]$. We refer to any such formula as an *absurdity*; and we use the (possibly decorated) symbol $\bot$ to denote, ambiguously, any absurdity. Note that all absurdities are actually $\mathcal{N}$-formulas.

If θ is an $\mathcal{N}^\dagger$-formula, we define the $\mathcal{N}^\dagger$-formula $\bar{\theta}$ to be the result of exchanging the symbols $\leq$ and $>$ in θ. That is:

$$\bar{\theta} = \begin{cases} (> C)[l, m] & \text{if } \theta = (\leq C)[l, m] \\ (\leq C)[l, m] & \text{if } \theta = (> C)[l, m]. \end{cases}$$

It is easy to see that, for any structure $\mathfrak{A}$, $\mathfrak{A} \models \theta$ if and only if $\mathfrak{A} \not\models \bar{\theta}$. Moreover, for any $\mathcal{N}^\dagger$-formula θ, we have $\bar{\bar{\theta}} = \theta$, and if θ is an $\mathcal{N}$-formula, then so is $\bar{\theta}$. Informally, we may think of $\bar{\theta}$ as the *negation* of θ. Thus, the languages $\mathcal{N}^\dagger$ and $\mathcal{N}$ are, in essence, closed under negation.

The *satisfiability problem* for $\mathcal{N}^\dagger$ is the following problem: given a finite set of $\mathcal{N}^\dagger$-formulas Θ, determine whether Θ is satisfiable. The *validity problem* for $\mathcal{N}^\dagger$ is the following problem: given a finite set of $\mathcal{N}^\dagger$-formulas Θ and an $\mathcal{N}^\dagger$-formula θ, determine whether $\Theta \models \theta$. The satisfiability and validity problems for $\mathcal{N}$ are defined analogously. Since $\mathcal{N}^\dagger$ and $\mathcal{N}$ are, in effect, closed under negation, satisfiability and validity are dual notions in the usual sense. It is known [21,14] that the satisfiability problems for $\mathcal{N}^\dagger$ and $\mathcal{N}$ are both NPTIME-complete; hence the corresponding validity problems are both CO-NPTIME-complete.

3   Proof Theory for $\mathcal{N}^\dagger$ and $\mathcal{N}$

This section develops a framework for formalizing systems of syllogism-like rules in $\mathcal{N}$ and $\mathcal{N}^\dagger$. Because we shall be deriving negative results about such systems, and wish these results to be as general as possible, our presentation will be in some respects rather abstract. However, we shall never stray far from the intuitions developed in Section 1.

We begin with some very general notions. Let $\mathcal{L}$ be any formal language, understood as a set of $\mathcal{L}$-formulas for which a truth-relation $\models$ is defined. By a *derivation relation* (*in $\mathcal{L}$*), we simply mean a subset of $\mathbb{P}(\mathcal{L}) \times \mathcal{L}$, where $\mathbb{P}(\mathcal{L})$ is the power set of $\mathcal{L}$. If $\vdash$ is a derivation relation, we write $\Theta \vdash \theta$ instead of $\langle \Theta, \theta \rangle \in \vdash$. We call $\vdash$ *sound* (*for $\mathcal{L}$*) if, for all sets of $\mathcal{L}$-formulas Θ and all $\mathcal{L}$-formulas θ, $\Theta \vdash \theta$ implies $\Theta \models \theta$. We call $\vdash$ *complete* (*for $\mathcal{L}$*) if, for all sets of $\mathcal{L}$-formulas Θ and all $\mathcal{L}$-formulas θ, $\Theta \models \theta$ implies $\Theta \vdash \theta$. In this paper, we are interested in derivation relations in $\mathcal{N}^\dagger$ and $\mathcal{N}$ generated by finite sets of syllogism-like rules. These we now proceed to define.

A *formula schema in $\mathcal{N}^\dagger$* is an expression of the form $(Q)[l, m]$ where Q is either of the symbols $\leq$ or $>$, and l and m are literals. A *formula schema in $\mathcal{N}$* is a formula schema in $\mathcal{N}^\dagger$ subject to the additional condition that at least one of l and m is positive. A *syllogistic rule in $\mathcal{N}^\dagger$* (*in $\mathcal{N}$*) is a pair $(\boldsymbol{\xi}, R)$, where, for some $k \geq 0$, $\boldsymbol{\xi}$ is a $(k+1)$-tuple of formula schemata in $\mathcal{N}^\dagger$ (respectively, $\mathcal{N}$), and R is a $(k+1)$-ary relation over $\mathbb{N}$. A *substitution* is a function $f : \mathbf{P} \to \mathbf{P}$. Substitutions are applied to negative literals in the expected way: $f(\bar{p}) = \overline{f(p)}$. An *instance* of the syllogistic rule

$$(\langle (Q_1)[l_1, m_1], \ldots, (Q_k)[l_k, m_k], (Q)[l, m] \rangle, R), \tag{4}$$

is any $(k+1)$-tuple

$$\langle (Q_1\ C_1)[f(l_1), f(m_1)], \ldots, (Q_k\ C_k)[f(l_k), f(m_k)], (Q\ C)[f(l), f(m)] \rangle \tag{5}$$

where f is a substitution and $C_1, \ldots, C_k, C$ are integers such that $\langle C_1, \ldots, C_k, C \rangle \in R$. It is easy to see that, if (4) is a syllogistic rule in $\mathcal{N}^\dagger$ (or in $\mathcal{N}$), then the elements of (5) are $\mathcal{N}^\dagger$-formulas (respectively, $\mathcal{N}$-formulas). The intuitive meaning of any instance $\langle \theta_1, \ldots, \theta_k, \theta \rangle$ of a syllogistic rule is that θ may be inferred from $\theta_1, \ldots, \theta_k$. Officially, no restrictions at all are placed on the relation R. In practice, however, R will usually be defined as $\{\langle x_1, \ldots, x_{k+1} \rangle \in \mathbb{N}^{k+1} \mid \pi(x_1, \ldots, x_{k+1})\}$, for some (arithmetic) expression π. In that case, we may display the syllogistic rule (4) in a more readable way as:

$$\frac{(Q_1\ C_1)[l_1, m_1] \quad \cdots \quad (Q_k\ C_k)[l_k, m_k]}{(Q\ C)[l, m]} \ (\pi(C_1, \ldots, C_k, C)). \tag{6}$$

A syllogistic rule is *valid* if, for any instance $\langle \theta_1, \ldots, \theta_k, \theta \rangle$ of that rule, we have $\{\theta_1, \ldots, \theta_k\} \models \theta$—that is to say, if all the inference steps it licenses are entailments.

Some examples will help to motivate the rather austere definitions just given. Consider the following syllogistic rules, displayed in the style of (6):

$$\frac{(\leq C)[l,\bar{m}] \qquad (\leq D)[m,n]}{(\leq E)[l,n]} \ (E \geq C + D) \tag{7}$$

$$\frac{(\leq C)[m,\bar{n}] \qquad (> D)[l,m]}{(> E)[l,n]} \ (0 \leq E \leq D - C). \tag{8}$$

These syllogistic rules are easily seen to be valid. We encountered (8), in a slightly different guise (and with all literals positive), as Rule (2) in Section 1.

If X is a set of syllogistic rules in $\mathcal{N}^{\dagger}$, we define the relation of *direct derivation relative to* X, denoted $\vdash_{\mathsf{X}}$, to be the smallest subset of $\mathbb{P}(\mathcal{N}^{\dagger}) \times \mathcal{N}^{\dagger}$ satisfying the following conditions:

1. if $\theta \in \Theta$, then $\Theta \vdash_{\mathsf{X}} \theta$;
2. if $\langle \theta_1, \ldots, \theta_k, \theta \rangle$ is an instance of some syllogistic rule in X, and $\Theta \vdash_{\mathsf{X}} \theta_i$ for all i $(1 \leq i \leq k)$, then $\Theta \vdash_{\mathsf{X}} \theta$.

Instances of the relation $\vdash_{\mathsf{X}}$ can be established by *derivations* in the form of finite trees in the usual way. For instance, from the premises of Argument (1), two applications of Rule (8) yield the derivation

$$\frac{(\leq 1)[\text{crpntr}, \overline{\text{dntst}}] \quad \dfrac{(\leq 3)[\text{bkpr}, \overline{\text{crpntr}}] \quad (> 12)[\text{artst}, \text{bkpr}]}{(> 9)[\text{artst}, \text{crpntr}]}}{(> 8)[\text{artst}, \text{dntst}]},$$

which, again, we encountered in Section (1). Thus, for any set of syllogistic rules X containing (8), we have:

$$\{(> 12)[\text{artst}, \text{bkpr}] \ , \ (\leq 3)[\text{bkpr}, \overline{\text{crpntr}}]$$
$$(\leq 1)[\text{crpntr}, \overline{\text{dntst}}]\} \vdash_{\mathsf{X}} (> 8)[\text{artst}, \text{dntst}]. \tag{9}$$

We remark in passing that, if X contains both Rules (7) and (8), we have an alternative derivation showing (9):

$$\frac{\dfrac{(\leq 3)[\text{bkpr}, \overline{\text{crpntr}}] \quad (\leq 1)[\text{crpntr}, \overline{\text{dntst}}]}{(\leq 4)[\text{bkpr}, \overline{\text{dntst}}]} \ \text{Rule (7)} \quad (> 12)[\text{artst}, \text{bkpr}]}{(> 8)[\text{artst}, \text{dntst}]} \ \text{Rule (8)}.$$

Classical treatments of the syllogistic actually recognize a slightly more liberal notion of derivation than that presented above. Suppose we have derived $\bot$ from a set of premises $\Theta \cup \{\theta\}$, where $\bot$ is some absurdity. The rule of *reductio ad absurdum* allows us then to infer the formula $\bar{\theta}$ (semantically: the negation of θ) from Θ alone. *Reductio* is not a syllogistic rule, in the technical sense employed in this paper: for one thing it decreases the set of premises in a derivation—something no syllogistic rule can do. Nevertheless, it evidently preserves entailment: if $\Theta \cup \{\theta\} \models \bot$, then $\Theta \models \bar{\theta}$. For these reasons, we might wish to take account of this rule in our analysis of the numerical syllogistic.

We do so as follows. If X is a set of syllogistic rules in $\mathcal{N}^\dagger$, we define the relation of *indirect derivation relative to* X, denoted $\Vdash_\mathsf{X}$, to be the smallest subset of $\mathbb{P}(\mathcal{N}^\dagger) \times \mathcal{N}^\dagger$ satisfying the following conditions:

1. if $\theta \in \Theta$, then $\Theta \Vdash_\mathsf{X} \theta$;
2. if $\Theta \Vdash_\mathsf{X} \theta_1, \dots, \Theta \Vdash_\mathsf{X} \theta_k$, and $\langle \theta_1, \dots, \theta_k, \theta \rangle$ is an instance of some syllogistic rule in X, then $\Theta \Vdash_\mathsf{X} \theta$;
3. if $\Theta \cup \{\theta\} \Vdash_\mathsf{X} \bot$, then $\Theta \Vdash_\mathsf{X} \bar{\theta}$ (the rule of *reductio ad absurdum*).

Instances of the indirect derivation relation $\Vdash_\mathsf{X}$ may be established by constructing proof-trees similar to those for direct derivations, except that we need a little more machinery to keep track of premises. This may be done as follows. Suppose we have a derivation (direct or indirect) showing that $\Theta \cup \{\theta\} \Vdash_\mathsf{X} \bot$, for some absurdity $\bot$. Let this derivation be displayed as

$$\Theta \quad \cdots \quad \cdots \quad \theta$$
$$\vdots$$
$$\bot.$$

Applying Clause 3 of the definition of $\Vdash_\mathsf{X}$, we have $\Theta \Vdash_\mathsf{X} \bar{\theta}$, which we take to be established by the derivation

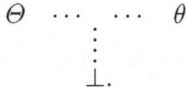

The premise θ of the original derivation no longer counts as a premise in the new derivation. As we say, the premise in question has been *discharged*. In displaying derivations, we enclose any discharged premise in brackets, and co-index it with the application of *reductio* which discharges it. Two minor complications regarding indirect derivations should be noted at this point. The first concerns the case where a derivation of $\bot$ involves multiple instances of some premise θ. In that case, the rule of reductio should be understood as allowing us to discharge *any number* (including zero) of those occurrences. The second complication concerns the case where a derivation of $\bot$ involves no instances of some premise θ. In that case, the rule of reductio should be understood as still allowing us to discharge (zero occurrences of) θ. Put another way: we do not have to discharge occurrences of premises if we do not want.

If X is a set of syllogistic rules in $\mathcal{N}$, we define the derivation relations $\vdash_\mathsf{X}$ and $\Vdash_\mathsf{X}$ analogously.

Thus, given a set of syllogistic rules X (in either $\mathcal{N}^\dagger$ or $\mathcal{N}$), we have two derivation relations of interest: $\vdash_\mathsf{X}$ (direct derivation) and $\Vdash_\mathsf{X}$ (indirect derivation), with the latter always including the former. The following are evidently equivalent:

(i) $\vdash_X$ is sound; (ii) $\Vdash_X$ is sound; (iii) every rule in X is valid. Moreover, if $\vdash_X$ is complete, then, trivially, so is $\Vdash_X$.

The following questions now arise. Does there exist a finite set X of syllogistic rules in $\mathcal{N}^\dagger$ such that the direct derivation relation $\vdash_X$ is sound and complete? If not, does there at least exist a finite set X of syllogistic rules in $\mathcal{N}^\dagger$ such that the indirect derivation relation $\Vdash_X$ is sound and complete? And is the situation any different for the smaller language $\mathcal{N}$? The main result of this paper is that the answer to all of these questions is no.

We close this section with a simple observation on derivations. Suppose Θ is a set of $\mathcal{N}^\dagger$-formulas and θ an $\mathcal{N}^\dagger$-formula such that $\Theta \Vdash_X \theta$; and let P be the signature of atoms occuring in $\Theta \cup \{\theta\}$. Consider any (indirect) derivation of θ from Θ (via the syllogistic rules X). If that derivation involves any atoms not in P, we may evidently uniformly replace them by atoms in P, obtaining another derivation of θ from Θ. The same holds for direct derivations, and also for the language $\mathcal{N}$. Thus, when considering derivations from Θ to θ, we may limit ourselves entirely to the languages $\mathcal{N}^\dagger(P)$ or $\mathcal{N}(P)$.

4   Main Result

Let n be an integer ($n \geq 4$), and let $P^{(n)}$ be a signature of cardinality $n + 1$— say $\{p_1, \ldots, p_n, q\}$. We denote by $\Gamma^{(n)}$ the following (infinite) set of $\mathcal{N}^\dagger(P^{(n)})$-formulas, where i, j range over all *distinct* integers in the interval $1, \ldots, n$, C ranges over all natural numbers in the intervals indicated, and o ranges over $P^{(n)}$.

1. There are exactly $n - 1$ objects in the domain, all satisfying q:

$$(\leq C)[q, q] \quad (C \geq n - 1) \qquad (\leq C)[\bar{q}, \bar{q}] \quad (C \geq 0)$$
$$(> C)[q, q] \quad (C \leq n - 2).$$

2. Each p_i is realized exactly once; and its complement is realized exactly $n - 2$ times:

$$(\leq C)[p_i, p_i] \quad (C \geq 1) \qquad (\leq C)[\bar{p}_i, \bar{p}_i] \quad (C \geq n - 2)$$
$$(> 0)[p_i, p_i] \qquad\qquad (> C)[\bar{p}_i, \bar{p}_i] \quad (C \leq n - 3).$$

3. All the p_i and all the non-p_i are q:

$$(\leq C)[p_i, q] \quad (C \geq 1) \qquad\qquad (\leq C)[p_i, \bar{q}] \quad (C \geq 0)$$
$$(> 0)[p_i, q]$$
$$(\leq C)[\bar{p}_i, q] \quad (C \geq n - 2) \qquad (\leq C)[\bar{p}_i, \bar{q}] \quad (C \geq 0)$$
$$(> C)[\bar{p}_i, q] \quad (C \leq n - 3).$$

4. No p_i is a p_j (remember that $i \neq j$):

$$(\leq C)[p_i, p_j] \quad (C \geq 0)$$
$$(\leq C)[p_i, \bar{p}_j] \quad (C \geq 1) \qquad (\leq C)[\bar{p}_i, \bar{p}_j] \quad (C \geq n - 3)$$
$$(> 0)[p_i, \bar{p}_j] \qquad\qquad\qquad (> C)[\bar{p}_i, \bar{p}_j] \quad (C \leq n - 4)$$

5. The logical truths of $\mathcal{N}^\dagger(P^{(n)})$:

$$(\le C)[o, \bar{o}] \qquad (C \ge 0).$$

When considering derivations from $\Gamma^{(n)}$, we limit ourselves entirely to the language $P^{(n)}$. Where n can be regarded as a constant, we omit it, and write Γ for $\Gamma^{(n)}$.

Lemma 1. Γ *is unsatisfiable.*

Proof. The formulas $(\le n - 1)[q, q]$, $(> 0)[p_i, q]$ $(1 \le i \le n)$ and $(\le 0)[p_i, p_j]$ $(1 \le i < j \le n)$ together violate the pigeonhole principle.

Lemma 2. *For every* $\theta \in \mathcal{N}^\dagger(P^{(n)})$, *either* $\theta \in \Gamma$ *or* $\bar{\theta} \in \Gamma$.

Proof. Exhaustive check.

For all i, $(1 < i \le n)$, define

$$\gamma_i = (\le 0)[p_1, p_i] \qquad\qquad \delta_i = (> 0)[p_1, \bar{p}_i]$$
$$\epsilon_i = (> 0)[\bar{p}_1, p_i] \qquad\qquad \zeta_i = (\le n - 3)[\bar{p}_1, \bar{p}_i],$$

so that

$$\bar{\gamma}_i = (> 0)[p_1, p_i] \qquad\qquad \bar{\delta}_i = (\le 0)[p_1, \bar{p}_i]$$
$$\bar{\epsilon}_i = (\le 0)[\bar{p}_1, p_i] \qquad\qquad \bar{\zeta}_i = (> n - 3)[\bar{p}_1, \bar{p}_i].$$

And for all i, $(1 < i \le n)$, define

$$\Theta_i = \{\gamma_i, \delta_i, \epsilon_i, \zeta_i\} \qquad\qquad \bar{\Theta}_i = \{\bar{\gamma}_i, \bar{\delta}_i, \bar{\epsilon}_i, \bar{\zeta}_i\}.$$

Note that $\Theta_i \subseteq \Gamma$; indeed, all the Θ_i are given in Clause 4 of the definition of Γ. (Remember: the order of literals in $\mathcal{N}^\dagger$-formulas is not significant.) In the presence of the formulas given in Clauses 1–3 of the definition of Γ, any formula in Θ_i is equivalent to any other, and states that the interpretations of p_1 and p_i are disjoint. Similarly, any formula in $\bar{\Theta}_i$ states that the interpretations of p_1 and p_i coincide. It is not hard to see that any set $(\Gamma \setminus \Theta_i) \cup \bar{\Theta}_i$ is satisfiable. For let $A = \{2, \ldots, n\}$, and, for all i $(1 < i \le n)$, let $\mathfrak{A}_i$ be the structure with domain A and interpretations

$$q^{\mathfrak{A}_i} = A \qquad p_1^{\mathfrak{A}_i} = \{i\} \qquad p_j^{\mathfrak{A}_i} = \{j\} \qquad (2 \le j \le n).$$

Thus, each $\mathfrak{A}_i$ distributes the interpretations of $p_2, \ldots, p_n$ disjointly over the universe $\{2, \ldots, n\}$, and makes the interpretations of p_1 and p_i coincide.

Lemma 3. *For all* i $(1 < i \le n)$, $\mathfrak{A}_i \models (\Gamma \setminus \Theta_i) \cup \bar{\Theta}_i$.

Proof. Routine check.

For all i, j, $(1 < i < j \le n)$, define

$$\Delta_{i,j}^{(n)} = \Gamma^{(n)} \setminus (\Theta_i \cup \Theta_j).$$

Again, for clarity, the superscript (n) is omitted where n can be regarded as a constant. Thus, $\Delta_{i,j}$ removes from Γ the formulas stating that the interpretation of p_1 is disjoint from those of both p_i and p_j.

Lemma 4. *Let θ be a formula of $\mathcal{N}^\dagger(P^{(n)})$, and let $1 < i < j \le n$. If $\Delta_{i,j} \models \theta$, then $\theta \in \Delta_{i,j}$.*

Proof. From Lemma 2, either $\theta \in \Gamma$ or $\bar{\theta} \in \Gamma$. Hence, if $\theta \notin \Delta_{i,j}$, then one of the following possibilities holds: (i) $\theta \in \Theta_i$; (ii) $\theta \in \Theta_j$; (iii) $\bar{\theta} \in \Delta_{i,j} \cup \Theta_i$; or (iv) $\bar{\theta} \in \Delta_{i,j} \cup \Theta_j$. From Lemma 3, we see that, in cases (i) and (iv), the fact that $\mathfrak{A}_i \models \Delta_{i,j} \cup \bar{\Theta}_i \cup \Theta_j$ contradicts $\Delta_{i,j} \models \theta$, and that, in cases (ii) and (iii), the fact that $\mathfrak{A}_j \models \Delta_{i,j} \cup \bar{\Theta}_j \cup \Theta_i$ contradicts $\Delta_{i,j} \models \theta$.

Lemma 5. *Let X be a finite set of valid syllogistic rules in $\mathcal{N}^\dagger$, and let r be the maximum number of antecedents in any syllogistic rule of X. If $\theta \in \mathcal{N}^\dagger(P^{(n)})$, and $\Gamma^{(n)} \vdash_\mathsf{X} \theta$, where $n \ge r + 3$, then $\theta \in \Gamma^{(n)}$.*

Proof. We proceed by induction on the lengths of (direct) derivations. If a derivation of θ from $\Gamma^{(n)}$ employs no syllogistic rules, then, trivially, $\theta \in \Gamma^{(n)}$. For the inductive step, consider the last rule-instance $\langle \theta_1, \ldots, \theta_k, \theta \rangle$ in the derivation. By inductive hypothesis, $\{\theta_1, \ldots, \theta_k\} \subseteq \Gamma^{(n)}$. But because $k \le r \le n - 3$, we in fact have, for some i, j $(1 < i < j \le n)$, $\{\theta_1, \ldots, \theta_k\} \subseteq \Delta_{i,j}^{(n)}$. Since every rule in X is valid, $\Delta_{i,j}^{(n)} \models \theta$. By Lemma 4, $\theta \in \Delta_{i,j}^{(n)} \subseteq \Gamma^{(n)}$. This completes the induction.

Note that, from Lemma 5, we see immediately that there is no finite set X of syllogistic rules for $\mathcal{N}^\dagger$ such that the *direct* derivation relation $\vdash_\mathsf{X}$ is sound and complete. For suppose r is the maximum number of antecedents in any of the syllogistic rules in X, and let $n \ge r + 3$. If $\bot = (> 0)[l, \bar{l}]$ is any absurdity, we have $\Gamma^{(n)} \models \bot$, by Lemma 1. But, by inspection, $\bot \notin \Gamma^{(n)}$. Of course, Lemma 5 does not by itself establish the incompleteness of the *indirect* system $\Vdash_\mathsf{X}$, which includes the rule of *reductio*. However, Lemma 2 ensures that *reductio* actually does no useful work in the present case, as we now proceed to show.

Theorem 1. *There is no finite set X of syllogistic rules in $\mathcal{N}^\dagger$ such that $\Vdash_\mathsf{X}$ is sound and complete for $\mathcal{N}^\dagger$.*

Proof. We assume otherwise and derive a contradiction. Suppose X is a finite set of syllogistic rules for the numerical syllogistic with $\Vdash_\mathsf{X}$ sound and complete. Let r be the maximum number of antecedents in any of the syllogistic rules in X, and let $n \ge r + 3$. For any absurdity $\bot = (> 0)[l, \bar{l}]$, we have $\Gamma^{(n)} \models \bot$, by Lemma 1. By the (assumed) completeness of $\Vdash_\mathsf{X}$, we have $\Gamma^{(n)} \Vdash_\mathsf{X} \bot$. Let k be the smallest integer with the property that there is a derivation of some absurdity in $\Vdash_\mathsf{X}$ from $\Gamma^{(n)}$ employing at most k applications of the rule of *reductio*. Since $\Gamma^{(n)}$ contains no absurdities at all, it follows from Lemma 5 that $k > 0$. Now take any such derivation employing the minimal number k of applications of *reductio*, and consider the last such application, which, we may suppose, discharges (more than zero occurrences of) a premise θ as a result of deriving some absurdity $(> 0)[m, \bar{m}]$:

$$\Gamma \quad \cdots \quad \cdots \quad [\theta]^1$$
$$\vdots$$
$$\frac{(> 0)[m, \bar{m}]}{\bar{\theta}} \; (\text{RAA})^1 \quad \cdots \quad \cdots \quad \Gamma$$
$$\vdots$$
$$(> 0)[l, \bar{l}].$$

By Lemma 2, either $\theta \in \Gamma^{(n)}$ or $\bar{\theta} \in \Gamma^{(n)}$. But then either one of the smaller derivations

$$\Gamma \quad \cdots \quad \cdots \quad \theta \qquad\qquad \bar{\theta} \quad \cdots \quad \cdots \quad \Gamma$$
$$\vdots \qquad\qquad\qquad\qquad \vdots$$
$$(> 0)[m, \bar{m}] \qquad\qquad (> 0)[l, \bar{l}]$$

is a derivation of an absurdity from $\Gamma^{(n)}$ involving fewer than k applications of *reductio*, which is impossible.

By restricting all formulas in the above proof to be $\mathcal{N}$-formulas, we obtain, by identical reasoning:

Theorem 2. *There is no finite set* X *of syllogistic rules in* $\mathcal{N}$ *such that* $\Vdash_X$ *is sound and complete for* $\mathcal{N}$.

The details are left to the reader.

Acknowledgments

This research was supported by the EPSRC, grant number EP/F069154/1. The author would like to thank Lawrence S. Moss for his helpful comments on a draft of this paper.

References

1. Corcoran, J.: Completeness of an ancient logic. Journal of Symbolic Logic 37(4), 696–702 (1972)
2. Martin, J.N.: Aristotle's natural deduction revisited. History and Philosophy of Logic 18(1), 1–15 (1997)
3. Pratt-Hartmann, I., Moss, L.S.: Logics for the relational syllogistic (2008), ArXiv preprint server, http://arxiv.org/abs/0808.0521
4. Smiley, T.: What is a syllogism? Journal of Philosophical Logic 2, 135–154 (1973)
5. de Morgan, A.: Formal Logic: or, the calculus of inference, necessary and probable. Taylor and Walton, London (1847)
6. Boole, G.: Of propositions numerically definite. Transactions of the Cambridge Philosophical Society XI(II) (1868)
7. Boole, G.: Collected Logical Works: Studies in Logic and Probability, vol. 1. Open Court, La Salle, IL (1952)

8. Jevons, W.: On a general system of numerically definite reasoning. Memoirs of the Manchester Literary and Philosophical Society (3rd Series) 4, 330–352 (1871)
9. Jevons, W.: Pure Logic and Other Minor Works. Macmillan, London (1890)
10. Grattan-Guinness, I.: The Search for Mathematical Roots, 1870-1940: logics, set theories and the foundations of mathematics from Cantor through Russell to Gödel. Princeton University Press, Princeton (2000)
11. Murphree, W.: The numerical syllogism and existential presupposition. Notre Dame Journal of Formal Logic 38(1), 49–64 (1997)
12. Murphree, W.: Numerical term logic. Notre Dame Journal of Formal Logic 39(3), 346–362 (1998)
13. Hacker, E., Parry, W.: Pure numerical Boolean syllogisms. Notre Dame Journal of Formal Logic 8(4), 321–324 (1967)
14. Pratt-Hartmann, I.: On the computational complexity of the numerically definite syllogistic and related logics. Bulletin of Symbolic Logic 14(1), 1–28 (2008)
15. Fitch, F.B.: Natural deduction rules for English. Philosophical Studies 24, 89–104 (1973)
16. Suppes, P.: Logical inference in English: a preliminary analysis. Studia Logica 38(4), 375–391 (1979)
17. Purdy, W.C.: A logic for natural language. Notre Dame Journal of Formal Logic 32(3), 409–425 (1991)
18. Purdy, W.C.: Surface reasoning. Notre Dame Journal of Formal Logic 33(1), 13–36 (1992)
19. Purdy, W.C.: A variable-free logic for mass terms. Notre Dame Journal of Formal Logic 33(3), 348–358 (1992)
20. Fyodorov, Y., Winter, Y., Francez, N.: Order-based inference in natural logic. Logic Journal of the IGPL 11(4), 385–416 (2004)
21. Kuncak, V., Rinard, M.: Towards efficient satisfiability checking for Boolean algebra with Presburger arithmetic. In: Pfenning, F. (ed.) CADE 2007. LNCS, vol. 4603, pp. 215–230. Springer, Heidelberg (2007)

Formal Grammars of Early Language

Shuly Wintner[1], Alon Lavie[2], and Brian MacWhinney[3]

[1] Department of Computer Science, University of Haifa
shuly@cs.haifa.ac.il
[2] Language Technologies Institute, Carnegie Mellon University
alavie@cs.cmu.edu
[3] Department of Psychology, Carnegie Mellon university
macw@cmu.edu

Abstract. We propose to model the development of language by a series of formal grammars, accounting for the linguistic capacity of children at the very early stages of mastering language. This approach provides a testbed for evaluating theories of language acquisition, in particular with respect to the extent to which innate, language-specific mechanisms must be assumed. Specifically, we focus on a single child learning English and use the CHILDES corpus for actual performance data. We describe a set of grammars which account for this child's utterances between the ages 1;8.02 and 2;0.30. The coverage of the grammars is carefully evaluated by extracting grammatical relations from induced structures and comparing them with manually annotated labels.

1 Introduction

1.1 Language Acquisition

Two competing theories of language acquisition dominate the linguistic and psycho-linguistic communities [59, pp. 257-258]. One, the *nativist* approach, originating in [14,15,16] and popularized by [47], claims that the linguistic capacity is innate, expressed as dedicated "language organs" in our brains; therefore, certain linguistic universals are given to language learners for free, requiring only the tuning of a set parameters in order for language to be fully acquired. The other, *emergentist* explanation [2,56,37,38,40,41,59], claims that language emerges as a result of various competing constraints which are all consistent with general cognitive abilities, and hence no dedicated provisions for universal grammar are required. Consequently,

> "[linguistic universals] do not consist of specific linguistic categories or constructions; they consist of general communicative functions such as reference and predication, or cognitive abilities such as the tendency to conceptualize objects and events categorically, or information-processing skills such as those involved with rapid vocal sequences." [58, p. 101].

Furthermore, language is first acquired in an *item-based* pattern:

O. Grumberg et al. (Eds.): Francez Festschrift, LNCS 5533, pp. 204–227, 2009.
© Springer-Verlag Berlin Heidelberg 2009

"[young children] do not operate on the basis of any linguistic abstractions, innate or otherwise. Fairly quickly, however, they find some patterns in the way concrete nouns are used and form something like a category of a noun, but schematization across larger constructions goes more slowly... The process of how children find patterns in the ambient language and then construct categories and schemas from them is not well understood at this point." [58, pp. 106-107].

Our ultimate goal in this work is to provide a formal environment in which these two hypotheses can be rigorously evaluated.

Children learn language gradually and inductively, rather than abruptly and deductively [37]. Still, it is possible to isolate "snapshots" of language development which are qualitatively distinct from previous snapshots, e.g., by moving from one word utterances to two words, exhibiting a new possibility for word order, introducing embedded clauses, etc. It should therefore be possible to construct grammars for each of these snapshots of language development in isolation, and then inspect the differences among those grammars. The underlying assumption of this work is that such snapshot grammars exhibit a gradual, smooth development which can be captured formally and accounted for mathematically.

As an example of "snapshots" of language acquisition, consider the first five *stages* defined by [11], namely:

1. Acquisition of *semantic roles* (e.g., agent, patient, instrument etc.) in simple sentences expressed by linear order and syntactic relations;
2. Acquisition of *semantic modulations* such as number, specificity, tense, etc., expressed morphologically or lexically;
3. *Modalities* of the simple sentences, e.g., interrogative, negative and imperative sentences;
4. *Embedding*, as in relative clauses or VP objects;
5. *Coordination* of full sentences or partial phrases with deletion.

[11] shows that these five stages correspond nicely to measures of grammar development, in particular mean length of utterance (MLU). However, there is nothing sacred in this number of stages and a finer granularity of representation is certainly possible. There is also evidence that in some languages (e.g., Hungarian) the order of the stages may be somewhat different [36]. A similar definition of "a phase-based model of language acquisition" is offered by [4], although the five stages she delineates are not identical to those of [11]. We will focus in this work only on the first 3–4 phases of [4], again dividing them up to a finer resolution.

In order to investigate the development of child language, corpora which document linguistic interactions involving children are needed. The CHILDES database [39], containing transcripts of spoken interactions between children at various stages of language development with their parents, provides vast amounts of useful data for linguistic, psychological, and sociological studies of child language development. Two developments make CHILDES an attractive experimental setup for investigating language acquisition. First, similar databases of child-adult linguistic interactions are constantly being collected and developed for a variety of languages. Furthermore, many of the corpora are morphologically analyzed and annotated in a compatible manner, which makes it possible to compare language development across different languages. Second, the English CHILDES database has recently been annotated with grammatical relations [51]. This is useful for various practical applications (e.g., assessing the syntactic development of children, as in [52]), but is particularly attractive as it provides a way for automatically evaluating the coverage and, to some extent, the correctness of formal grammars which attempt to generate child language (see, e.g., [7]).

This paper presents some preliminary results. Following a discussion of related work below, we describe in section 3 a set of formal grammars accounting for the early utterances of one English speaking child, *Seth* [44,45], as reflected in the CHILDES database [39]. Faithful to the item-based model of language acquisition, the grammars that we develop are highly lexicalized (typed) unification grammars [13], inspired by Head-Driven Phrase Structure Grammar (HPSG, [48]). Such grammars provide means for expressing deep linguistic structures in an integrated way. We use LKB [19] to implement the grammars and apply them to the data in the corpus, producing deep linguistic structures which describe the syntax, but also some aspects of the morphology and the semantics, of child utterances. In particular, we convert the structural descriptions produced by the grammars to the functional annotation of [51]. This facilitates an automatic evaluation of the coverage of the grammars (section 4). On our development set, the error rate is lower than 10%.

1.2   Formal Grammars

Formal grammars are mathematical systems for describing the structure of languages, both formal and natural. In this work we employ *unification grammars* [65,66] and in particular unification grammars that are based on *typed* feature structures [13]. Such grammars underly linguistic theories such as Lexical Functional Grammar [28] and Head-Driven Phrase-Structure Grammars (HPSG) [48].

Typed unification grammars are based on a *signature* consisting of a partial order of *types*, along with an *appropriateness specification* which lists the *features* which are appropriate for each type and the types of their values. The signature facilitates the specification of an ontology of linguistic entities, from basic signs to rules and grammars. The building blocks of unification grammars are *feature structures*, which are nested data structures licensed by the signature. Grammar rules can then refer to and manipulate various pieces of the data encoded in feature structures, providing the linguist with powerful means for expressing linguistic theories in various strata: phonology, morphology, syntax, semantics and information structure. Unification grammars, and in particular HPSG grammars, have been used successfully for describing the structure of a variety of natural languages [3,26,29].

Typed unification grammars are attractive as a grammatical formalism, and in particular as an environment for investigating language acquisition, for several reasons. First, grammars are multi-stratal, allowing for the expression of various types of linguistic information and interactions thereof in the same sort of object. Second, grammar rules are basically constraints which limit the domain of grammatical utterances rather than procedures for generating linguistic structures. Third, such grammars are highly lexical: most of the information is encoded in the lexicon and rules tend to be few and very general. This goes hand in hand with item-basd theories of language acquisition. Finally, the unification operation which is so essential to the formalism provides a natural implementation of unifying the effect of various competing constraints, although it currently does not clearly support the kind of violable constraints that are known in, e.g., optimality theory [49]. While unification grammars, as a constraint-based formalism, are adequate for characterizing *structure*, they may not be ideal for characterizing *processes*; as will be made clear in section 2, we focus in this work on the formal representation of linguistic knowledge, rather than on the processes which lead to its acquisition. Unification grammars, therefore, are a natural formalism to use for this task.

One task that unification grammars may fail in is the truthful modeling of *competition* among different, incompatible constraints. This seems to be one of the core mechanisms in language acquisition, and is probably best modeled with some kind of statistical or probabilistic framework. We conjecture that stochastic unification grammars [1,27] may be adequate for this, but we leave this direction for future research.

1.3   Formal Approaches to Language Acquisition

Many existing theories of language acquisition focus on exploring *how* language develops, applying computational learning theory to the problem of language acquisition [62,5]. They assume a class of formal languages and an algorithm for inducing a grammar from examples, and differ in the criterion of success (e.g., identification in the limit [22] vs. PAC learning [60]), the class of algorithms the learner is assumed to apply (accounting for memory limitations, smoothness etc.) and the class of formal languages that can be learned.

A different kind of formal approaches to language acquisition utilizes computer simulations; this line of research is sometimes called *computational approaches* to language acquisition [10]. For a discussion of such approaches and their interrelations with behavioral, experimental methods, see [9]. Of particular interest is the *What and why vs. how* dichotomy that [9] introduces: much of the research conducted under the computational framework seems to be interested in explaining *how* language (in any case, certain aspects of language) is acquired. In this work we are interested in how linguistic knowledge is organized and how this organization develops as the child matures; in the terminology of [9], this is a *what* question, which we nevertheless address formally.

The dominant nativist approach to language acquisition attempts to place it in the context of *principles and parameters* theories of language [14,17]. Assuming some form of Universal Grammar with a set of parameters that have to be finely tuned as a particular language is acquired, [21] propose a *triggering learning algorithm* (TLA) which is aimed at setting the parameters correctly given positive examples. Some shortcomings of this algorithm were pointed out, e.g., by [43], who propose a model with fewer assumptions and better convergence performance.

In the main stream of generative approaches to syntax, a representative work on language acquisition is [31], which attempts to answer the question "What is the best way to structure a grammar?" and in particular to investigate "the relation between the sequence of grammars that the child adopts, and the basic formation of the grammar." One of the answers of [31] is very similar to our approach here; namely, that "the grammar is arranged along the lines of *subgrammars*... so that the child passes from one to the next." However, [31] suffers from two major shortcomings. First, this work is not informed by actual data: the grammar fragments it provides are supposed to account for competence rather than performance. In this respect, it is interesting to quote [11, p. 56–58], who says:

> "not many people know how much can be milked from mere performance in the case of small children... [This work] is about knowledge; knowledge concerning grammar and the meanings coded by grammar. Knowledge inferred, of course, from performance..."

Second, [31] is, to use the terminology of [56, p. 136], "described in linguistically specific terms such that it is very difficult to relate them to cognition in other psychological domains"; in other words, this work is only accessible to experts in contemporary generative approaches to syntax.

These two issues are remedied in [61]. Here, the Universal Grammar is implemented as a Unification-Based Generalised Categorial Grammar, embedded in a default inheritance network of lexical types. The learning algorithm receives input from a corpus of spontaneous child-directed transcribed speech, annotated with logical forms, and sets the parameters based on this input. This framework is used as a basis for investigating several aspects of language acquisition from data, focusing on the acquisition of subcategorization frames and word order information. However, this work is still embedded in the principles and parameters framework, assuming a (very elaborate) universal grammar which is expressed in unification-based categorial grammar and a large set of parameters, including 89 categorial parameters, 18 word order parameters etc. As the work focuses on the acquisition of verb arguments, only 1517 carefully selected sentences were selected and annotated (out of a much larger corpus). Most importantly, this work concentrates on the "how" of language acquisition, discussing an architecture for learning the lexicon and the parameters of a grammar. The focus of our work, in contrast, is exactly on the component which [61] assumes given, namely the internal representation of the grammar, and how this grammar changes as language develops. Our work is not committed to the full details of Government-Binding theory (or any other linguistic theory, for that matter); rather, we expect child grammars to naturally emerge from actual data rather than comply with predefined principles. For a detailed, well-argued criticism of the nativist approach see [54].

Building on the work of [61], but assuming very limited innate linguistic mechanisms (e.g., the ability to recognize objects, segment words or combine constituents), and not subscribing to the principles and parameters discipline, [12] proposes a computational model for first language acquisition. The model, which infers linguistic structure from data, is statistical and can cope with noisy input. The model is shown to outperform TLA on simple learning tasks. Again, this works focuses on the "how" rather than the "what".

Surprisingly, very few works address the issue of accounting for child language data by means of formal grammars. It is worth mentioning that [11] himself developed formal grammars for describing the utterances in his corpus. He says (p. 56):

> "...the requirement to be fully explicit and develop rules that will really derive the sentences you have obtained forces a kind of intense examination and continuing re-examination of the data, and so is a good way to get to know it very well...."

We do not stop at writing the grammars, and do not limit ourselves to the goal of getting to know the data. Rather, we investigate the possible changes in grammar as it develops and the constraints imposed on such changes. Perhaps the closest approach to the one we propose here is [25], which sketches

> "...a solution to the string-to-structure problem in first language acquisition within a set of emergentist assumptions that minimizes the need to assume innate linguistic knowledge, minimizes demands for linguistic analysis by the language learner, and exploits the projection of lexical properties of words. These conceptual constraints minimize the number, complexity, and diversity of principles that have to develop in learning a grammar."

[25] sketches, very briefly, some milestones in language acquisition, and models them by means of an HPSG grammar which is gradually being developed to account for more data. The two most important principles for [25] are assuming no innate linguistic knowledge (rather, she assumes general cognitive capabilities such as "the ability to discriminate kinds of things", "the capacity to add to a store of propositional knowledge", or "the ability to identify what is missing when something is missing"); and proposing a *monotonic* theory of grammar development, although exactly what is meant by monotonicity is not explicitly defined.

[25] advocates the use of HPSG as the framework in which to model language acquisition. HPSG is chosen to demonstrate the emergentist theory mainly because of its underlying architecture, in particular the use of a type signature and, separately, a set of well-defined formal principles. The signature provides an ontology of entities (e.g., linguistic signs), and the principles constrain and restrict those entities. [25] shows that it is possible to account for several stages in child language development by means of a single mechanism: "incremental and largely monotonic changes to a type hierarchy that constitutes an increasingly less skeletal constraint-based grammar".

The main drawbacks of this work are two. First, while [25] lists several fragments of the proposed HPSG grammar, those fragments are never tested and their predictive power is never contrasted with actual corpora recording child language use. Second, the possible changes that grammars may undergo in the process of language development are never spelled out. Two of our goals in this work are to remedy these problems, by providing actual grammars that can (and are) tested against the CHILDES database corpora; and by formally defining a *grammar refinement* operator that will account for the possible changes in grammar development.

2   Research Objectives

This work can be divided into two main sub-tasks: developing grammars for representative stages of child language data; and developing grammar refinement operators that will explain the possible changes in grammars as language develops.

2.1   Grammar Development

Our departure point is the item-based, functional, emergentist theory of grammar acquisition; our grammars are therefore highly lexicalized, and assume few innate components. However, we do expect certain universal patterns to emerge from the multilingual data that we explore, and these patterns indeed inform and guide the development of the grammars. Faithful to the item-based model of language acquisition, the grammars that we develop are highly lexicalized (typed) unification grammars. As pointed out above, such grammars provide means for expressing deep linguistic structures in an integrated way. We use the LKB system [19] as our grammar development environment.

In the first phase we manually developed a few grammars for some representative corpora, focusing on the *Eve* corpus [11,39] and the *Seth* corpus [44,45]. In the next phase we will extend the coverage of the grammars to more corpora, longitudinally accounting for a small number of individuals along the first few years of language acquisition. We will consequently extend the coverage of the grammars to cover also data in other languages, for similar language development stages.

The grammars are all applied to the data in the corpus, producing deep linguistic structures describing the syntax, but also some aspects of the morphology and the semantics, of the utterances observed in the data. In particular, it is possible to automatically and deterministically convert the structural descriptions produced by the grammars to the functional annotation of [51].

2.2   Grammar Refinement

A major research question that we leave for future research involves the possible changes that a grammar can undergo as language develops; we use the term *grammar refinement* to refer in general to any of several operators which may convert one grammar, accounting for some layer of language command, to a subsequent grammar, accounting for more developed linguistic capacity. Grammar refinement must have access to all the components of a grammar, and in particular to the type signature on which a grammar is based and to the lexicon.

Current work in unification grammars focuses on issues of grammar engineering. In a recent work, [18] define the concept of signature *modules* for unification grammars and show how several modules can combine through a *merge* operation. They introduce the concept of *partially specified signature (PSSs)* to abstract away from specific type- and feature-names in grammars. This work is extended in [55], where PSSs are extended to modules, facilitating the combination of grammar fragments, which interact via an abstract interface, and which can be combined as needed to form full grammars. While this work is preliminary, it sets the stage for several possible developments in collaborative grammar engineering. It is also a very promising starting point for defining "core" grammars which embody universal principles, and which can then be combined with more language specific and parochial modules and constraints.

We believe that these results can provide the infra-structure for defining grammar refinement operators. Operations such as splitting an existing type to several subtypes (reflecting the child's ability to make finer distinctions), or adding appropriate features to types (reflecting a more complex representation of a given sign), are easily definable with PSSs. Lexical acquisition can be accounted for by means of *grammar union* [63,64]. Acquisition and refinement of constructions can also be defined as changes to the type signature, following the encoding of constructions advocated by, e.g., [50]. In sum, unification grammars in general and those that are based on typed feature structures in particular provide suitable means for modeling language development; we hope to make significant contributions to such modeling in future extensions of this work.

3 Formal Grammars of Children's Language

In the preparation of the successive series of grammars for a single child's language, we are constrained by three requirements. First, we are bound to be faithful to the data: the grammars must adequately account for the kind of utterances observed in the CHILDES data that drive this research, and the structures they induce on utterances must be compatible with the existing syntactic annotation of [51] (henceforth, GRASP annotation). Second, our grammars must be consistent with existing psycholinguistic theories which ground the formal description in well-founded experimental and theoretical results. Formally speaking, it is always possible to provide infinitely many different grammars for any given finite language; it is the predictions of the grammar and their theoretical adequacy which distinguish a good grammar from a better one. Finally, the rules that we stipulate must be *learnable*: one can imagine a postulated rule that is generatively adequate and psycholinguistically plausible, but unlearnable. We will therefore strive to define grammar refinement operators that are compatible with existing theories of learnability.

The forces that drive grammar development are manifold; a good grammar must simultaneously account for and generate phonological, intonational, morphological, syntactic, semantic and pragmatic structures. Since this work is preliminary, we are unable to provide such wealth of information, although the framework in which we work certainly facilitates it. Two examples suffice to delineate the types of linguistic structure that our grammars abstract over.

It is clear that intonation plays a major part in (some) children's first utterances [44]. The early speech of some children is interspersed with *fillers*, and "One function underlying production of these early fillers seems to be preservation of the number of syllables in and/or the prosodic rhythm of the target" [46, p. 234]. Along the same lines, [20] proposes that "much of the null and variable occurrence of functional categories in early speech can be more accurately explained by appealing to *phonological* rather than *syntactic* aspects of children's developing grammars". However, since a formal theory of phonological and intonational structures and their interaction with other components of the grammar, even in adult language, is still unavailable, we suppress a discussion of these aspects in the sequel.

Similarly, [42] demonstrate that the very first utterances of children are better understood from a functional (i.e., pragmatic) point of view: "children's earliest words express specific communicative function" that a focus on syntax and semantics fails to analyze correctly. We believe that information structure, speech-act and pragmatics in general are essential to understanding child language, but our current grammars do not specify such information, which is not sufficiently well understood to be formalized.

3.1 Preliminaries

LKB grammars consist of three main components: a signature, specifying the type hierarchy along with constraints on types; grammar rules, which specify phrase structure combinatorics; and a lexicon, associating feature structures with words. The examples below use the LKB syntax, which is described in detail in [19].

We begin with an inventory of general-purpose types, such as lists and difference lists (figure 1). These are needed mostly for technical reasons at this preliminary stage of the grammar, so we only mention them in passing.

Next, we associate words with *part of speech (POS)* categories. It is clear that children can distinguish among some different POS categories from a very early age [30]. Several works address the question of the acquisition of POS categories, and the psychological, neurological and computational underpinnings of this process have been extensively investigated and described [33,34,35]. A classification of POS categories is manifested in our grammar as in figure 2; *co* is a communicator, e.g., *hi, bye, okay, please*, and *fil* is a filler, e.g., *uh*.

Similarly, we assume a set of *grammatical relations (GRs)* of the same inventory that is used by GRASP [51] to annotate the corpora. In this early stage of the grammar, we cannot assume that the child had learned any of those, so we resort to the general specification:

```
string := *top*.
*list* := *top*.
*ne-list* := *list* & [ FIRST *top*, REST *list* ].
*null* := *list*.
*diff-list* := *top* & [ LIST *list*, LAST *list* ].
*empty-diff-list* := *diff-list* & [ LIST #list, LAST #list ].
```

Fig. 1. General purpose types

```
cat := *top*.
  adj := cat.
  adv := cat.
  co := cat.
  fil := cat.
  nominal := cat.
    n := nominal.
    prop := nominal.
  prep := cat.
  v := cat.
```

Fig. 2. Parts of speech

```
gr := *top*.
```

Furthermore, we define a structure for dependency relations. A dependency is a triple consisting of a type, which is a GR, and a head and a dependent, which are lexical structures:

```
dep := *top* &
[ GRTYPE gr,
  GRDEP lex,
  GRHEAD lex ].
```

Lexical structures consist of a string, representing the standard orthography of the lexical item, and a GR. The string is merely a convenient abstraction over the actual phonological structure of the word, which should be properly modeled as an extension of the present work. The reason for the GR is that oftentimes it is the lexical item which determines the grammatical relation of some construction; we return to this below.

```
lex := *top* &
[ ORTH string,
  DEP gr ].
```

Syntactic structures consist of a part of speech CATegory; a sub-category, which we take to be a list of complements; and a list of specifiers, as in HPSG. The SPR list will be used by non-heads to specify the heads they modify; it will either have a single element or be empty. In contrast, SUBCAT lists are used by heads to specify their complements and may have zero or more elements.

```
syn := *top* &
[ CAT cat,
  SPR *list*,
  SUBCAT *list* ].
```

Finally, we define *constructions*. A construction [23,24,57] is a collection of linguistic information pertaining to the form, combinatorics and meaning of an utterance, similarly to HPSG's *signs*. In our grammars, we roughly approximate constructions by specifying the lexical and syntactic information structures of words and phrases, as well as a list of grammatical relations GRS in lieu of semantics. Additionally, the feature STRING lists the surface form of the construction, as in figure 3.

```
construction := *top* &
[ LEX lex,
  SYN syn,
  GRS *diff-list*,
  STRING *diff-list* ].
```

Fig. 3. Constructions

Of course, figure 3 only lists the very basic constructions, and sub-types of the type *construction* are defined and refined as the grammar develops. For our first grammar we can assume that two sub-types are defined, namely *lexeme* and *phrase*; the former is associated with lexical items, whereas the latter has internal structure, expressed via (at least) the feature DTR1, standing for 'daughter 1', as in figure 4.

```
lexeme := construction &
[ LEX #lex,
  GRS *empty-diff-list*,
  STRING [ LIST [ FIRST #lex , REST #last ], LAST #last ] ].

phrase := construction &
[ DTR1 construction ].
```

Fig. 4. Lexemes and phrases

3.2 Earliest Language

The first grammar reflects the very first utterances recorded in the corpus; these are usually one-word utterances and *holophrases* [59, p. 261]. We constructed a lexicon of all such items which occur in the corpus; each item is associated with a

```
Julie   := [ LEX [ ORTH "Julie" ] ] & lex_prop .
back    := [ LEX [ ORTH "back" ] ] & lex_adv .
byebye  := [ LEX [ ORTH "byebye" ] ] & lex_co .
cake    := [ LEX [ ORTH "cake" ] ] & lex_n .
came    := [ LEX [ ORTH "came" ] ] & lex_v .
```

Fig. 5. Some lexical items

```
lex_adj := lexeme & [ SYN [ CAT adj ] ].
lex_adv := lexeme & [ SYN [ CAT adv ], LEX [ DEP jct-rel ] ].
lex_co := lexeme & [ SYN [ CAT co ] ].
lex_fil := lexeme & [ SYN [ CAT fil ] ].
lex_n := lexeme & [ SYN [ CAT n ] ].
lex_prop := lexeme & [ SYN [ CAT prop ] ].
lex_prep := lexeme & [ SYN [ CAT prep ] ].
lex_v := lexeme & [ SYN [ CAT v ] ].
```

Fig. 6. Sub-types of *lexeme*

feature structure of type *lexeme*, reflecting at least its orthography and syntactic (POS) category. Some examples are listed in figure 5. The types *lex_prop*, *lex_adv* etc. are defined as in figure 6.

These definitions are sufficient for handling the one-word utterances that are observed in the first files of the Seth corpus.

3.3   The Emergence of Constructions

The next stage of language development allows for combinations of words and holophrases. For example, the first multi-word utterances in the Seth corpus are *came in; come in; had cake; had chocolate; orange juice;* and *right back*. We view the latter two as holophrases, but the first four, which are augmented in subsequent data by utterances such as *Dabee toast* and *swing high*, justify the introduction of item-based constructions to the grammar.

Exactly how constructions are learned by children remains largely unknown [32], and we do not attempt to address this question directly here. However, at least three types of constructions must be supported by a formal account of grammar development. First, basic items can be concatenated, yielding "successive single-word utterances" [6]; according to [8], these are the first indications of syntax in speech. Second, *pivot-schemas* allow the combination of one highly frequent "event-word" (e.g., *more, It's, I*) with a variety of items. Third, *item-based constructions* allow more complex and more abstract combination, in which the construction itself adds, for the first time, to the semantics of the full utterance [59, pp. 263-264].

Let us focus first on utterances such as *came in; come in* and *had cake; had chocolate*. By the end of the language learning process we may want to say that such utterances reflect *verb–adjunct* and *verb–object* relations, respectively. This is indeed how these utterances are annotated in the corpus. Therefore, during the learning process the child must somehow acquire these constructions. A possible way of stepping through this process is outlined below.

First, the grammar must allow binary combinations of words; this is captured by the type *binary-phrase*, which is a sub-type of *phrase*. A *binary-phrase* has two daughters, DTR1 and DTR2. Furthermore, the list of GRS of the mother is obtained by concatenating the same lists in the two daughters, and the same holds for STRING. In order to distinguish between the two daughters, we introduce the type *headed-phrase*, in which an additional feature, HEAD-DTR, reflects (and is reentrant with) the head daughter, which can be either DTR1 or DTR2. The head daughter shares its LEX structure and its main CATegory, but not necessarily its SUBCATegory, with its mother. We next define a *binary-headed-phrase*, a sub-type of both *binary-phrase* and *headed-phrase*, with an additional feature, NON-HEAD-DTR. These definitions are given in figure 7.

```
binary-phrase := phrase &
[ DTR1 [ GRS [ LIST #first, LAST #middle ],
         STRING [ LIST #str1, LAST #str3 ] ],
  DTR2 [ GRS [ LIST #middle, LAST #last ],
         STRING [ LIST #str3, LAST #str2 ] ],
  GRS [ LIST [ FIRST dep, REST #first ], LAST #last ],
  STRING [ LIST #str1, LAST #str2 ] ].

headed-phrase := phrase &
[ LEX #lex,
  SYN [ CAT #cat ],
  HEAD-DTR construction & [ LEX #lex, SYN [ CAT #cat ] ] ].

binary-headed-phrase := headed-phrase & binary-phrase &
[ NON-HEAD-DTR construction ].
```

Fig. 7. Binary phrases

In both verb–adjunct and verb-object constructions, the head is the first daughter; furthermore, in both the second daughter is a dependent of the first. We therefore define two sub-types of *binary-phrase*, namely *head-first* and *dep-second*, which reflect these observations. A major difference between adjuncts and objects is that the latter are subcategorized for by the head verb; we define two subtypes of *binary-phrase* which account to this distinction: in *subcat-unchanged*, the SUBCATegorization of the mother and the head daughter are identical, whereas in *subcat1* constructions the head daughter specifies its object argument, which is removed from the SUBCATegorization list that is propagated to the mother. With these definitions in place, *obj-construction* and *jct-construction* are defined as in figure 8.

The only grammar rules that are needed for this elementary stage are two: one for each type of constructions. These rules are depicted in figure 9. While the

```
head-first := binary-headed-phrase &
[ HEAD-DTR #head-dtr,
  DTR1 #head-dtr,
  NON-HEAD-DTR #non-head-dtr,
  DTR2 #non-head-dtr ].

dep-second := binary-phrase &
[ GRS [ LIST [ FIRST [ GRHEAD #dep1, GRDEP #dep2 ] ] ],
  DTR1 [ LEX #dep1 ],
  DTR2 [ LEX #dep2 ] ].

subcat-unchanged := binary-headed-phrase &
[ SYN [ SUBCAT #subcat, SPR #spr ],
  HEAD-DTR [ SYN [ SUBCAT #subcat, SPR #spr ] ] ].

subcat1 := binary-headed-phrase &
[ SYN [ SUBCAT #rest, SPR #spr ],
  HEAD-DTR [ SYN [ SPR #spr ] ],
  DTR2 #head-dtr,
  DTR1 [ SYN [ SUBCAT [ FIRST #head-dtr, REST #rest ] ] ] ].

obj-construction := head-first & dep-second & subcat1 &
  [ GRS.LIST.FIRST.GRTYPE obj-rel ].
jct-construction := head-first & dep-second & subcat-unchanged &
  [ GRS.LIST.FIRST.GRTYPE jct-rel ].
```

Fig. 8. Types of binary phrases

rules induce isomorphic trees on the utterances *had cake* and *come in* (figure 10), the feature structures induces by the grammar on the two strings are significantly different (figure 11). In fact, words such as *in* are categorized as *adv*erbs for convenience only. We do not claim that the child had acquired the category of adverbs yet; rather, it is clear that he uses some verbs with some adjuncts, as in *come in* or *swing high*. It is also clear that these constructions are different from, say, *had cake* or *had chocolate*. We therefore classify "words which can follow verbs" as adjuncts, and "nouns which can follow the verb *have*" as objects. Also, the rule which allows objects to combine with verbs is general: it can apply to any verb. It is well known that only few verbs are used in early language, and that argument structure is acquired in an item-based fashion [24]. The rule could therefore have been stated in terms of a particular verb, by specifically referring to the LEX feature of the verb, or to a group of verbs, by specifying a sub-type of *verb*.

The first utterance in the Seth corpus which is clearly not an instance of either the verb–object or the verb-adjunct constructions is *Dabee toast* (Dabee is the child's grandmother). We treat such utterances as *vocative* constructions: the first word of the construction is a proper name, and the second is unconstrained. Few additions to the grammar are required in order to account for such

```
vp_v_np_rule := obj-construction &
[ SYN [ CAT v ],
  DTR2 [ SYN np ] ] .

vp_v_advp_rule := jct-construction &
[ SYN [ CAT v ],
  DTR2 [ SYN [ CAT adv ] ] ] .
```

Fig. 9. Two phrase-structure rules

Fig. 10. Two trees

utterances. In a vocative construction, the head is the second daughter, and the first word is a dependent of the second; we therefore add the types *head-second* and *dep-first*, analogously to *head-first* and *dep-second*. Also, the resulting phrase is saturated, as no further modification is allowed. We subsequently add a *sub-cat0* construction which specifies that the SUBCATegorization list of the mother is empty. A *voc-construction* is a sub-type of both. One rule is added, too, as shown in figure 12.

3.4   Towards Full Syntax

The infra-structure set up above facilitates a formal account of the various types of constructions learned at this early stage. In the first files of the Seth corpus different kinds of multi-word utterances are observed, including *Dabee toast; swing high; Uncle Bob; bird outside; I bet; kiss it; come see me; put shoes on; take a bite* and *who loves you*. These phrases represent various grammatical relations, and while some of them are pivot-based or item-based, others may be the beginning of the emerging syntax. We describe below the extensions to the grammar which are required to account for these new data.

Let us first account for the emergence of subjects, as in *I bet*. We add a sub-type of *nominal*, called *pronoun*, of which *I* is an instance. We also add a sub-type of *gr*, namely *subj-gr*. A subject construction is a special case of the newly introduced *specifier–head* construction, in which the first daughter is the specifier of the head of the second daughter, and SUBCAT is empty and shared between the mother and the head daughter. Also, the first constituent depends on the second, hence the type specification and grammar rule listed in figure 13.

```
 ○ ○ ○          Edge 3 P – Tree FS              ○ ○ ○          Edge 3 P – Tree FS

 Edge                                            Edge

 obj-construction                                jct-construction
 LEX: <0> = [lex                                 LEX: <0> = [lex
            ORTH: had                                       ORTH: come
            DEP: gr]                                        DEP: gr]
 SYN: [syn                                       SYN: [syn
       CAT: <1> = v                                    CAT: <1> = v
       SPR: <2> = "list"                               SPR: <2> = "list"
       SUBCAT: <3> = "list"]                           SUBCAT: <3> = "list"]
 GRS: ["diff-list"                               GRS: ["diff-list"
       LIST: ["ne-list"                                LIST: ["ne-list"
              FIRST: [dep                                      FIRST: [dep
                      GRTYPE: obj-rel                                  GRTYPE: jct-rel
                      GRDEP: <4> = [lex                              GRDEP: <4> = [lex
                                    ORTH: cake                                     ORTH: in
                                    DEP: gr]                                       DEP: jct-rel]
                      GRHEAD: <0>]                                   GRHEAD: <0>]
              REST: <5> = "list"]                              REST: <5> = "list"]
       LAST: <5>]                                       LAST: <5>]
 STRING: ["diff-list"                            STRING: ["diff-list"
       LIST: <6> = ["ne-list"                           LIST: <6> = ["ne-list"
                     FIRST: <0>                                       FIRST: <0>
                     REST: <7> = ["ne-list"                          REST: <7> = ["ne-list"
                                  FIRST: <4>                                      FIRST: <4>
                                  REST: <8> = "list"]]                           REST: <8> = "list"]]
       LAST: <8>]                                       LAST: <8>]
```

Fig. 11. Feature structures of two two-word utterances

```
head-second := binary-headed-phrase &
[ HEAD-DTR #head-dtr,
  DTR2 #head-dtr,
  NON-HEAD-DTR #non-head-dtr,
  DTR1 #non-head-dtr ].

dep-first := binary-phrase &
[ GRS [ LIST [ FIRST [ GRHEAD #dep2, GRDEP #dep1 ] ] ],
  DTR1 [ LEX #dep1 ],
  DTR2 [ LEX #dep2 ] ].

subcat0 := binary-phrase &
[ SYN [ SUBCAT *null*, SPR *null* ] ].

 voc-construction := dep-first & subcat0 & subcat1 &
 [ GRS.LIST.FIRST.GRTYPE voc-rel ].

vocative_rule := voc-construction &
[ DTR1 [ SYN [ CAT prop ] ] ] .
```

Fig. 12. Accounting for vocative constructions

We now move on to the emergence of determiners, as in *take a bite*. We need an additional sub-type of *cat*, namely *det*erminer, and an additional sub-type of *gr*, namely *det-rel*. Like subjects, determiner are specifiers of their heads, but unlike subjects they specify the category of their heads in their lexical types, as shown in figure 14.

```
spr-head := binary-headed-phrase & dep-first &
[ SYN [ SUBCAT #subcat & *null*, SPR *null* ],
  HEAD-DTR #head-dtr & [ SYN [ SUBCAT #subcat ] ],
  DTR1 #spr-dtr,
  DTR2 #head-dtr & [ SYN [ SPR [ FIRST #spr-dtr ] ] ] ].

subj-construction := spr-head & [ GRS.LIST.FIRST.GRTYPE subj-rel ].

s_np_vp_rule := subj-construction &
[ SYN clause,
  DTR1 [ SYN np ],
  DTR2 [ SYN [ CAT v ] ] ] .
```

Fig. 13. Accounting for subject constructions

```
det-construction := spr-head & [ GRS.LIST.FIRST.GRTYPE det-rel ].

lex_det := lexeme &
           [ SYN [ CAT det, SPR [ FIRST [ SYN np ], REST *null* ] ] ].

np_det_n_rule := det-construction &
[ SYN np,
  DTR1 lex_det,
  DTR2 lex_n ] .
```

Fig. 14. Accounting for determiner constructions

The trees induced by the grammar on *I bet* and *take a bite* are depicted in figure 15. Of course, with the available constructions the grammar can generate more complex structures, such as *I take a bite*, which are not observed in the data at this early stage. We take this to be a performance constraint on grammatical competence, which is better realized by limiting, e.g., the length of the utterances or the degree of tree nesting.

Immediately after the appearance of determiners, quantifiers are observed in the data in very similar constructions: *some toast, some juice*. In addition to the introduction of the types *qn* (a sub-type of *cat*) and *quant-rel* (a sub-type of *gr*), no additional construction type is needed. Rather, we specify quantifiers lexically analogously to determiners, and add a designated rule which is almost identical to the determiner rule, see figure 16.

Another construction that is observed in this early stage involves *communicators*: words such as *hi, oh, please* or *yes*, followed by some other short utterance, as in *oh good* or *please Daddy*. These are accounted for by the introduction of a new sub-type of *gr*, namely *com-rel*; a new construction type, *com-construction*; and an additional rule, depicted in figure 17.

At a very early age children begin to develop more complex syntactic structures. Seth, for example, produces an interrogative construction at the age of

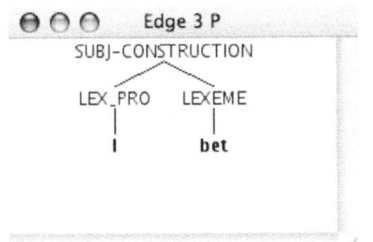
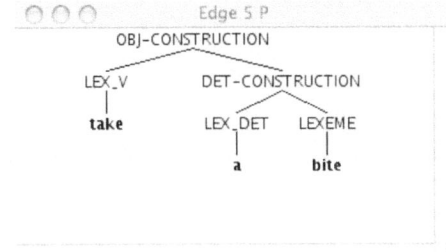

Fig. 15. Two more trees

```
lex_qn :=  lexeme &
           [ SYN [ CAT qn,  SPR [ FIRST [ SYN np ], REST *null* ] ] ].

np_qn_n_rule := det-construction &
[ SYN np,
  DTR1 lex_qn,
  DTR2 lex_n ] .
```

Fig. 16. Accounting for quantifiers

```
com-construction := head-second & dep-first & subcat1 &
                    [ GRS.LIST.FIRST.GRTYPE com-rel ].

communicator_rule := com-construction &
[ DTR1 lex_co
] .
```

Fig. 17. Accounting for communicator constructions

twenty months. It is *what geese say*, whose meaning can be interpreted from the following utterance in the corpus: his father repeating *what do the geese say?* In what follows we provide an account of such constructions, inspired by the HPSG treatment of "movement", or transformations, using *slash* categories.

First, we split the type *pronoun* into two sub-types, *pers-pro* and *wh-pro*, distinguishing between personal and interrogative pronouns, respectively. We similarly split *lex-pro* into *lex-pers-pro* and *lex-wh-pro*. Also, in order to limit the extraction of the object in utterances like *what geese say* to transitive verbs only, we define the type *lex-trans*, which is a sub-type of *lex-v* which has a single, noun phrase element on its SUBCAT list.

Next, we augment the type *construction* by adding a feature SLASH, whose value is a list. We expect at most one element on slash lists, an instance of the type *slash*:

```
slash := *top* &
[ CONST construction,
  SLASH-DEP dep ].
```

When an expected argument is "moved", as in the case of the object of *say* in the running example, it is removed from the SUBCAT list of its mother and stored in the SLASH list. This is accounted for by *gap* constructions. Such elements are expected to be realized higher up the tree as *wh-pro*nouns; they are then matched against the stored element in the SLASH list of the second daughter, through the *filler* construction. These two constructions, along with some additional supporting types, are depicted in figure 18 (some additional, minor modifications are required in the grammar in order to propagate the value of the SLASH feature along the derivation). The resulting tree is depicted in figure 19.

4   Evaluation

In order to evaluate the plausibility of the grammars, we extracted from the structures that they induce on child utterances the sequence of grammatical relations, and compared them to the manual GRASP annotations. We briefly describe the methodology and discuss the results of the evaluation in this section.

```
unary-phrase := phrase &
[ DTR1 [ GRS #grs, STRING #str ],
  GRS #grs,
  STRING #str ].

unary-headed-phrase := headed-phrase & unary-phrase &
[ HEAD-DTR #dtr,
  DTR1 #dtr ].

gap := unary-headed-phrase &
[ SYN [ SUBCAT #rest, SPR #spr ],
  SLASH [ FIRST [ CONST #missing-dtr,
                  SLASH-DEP [ GRTYPE obj-rel ] ] ],
  DTR1 [ SYN [ SUBCAT [ FIRST #missing-dtr, REST #rest ], SPR #spr ] ] ].

filler := head-second & dep-first &
[ DTR2 [ SYN #syn, SLASH [ FIRST [ CONST #dtr,
                                   SLASH-DEP [ GRTYPE #grtype ] ] ] ],
  DTR1 #dtr & [ SYN [ CAT wh-pro ] ],
  SYN #syn,
  GRS.LIST.FIRST.GRTYPE #grtype ].
```

Fig. 18. Gaps and fillers

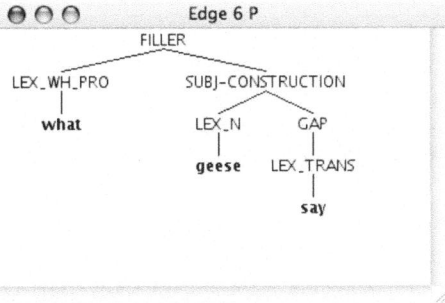

Fig. 19. Filler-gap constructions

First, we manually annotated a subset of the Seth corpus, focusing only on child utterances and ignoring any adult ones, using the GRASP scheme. We then extracted from this subset only those utterances that are *well-formed*, i.e., include only words which occur in the lexicon (as opposed to filler syllables, unrecognized sounds etc.) These 516 utterances constitute our development corpus. We then parsed the utterances with the set of grammars described above. From the resulting feature structures we extracted the values of the GRS feature, and converted them to GRASP format.

Table 1 lists the number of errors of the parser, in terms of unlabeled and labeled dependencies, with respect to the manually annotated corpus. Since the corpus contains many repetitions we present both the number of utterance tokens (total) and the number of utterance types (unique occurrences).

Table 1. Evaluation results, development set

File	# utterances	# of GRs	Grammar Errors unlabeled		labeled	
			#	%	#	%
0a tokens	107	234	48	21	48	21
0a types	38	82	6	7	6	7
1a tokens	108	223	3	1	7	3
1a types	38	79	3	4	5	6
2a tokens	115	247	16	6	18	7
2a types	45	105	13	12	15	14
2b tokens	68	142	12	8	12	8
2b types	30	64	6	9	6	9
3a tokens	118	240	0	0	0	0
3a types	34	72	0	0	0	0
Total tokens	516	1086	79	7.9	85	7.8
Total types	185	402	28	6.9	32	7.9

5   Conclusion

We presented a sequence of grammars which adequately cover the first stages of the emergence of syntax in the language of one child. The structures produced by the grammars were evaluated against a manually annotated corpus, revealing a low error rate (below 10%). These are preliminary results, and the evaluation is very basic, but we believe that the results are promising.

Space considerations prevent us from listing the complete grammars. However, we note that the changes introduced in each grammar, compared to the previous one, consist only of two operations: adding types, by splitting a single type to two or more sub-types; and adding constraints on types, in the form of additional features or reentrancy constraints. This is an encouraging outcome: in a very well-founded, mathematical sense, our grammars are monotonically increasing.

This preliminary work will be continued in three main tracks. First, we intend to continue the development of the grammars, accounting for more constructions and evaluating the accuracy on the entire Seth corpus, as well as on other manually annotated child language corpora. Second, we will develop mechanisms of grammar engineering that will facilitate both the definition of refinement operators and the modular development of sequences of grammars. Finally, we will account for competition among constraints, setting the stage for a theory which could also explain *how* these grammars can be learned from data.

Acknowledgements

This research was supported by Grant No. 2007241 from the United States-Israel Binational Science Foundation (BSF). This work was supported in part by the National Science Foundation under grant IIS-0414630. We are grateful to Eric Davis for his invaluable help.

References

1. Abney, S.P.: Stochastic attribute-value grammars. Computational Linguistics 23(4), 597–618 (1997)
2. Bates, E., MacWhinney, B.: Competition, variation, and language learning. In: MacWhinney, B. (ed.) Mechanisms of language acquisition, ch. 6, pp. 157–193. Lawrence Erlbaum Associates, Hillsdale (1987)
3. Bender, E.M., Flickinger, D., Fouvry, F., Siegel, M.: Shared representation in multilingual grammar engineering. Research on Language and Computation 3, 131–138 (2005)
4. Berman, R.A.: Between emergence and mastery: The long developmental route of language acquisition. In: Berman, R.A. (ed.) Language development across childhood and adolescence. Trends in Language Acquisition Research, vol. 3, pp. 9–34. John Benjamins, Amsterdam (2004)
5. Bertolo, S.: Acquisition, formal theories of. In: Wilson, R.A., Keil, F.C. (eds.) The MIT Encyclopedia of the Cognitive Sciences. Bradford Books (September 2001)

6. Bloom, L.: One word at a time: The use of single word utterances before syntax. Mouton, The Hague (1973)
7. Borensztajn, G., Zuidema, W., Bod, R.: Children's grammars grow more abstract with age - evidence from an automatic procedure for identifying the productive units of language. In: Proceedings of CogSci (2008)
8. Branigan, G.: Some reasons why successive single word utterances are not. Journal of Child Language 6, 411–421 (1979)
9. Brent, M.R.: Advances in the computational study of language acquisition. In: Brent, M.R. (ed.) Computational Approaches to Language Acquisition, Cognition Special Issues, ch. 1, pp. 1–38. The MIT Press, Cambridge (1997)
10. Brent, M.R. (ed.): Computational Approaches to Language Acquisition. Cognition Special Issues. The MIT Press, Cambridge (1997)
11. Brown, R.: A first language: the Early stages. Harvard University Press, Cambridge (1973)
12. Buttery, P.J.: Computational models for first language acquisition. Technical Report UCAM-CL-TR-675, University of Cambridge, Computer Laboratory (November 2006)
13. Carpenter, B.: The Logic of Typed Feature Structures. Cambridge Tracts in Theoretical Computer Science. Cambridge University Press, Cambridge (1992)
14. Chomsky, N.: Aspects of the theory of syntax. MIT Press, Cambridge (1965)
15. Chomsky, N.: Language and Mind. Harcourt Brace Juvanovich, New York (1968)
16. Chomsky, N.: Rules and representations. Behavioral and Brain Sciences 3, 1–61 (1980)
17. Chomsky, N., Lasnik, H.: The theory of principles and parameters. In: Jacobs, J., von Stechow, A., Sternefeld, W., Vannemann, T. (eds.) Syntax: An International Handbook of Contemporary Research. Walter de Gruyter, Berlin, pp. 506–569 (1993); Reprinted in Chomsky, N.: The Minimalist Program, pp. 13–127. MIT Press, Cambridge (1995)
18. Cohen-Sygal, Y., Wintner, S.: Partially specified signatures: a vehicle for grammar modularity. In: Proceedings of Coling–ACL 2006, Sydney, Australia, July 2006, pp. 145–152 (2006)
19. Copestake, A.: Implementing Typed Feature Structure Grammars. CSLI Publications, Stanford (2002)
20. Demuth, K.: On the 'underspecification' of functional categories in early grammars. In: Lust, B., Suñer, M., Whitman, J. (eds.) Syntactic theory and first language acquisition: cross-linguistic perspectives, pp. 119–134. Lawrence Erlbaum Associates, Hillsdale (1994)
21. Gibson, E., Wexler, K.: Triggers. Linguistic Inquiry 25(4), 407–454 (1994)
22. Mark Gold, E.: Language identification in the limit. Information and Control 10(5), 447–474 (1967)
23. Goldberg, A.: Constructions. A Construction Grammar approach to argument structure. University of Chicago Press, Chicago (1995)
24. Goldberg, A.: Constructions at Work: the nature of generalization in language. Oxford University Press, Oxford (2006)
25. Green, G.M.: Modelling grammar growth; universal grammar without innate principles or parameters, Unpubl. ms., University of Illinois (July 2003)
26. Hinrichs, E.W., Meurers, W.D., Wintner, S.: Linguistic theory and grammar implementation. Research on Language and Computation 2, 155–163 (2004)
27. Johnson, M., Riezler, S.: Statistical models of language learning and use. Cognitive Science 26(3), 239–253 (2002)

28. Kaplan, R., Bresnan, J.: Lexical functional grammar: A formal system for grammatical representation. In: Bresnan, J. (ed.) The Mental Representation of Grammatical Relations, pp. 173–281. MIT Press, Cambridge (1982)
29. King, T.H., Forst, M., Kuhn, J., Butt, M.: The feature space in parallel grammar writing. Research on Language and Computation 3, 139–163 (2005)
30. Labelle, M.: The acquisition of grammatical categories: a state of the art. In: Cohen, H., Lefebvre, C. (eds.) Handbook of categorization in cognitive science, pp. 433–457. Elsevier, Amsterdam (2005)
31. Lebeaux, D.: Language acquisition and the form of the grammar. John Benjamins, Amsterdam (2000)
32. Levy, Y., Schlesinger, I.M.: The child's early categories: approaches to language acquisition theory. In: Levy, et al. (eds.) [33], ch. 9, pp. 261–276.
33. Levy, Y., Schlesinger, I.M., Braine, M.D.S. (eds.): Categories and Processes in Language Acquisition. Lawrence Erlbaum Associates, Hillsdale (1988)
34. Li, P.: Language acquisition in a self-organizing neural network model. In: Quinlan, P. (ed.) Connectionist models of development: Developmental processes in real and artificial neural networks. Psychology Press, Hove (2003)
35. Li, P., Farkas, I., MacWhinney, B.: Early lexical development in a self-organizing neural network. Neural Network 17, 1345–1362 (2004)
36. MacWhinney, B.: Rules, rote, and analogy in morphological formations by Hungarian children. Journal of Child Language 2, 65–77 (1975)
37. MacWhinney, B.: Models of the emergence of language. Annual Review of Psychology 49, 199–227 (1998)
38. MacWhinney, B.: The emergence of language. In: Carnegie Mellon Symposia on Cognition. Lawrence Erlbaum Associates, Mahwah (1999)
39. MacWhinney, B.: The CHILDES Project: Tools for Analyzing Talk, 3rd edn. Lawrence Erlbaum Associates, Mahwah (2000)
40. MacWhinney, B.: A multiple process solution to the logical problem of language acquisition. Journal of Child Language 31, 883–914 (2004)
41. MacWhinney, B.: A unified model of language acquisition. In: Kroll, J., De Groot, A. (eds.) Handbook of bilingualism: Psycholinguistic approaches. Oxford University Press, Oxford (2004)
42. Ninio, A., Snow, C.E.: Language acquisition through language use: the functional source of children's early utterances. In: Levy, et al. (eds.) [33], ch. 1, pp. 11–30
43. Niyogi, P., Berwick, R.C.: A language learning model for finite parameter spaces. Cognition 61(1-2), 161–193 (1996)
44. Peters, A.M.: The Units of Language Acquisition. In: Monographs in Applied Psycholinguistics. Cambridge University Press, New York (1983)
45. Peters, A.M.: Strategies in the acquisition of syntax. In: Fletcher, P., MacWhinney, B. (eds.) The handbook of child language, pp. 462–482. Blackwell, Oxford (1995)
46. Peters, A.M.: Filler syllables: what is their status in emerging grammar? Journal of Child Language 28, 229–242 (2001)
47. Pinker, S.: The Language Instinct. William Morrow and Company, New York (1994)
48. Pollard, C., Sag, I.A.: Head-Driven Phrase Structure Grammar. University of Chicago Press and CSLI Publications (1994)
49. Prince, A., Smolensky, P.: Optimality: from neural networks to universal grammar. Science 275, 1604–1610 (1997)
50. Sag, I.A.: English relative clause constructions. Journal of Linguistics 33(2), 431–484 (1997)

51. Sagae, K., Davis, E., Lavie, A., MacWhinney, B., Wintner, S.: High-accuracy annotation and parsing of CHILDES transcripts. In: Proceedings of the ACL 2007 Workshop on Cognitive Aspects of Computational Language Acquisition, Prague, Czech Republic, June 2007, pp. 25–32. Association for Computational Linguistics (2007)

52. Sagae, K., Lavie, A., MacWhinney, B.: Automatic measurement of syntactic development in child language. In: Proceedings of the 43rd Annual Meeting of the Association for Computational Linguistics (ACL 2005), Ann Arbor, Michigan, June 2005, pp. 197–204. Association for Computational Linguistics (2005)

53. Sagae, K., MacWhinney, B., Lavie, A.: Automatic parsing of parent-child interactions. Behavior Research Methods, Instruments, and Computers 36, 113–126 (2004)

54. Sampson, G.: The 'Language Instinct' Debate, revised edn. Continuum, London (2005)

55. Sygal, Y., Wintner, S.: Type signature modules. In: de Groote, P. (ed.) Proceedings of FG 2008: The 13th conference on Formal Grammar, August, pp. 113–128 (2008)

56. Tomasello, M.: Language is not an instinct. Cognitive Development 10, 131–156 (1995)

57. Tomasello, M.: Constructing a language. Harvard University Press, Cambridge (2003)

58. Tomasello, M.: On the different origins of symbols and grammars. In: Christiansen, M.H., Kirby, S. (eds.) Language Evolution, Studies in the Evolution of Language, ch. 6, pp. 94–110. Oxford University Press, Oxford (2003)

59. Tomasello, M.: Acquiring linguistic constructions. In: Kuhn, D., Siegler, R. (eds.) Handbook of Child Psychology, pp. 255–298. Wiley, New York (2006)

60. Valiant, L.G.: A theory of the learnable. Commun. ACM 27(11), 1134–1142 (1984)

61. Villavicencio, A.: The Acquisition of a Unification-Based Generalised Categorial Grammar. Technical report UCAM-CL-TR-533, Computer Laboratory, University of Cambridge (April 2001)

62. Wexler, K., Culicover, P.W.: Formal principles of language acquisition. The MIT Press, Cambridge (1980)

63. Wintner, S.: Modular context-free grammars. Grammars 5(1), 41–63 (2002)

64. Wintner, S.: On the semantics of unification grammars. Grammars 6(2), 145–153 (2003)

65. Wintner, S.: Introduction to unification grammars. In: Ésik, Z., Martín-Vide, C., Mitrana, V. (eds.) Recent Advances in Formal Languages and Applications, ch. 13. Studies in Computational Intelligence, vol. 25, pp. 321–342. Springer, Heidelberg (2006)

66. Wintner, S.: Unification: Computational issues. In: Brown, K. (ed.) Encyclopedia of Language and Linguistics, 2nd edn., vol. 13, pp. 238–250. Elsevier, Oxford (2006)

Hybrid BDD and All-SAT Method for Model Checking

Avi Yadgar, Orna Grumberg, and Assaf Schuster

Computer Science Department, Technion, Haifa, Israel

Abstract. We present a new hybrid BDD and SAT-based algorithm for model checking. Our algorithm is based on backward search, where each pre-image computation consists of an efficient All-SAT procedure. The All-SAT procedure exploits a graph representation of the model to dynamically prune the search space, thus preventing unnecessary search in large sub-spaces, and for identifying independent sub-problems. Apart from the SAT mechanisms, BDD structures are used for storing the input to, and output of the pre-image computation. In this way, our hybrid approach enjoys the benefits of both worlds: on the one hand, basing the pre-image computation on SAT technology avoids expensive BDD quantification operations and the corresponding state space blow up. On the other hand, our model checking framework still enjoys the advantages of symbolic space reduction in holding intermediate images. Furthermore, our All-SAT analyzes the model and avoids redundant exploration of sub-spaces that are completely full with solutions, paying in these cases for the instantiation of a single assignment only.

We implemented our algorithm using the zChaff SAT solver and the CUDD BDD library. Experimental results show a potential for substantial improvement over existing model checking schemes.

1 Introduction

This work presents a hybrid, symbolic model checking algorithm for temporal safety properties, composed of both BDD and All-SAT procedures. This algorithm exploits the strengths of both BDD-based and SAT-based approaches, while trying to avoid their respective drawbacks. In addition to the common representation of the model being checked as a CNF formula or as a BDD structure, we also make use of a graph representation, which we use to prune the search space and improve performance in several ways.

We first suggest an efficient implementation of the pre-image computation using the All-SAT procedure. We then use the pre-image computation in a backward search algorithm to perform full model checking of temporal safety properties. The resulting algorithm uses BDDs for all operations, except for the pre-image computation, where the All-SAT method is used instead.

SAT-based methods for image and pre-image computation [1,2,3] are based on All-SAT engines, which return the set of all the solutions to a given formula (all satisfying assignments). The All-SAT engine for pre-image computation receives

O. Grumberg et al. (Eds.): Francez Festschrift, LNCS 5533, pp. 228–244, 2009.

as input a propositional formula describing the application of a transition relation T to a set of *"next-states"* S'. The resulting set of solutions represents the pre-image of S', which is the set of all predecessors for states in S', also referred to as *"current-states"*.

Most modern SAT solvers implement the DPLL [4] backtrack search. These solvers learn and add conflict clauses to the formula in order to block searching in subspaces that are known to contain no solution. SAT solvers also implement efficient Boolean propagation procedures, as in [5]. However, when used for model checking, these algorithms do not make use of available knowledge about the structure of the model which is being checked. Thus, SAT-based model checking still suffers from the exponential complexity of search procedures that explore too many potential assignments.

In this paper we propose an All-SAT algorithm that makes use of two representations of the model's transition relation T: a propositional CNF formula and a graph of the hardware gates. We exploit the two representations for efficient search in the pre-image computation: the CNF representation is used for the usual backtrack search of SAT algorithms, whereas the graph representation is used to extract information about the structure of the design. We dynamically modify the graph representation according to the currently searched sub-space, thus exploiting more information than by static analysis of the model.

Our algorithm uses the information extracted from the structure of the model to do the following:

i Process whole sets of next states instead of processing them one by one, unlike other All-SAT based image/pre-image algorithms [1,2,6,3].
ii Each set of next-states is represented by a partial assignment to the next-state variables. The values of these variables are determined by only a subset of the current- state variables in the CNF representation of the model. In every iteration of the model checking, when pre-image computation is performed, our algorithm assigns values only to this significant subset of variables. This *Dynamic Transition Relation* is obtained by our algorithm without computational overhead.
iii Similar to [2,7,8,9,10,11], our algorithm uses the graph representation of the model to find partial assignments to the current state variables, instead of complete ones, thus saving time and space. However, unlike other works, the required analysis of the graph transition relation is carried on-the-fly, costing $O(1)$ operations for the branching procedure.
iv Detect independent sub-spaces and solve them independently.
v Detect sub-spaces where solving SAT instead of All-SAT problem is sufficient.

Built on top of a SAT algorithm, our All-SAT algorithm benefits from the mechanisms incorporated in the original SAT algorithm for learning conflict clauses. Moreover, when used for pre-image computation, our All-SAT algorithm is capable of learning conflict clauses incrementally. Thus, in each iteration of

the backward search model checking, the algorithm exploits the knowledge that
was gained in earlier iterations.

The set of states S' is given to the All-SAT engine in the form of a BDD.
Similar to [12], our All-SAT algorithm stores the negation of the solutions which
were already found in a BDD structure. Consequently, both S' and S, the input
and the output of our All-SAT algorithm, consist of BDD structures only. This
allows us to easily use the All-SAT algorithm in the BDD based model checking
algorithm.

The rest of the paper is organized as follows: In Section 2 we describe our
hybrid All-SAT algorithm for pre-image computation. In Section 3 we show how
to employ it for full model checking. In section 4 we present our experimental
results, and in Section 5 we discuss related work.

2 Hybrid BDD and All-SAT Method for Pre-image Computation

In this section we describe our algorithm for an All-SAT-based pre-image computation. This algorithm is based on the DPLL backtrack search [4] and uses
conflict clause learning. Apart from the CNF description of the problem, the
algorithm uses a graph representation of the model's transition relation. The
added information about the structure of the model is used to speed up the
search by reducing the number of solutions which are instantiated.

2.1 Pre-image Computation

In Section 3 we describe the model checking algorithm. In order to implement it
using SAT methods, we have to build an All-SAT engine which will perform the
pre-image computation. This engine should find all the assignments to $\overline{x}$ which
are solutions to the formula

$$\varphi(\overline{x}) = \exists \overline{x'} \exists \overline{I} [T(\overline{x}, \overline{I}, \overline{x'}) \wedge S'(\overline{x'}) \wedge \neg S^*(\overline{x})] \tag{2.1.1}$$

T is the transition relation, defined over the current-state variables $\overline{x}$, the inputs
$\overline{I}$, and the next-state variables $\overline{x'}$. S' is the given set of next-states, and S^* is
the set of previously found states. Each solution for this formula represents a
current-state $\overline{x}$, which is not in S^*, and is a predecessor of some state in S'. The
fact that $\overline{x}$ is not in S^* implies that the algorithm finds only states which were
not found before. In our model checking algorithm, S' and S^* are given to the
All-SAT algorithm as BDDs, and T is given both in CNF and as a graph (see
below). The set of solutions $S(\overline{x})$ is returned as a BDD.

2.2 Justification of Assignments

The input to our All-SAT algorithm is a transition relation of a hardware
model which is given in some hardware design language. This transition relation

describes the values of the next-state variables $\overline{x'}$, and is given as a *partitioned transition relation* [13]. That is, T is a set of functions f_i, such that, $x'_i = f_i(\widetilde{x}_i, \widetilde{I}_i)$ for some $\widetilde{x}_i \subseteq \overline{x}$ and $\widetilde{I}_i \subseteq \overline{I}$. For $x'_i \in \overline{x'}$, the *cone of influence* of x'_i, $C_{x'_i}$, is $C_{x'_i} = \widetilde{x}_i \cup \widetilde{I}_i$.

A partial assignment b' to $C'_{x'_i} \subseteq C_{x'_i}$ *justifies* an assignment a to x'_i if for every extension b of b' over all the variables in $C_{x'_i}$ it holds that, $a(x'_i) = f_i(b(C_{x'_i}))$. b' represents all the assignments to $C_{x'_i}$ that agree with b' on the assignment to $C'_{x'_i}$. We say that a partial assignment b' *maximally justifies* a if b' justifies a, and for any variable which is assigned by b', removing it from b' will make b' not justify a anymore. Note that such an assignment is not unique.

Recall that we are looking for all the assignments to $\overline{x}$ which can be extended to a solution (satisfying assignment) of the formula $T(\overline{x}, \overline{I}, \overline{x'}) \wedge S'(\overline{x'}) \wedge \neg S^*(\overline{x})$ (see Equation 2.1.1). Therefore, we are actually looking for all the assignments to $\overline{x}$ for which there is an assignment to $\overline{I}$ such that they justify some assignment to $\overline{x'} \in S'$. The assignments to $\overline{x}$ should also not conflict with $\neg S^*$. We discuss the last constraint later.

We find these assignments to $\overline{x}$ by finding all the justifying assignments for all the assignments in S' that differ on the assignment to $\overline{x}$. First we describe a SAT based algorithm for finding a single maximally justifying assignment to $\overline{x} \cup \overline{I}$, for a given assignment in S'. We then describe an All-SAT based algorithm which finds all the maximally justifying assignments to $\overline{x} \cup \overline{I}$ that differ on the assignment to $\overline{x}$, for a single assignment in S'. Last, we extend it to find all the maximally justifying assignments to $\overline{x} \cup \overline{I}$ that differ on their assignment to $\overline{x}$, for all of the assignments in S'.

2.3   Transition Relation Graph

The transition relation $T(\overline{x}, \overline{I}, \overline{x'})$ can be presented as a DAG, where the roots are the next-state variables $\overline{x'}$, and the terminal nodes are the current-state and input variables $\overline{x}$ and $\overline{I}$. Each internal node in the graph is associated with an auxiliary variable and a Boolean operator. It corresponds to a subexpression of the function that defines the transition relation. The value of a node is the result of applying it's corresponding operation on its successors. Such a graph is equivalent to the RTL description of the hardware model. We refer to this graph as Transition Relation Graph (TRG), and to a node in it by its variable. The graph in Figure 1(a) corresponds to the partitioned transition relation: $x'_1 = (x_1 \wedge (x_2 \wedge i_3)) \vee ((x_2 \wedge i_3) \vee i_2)$ and $x'_2 = i_3 \vee i_1$.

The CNF representation of the transition relation is constructed from its hardware design language, in a manner similar to the TRG [14]. For each node in the TRG, the CNF representation has a corresponding variable, and clauses which describe the node's Boolean operation. Thus, we have two representations for the transition relations, which are closely related and therefore it is easy to switch between them.

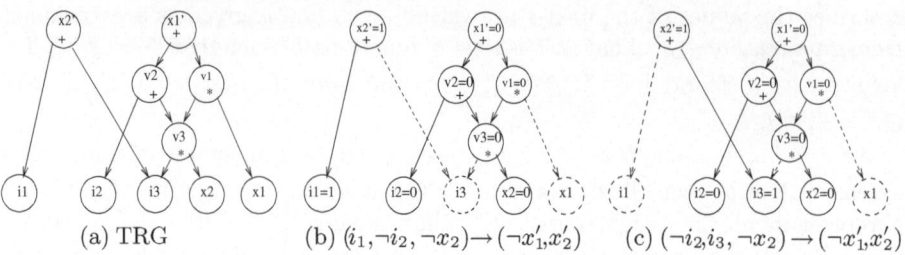

(a) TRG   (b) $(i_1, \neg i_2, \neg x_2) \rightarrow (\neg x'_1, x'_2)$ (c) $(\neg i_2, i_3, \neg x_2) \rightarrow (\neg x'_1, x'_2)$

Figure 1. Transition Relation Graph. (a) A TRG. $\overline{x} = \{x_1, x_2\}$, $\overline{I} = \{i_1, i_2, i_3\}$, $\overline{x'} = \{x'_1, x'_2\}$. v_1, v_2 and v_3 are auxiliary variables representing the sub-expressions of the transition relation. (b,c) Two maximally justifying assignments that differ only in the assignments to $\overline{I}$.

2.4   Maximally Justifying a Given Assignment to $\overline{x'}$

In this section we describe a SAT based algorithm which uses the TRG for pruning the search space for a SAT engine, such that it finds a single maximally justifying assignment to $\overline{x} \cup \overline{I}$, for a given assignment to $\overline{x'}$. The assignment to $\overline{x'}$ is given as an initial partial assignment to the SAT solver before the search begins.

We introduce a new branching procedure into a SAT solver. This procedure uses the TRG to decide which variable to assign next as part of the regular DPLL. The branching procedure also detects that a solution is found. The rest of the search algorithm is not changed, and uses the CNF representation of the problem. In particular, deduction of conflict clauses and detection of empty subspaces are performed as in [3], and the efficient *bcp()* procedure of the original SAT solver is used. Throughout the rest of this paper, we use the term *backtrack* to refer to the backtrack of the SAT algorithm in the search tree, as opposed to the operations performed on the TRG.

We generalize the notion of cone of influence of a node v in TRG by setting C_v to be the set of all descendants of v.

Our new branching procedure is a variation of [15]. It performs a traversal of the TRG, assigning the variables with values in a pre-order manner. Each $x' \in \overline{x'}$ is a root in the TRG. For a given root, our branching procedure only assigns values to the root's descendants that are needed in order to justify its value, leaving the other descendants unassigned. Each branch justifies the previous one until a maximally justifying assignment is found. The next branches are made such that they justify the next root and so on.

More specifically, assume that the branching level i gives a value to a node v in the TRG. In branching level $i + 1$, the branching procedure assigns a value to one of its successors v_1, such that it justifies the value of v, and removes the edges to the other successors from the graph. For example, if v is associated with the "AND" operator and its value is 0, we assign 0 to one of its successors and disconnect the others. When a new branch is required at level $i + 2$, we continue the traversal of the TRG over the connected successors only, trying to justify

v_1. When backtracking in the SAT search space, we also go backwards in the traversal of the TRG. If we backtrack to branching level $i + 1$, we change the value of v_1 as in regular DPLL. The next branch, at level $i + 1$, will assign a value to v_2, another successor of v, which justifies the value of v. In this manner, we can justify all the roots of the TRG.

We conclude that a justifying assignment was found when we complete the traversal of the connected components of the TRG, starting from each of the roots. This is opposed to satisfying all the clauses of the CNF formula, or assigning values to all of the variables, as in regular SAT procedures. The returned result is the partial assignment to $\overline{x} \cup \overline{I}$, obtained at the termination of the algorithm.

Consider a partial assignment to $\overline{x} \cup \overline{I}$ returned by the algorithm. The branching procedure only traverses the parts of the TRG that are required to justify the assignment to $\overline{x'}$. Therefore, each value in the partial assignment takes part in the justification, and removing it would make the partial assignment not justifying the assignment to $\overline{x'}$ anymore. Therefore, this partial assignment maximally justifies the assignment to $\overline{x'}$, and thus we have found the required result.

2.5   All Justifications for a Given Assignment to $\overline{x'}$

Let a solution for a given assignment a' to $\overline{x'}$ be an assignment to $\overline{x}$, for which there is some assignment to $\overline{I}$, such that they justify a'. In this section we describe an algorithm for finding all the solutions for a given a'.

Note that any extension of a maximally justifying assignment in which all the variables in $\overline{x}$ have a value, is a solution for a'. Thus, a maximally justifying assignment actually represents a set of solutions to a'.

We build an All-SAT algorithm on top of the SAT algorithm presented in the previous section, in order to find all the maximally justifying assignments to $\overline{x}$, and thus find all the solutions to our problem. This algorithm uses a blocking BDD $S^{not}(\overline{x})$, as in [12]. When a solution is found by the SAT algorithm, we consider its projection a over $\overline{x}$. The conjunction of the literals in a are negated and conjuncted with $S^{not}(\overline{x})$. The SAT algorithm backtracks one level, and the search is resumed. We use the procedure $BDD\_agree(S^{not}(\overline{x}))$ [12] in order to check the solutions during their creation, making sure that they agree with S^{not}, meaning that they were not produced before. Thus, this procedure blocks the solver from finding the same partial assignment to $\overline{x}$ again. The algorithm terminates when no new solution is found.

Next we explain the reason for using only the projections of the solutions over $\overline{x}$ in the blocking BDD. Each maximally justifying assignment which is found by the algorithm is a partial assignment to $\overline{x} \cup \overline{I}$. However, multiple maximally justifying assignments might give the same assignments to the variables in $\overline{x}$, and differ in the assignment to variables in $\overline{I}$ only. This is demonstrated in Figure 1(b,c). Moreover, we might reach the same maximally justifying assignment by choosing different values to the internal nodes in the TRG. However, we are only interested in all the maximally justifying assignments which differ on their assignment to $\overline{x}$. Therefore, we use only the assignments to variable in $\overline{x}$ for blocking future solutions.

Note that S^{not} is defined over $\overline{x}$, which typically constitute about 10% of the total number of variables [16]. This reduces the overhead of using $BDD\_agree(S^{not})$. Note, also, that by constructing S^{not}, we actually get for free the BDD $S(\overline{x})$ of the solutions to the problem. This is because $S = \neg S^{not}$, which is obtained by an O(1) operation. Thus, the algorithm results in $S(\overline{x})$, as required.

2.6   Justifying All the Assignments to $\overline{x'}$

In the previous section we showed how to find all the solutions for a given assignment to the next-state variables. In this section we find all the solutions to all of the partial assignments represented by a given BDD $S'(\overline{x'})$. The All-SAT algorithm then returns all the maximally satisfying assignments of all the assignments in S', which is the solution to our problem.

We exploit the fact that $S'(\overline{x'})$ is given as a BDD. The assignments to $\overline{x'}$ are represented by paths from the root of the BDD to the terminal node '1'. In Figure 2(a,b) we show a BDD and its corresponding assignments. For a path π in $S'(\overline{x'})$, $v(\pi)$, the set of variables in π, is a subset of $\overline{x'}$. We can refer to π as an assignment to $v(\pi)$. Note that the branching procedure presented in previous sections is applicable also to partial assignments to $\overline{x'}$.

We now show how to introduce all the possible π in S' into the All-SAT solver. If we reverse the direction of the edges in S', and traverse it in a DFS-like manner, starting from the terminal node '1', we will get all the paths from the root of S' to the terminal node '1' in reverse order. The value of a node on the path is '1' if we reached it from its 'right' successor, and '0' otherwise. A straightforward approach for finding the maximally justifying assignments for the paths in the BDD is to apply the algorithm given in the previous section for each π found in the DFS. However, this may cause unnecessary duplicated work. For example, consider two assignments to $\overline{x'}$, $\pi_1 = \{x'_1, x'_2, x'_3\}$ and $\pi_2 = \{x'_1, x'_2, \neg x'_4\}$. Applying this approach over these assignment would require justifying x'_1 and x'_2 twice, once for each assignment to $\overline{x'}$.

In order to avoid the repetition, we integrate the DFS of the BDD into the All-SAT branching procedure. At branching level $i + 1$, if the branch in level i is not justified yet, the branching procedure uses the TRG for choosing an assignment to a variable that will justify the branch at level i. This is done as described in the Section 2.4. Otherwise, if there is no unjustified branch, then a value for one of the (unassigned yet) roots of the TRG is chosen, based on the DFS over the BDD. When backtracking in the All-SAT procedure, we also backtrack in the DFS over the BDD of S' respectively. We demonstrate this procedure in Figure 2(c,d).

By integrating the DFS with the branching procedure, the All-SAT algorithm assigns $\overline{x'}$ with all possible assignments corresponding to some π in S', and them only. Since π corresponds to a partial assignment, some of the TRG roots will not be assigned and therefore will not have to be justified. As a result, only the parts of the TRG which are reachable from the assigned roots are traversed, and work is saved. This is similar to using dynamic transition relation,

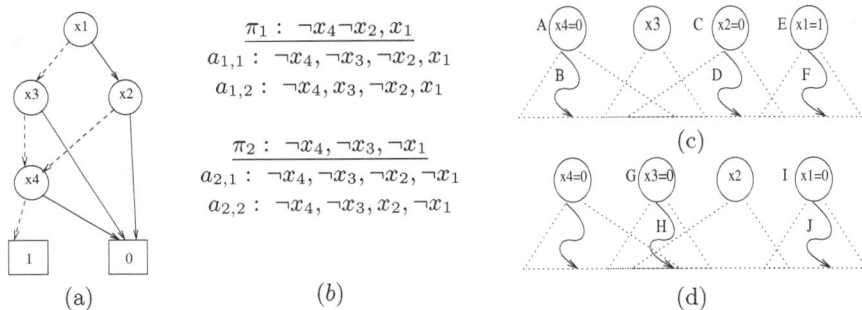

Figure 2. (a) A BDD representing $S'(\overline{x'})$. Solid lines represent the 'right' successor of a node, while dashed lines represent the 'left' successor of a node. (b) The assignments corresponding to the paths from the root to the terminal node '1'. (c) A,B...F are the branching sequence of the All-SAT. Branches A,C, and E are imposed on the All-SAT engine by the DFS over the BDD. B,D and F represent the branches which are made for justifying A,C and E respectively. (d) After we backtrack, branch G corresponds to the DFS search, followed by branching sequences H,I and J. The justification of x_4 is performed only once.

as explained in the next section. Altogether, the All-SAT algorithm receives partial assignments, representing sets of next-states (from S') and returns partial assignments, representing sets of current-states.

2.7 Optimizations

The incorporation of the TRG into the All-SAT solver allows us to apply additional optimizations on the search. In this section we describe how the information extracted from the TRG is used in order to detect independent subproblems, and to reduce All-SAT sub-problems to SAT problems.

Independent Roots

For a node v in the TRG, and for G, a sub-graph of the TRG, we say that v is the root of G if G consists of exactly v and all of its descendants. We say that v is an *Independent Root* of G if v is the root of G, and all the paths from outside of G into G pass through v. That is, For every v_1, an ancestor of a vertex $v_2 \in G$, v_1 is either a descendant of v, or v is on all the paths $v_1 \rightsquigarrow v_2$. This property can be decided statically before starting the All-SAT solving process. However, when using the TRG for branching during the search, we detach nodes that are not required for justifying the current assignment. When backtracking in the All-SAT procedure, we reattach these nodes. Therefore, independent roots should also be detected dynamically during the solving process. Terminal nodes in the TRG are, by their definition, independent roots. We call a maximally justifying assignment to the terminal nodes in G, for an assignment to v, a *solution in* G. In Figure 3(a), only the terminal nodes are independent roots.

For a node v which is an independent root of a subgraph G, all the nodes in C_v are in the cone of influence of no other node outside of G. Therefore, the assignment to these nodes have no effect outside of G that is not expressed in the value of v. For a given assignment to v, finding all the justification for it is independent of the rest of the nodes in the TRG. Therefore, when we reach an independent root, we can solve the All-SAT problem of its corresponding sub-graph independently of the rest of the TRG. The solutions of the complete problem are the product of the partial solutions. In Figure 3(b), the edge from x_2' to i_3 is detached, and v_3 becomes an independent root. Thus, for any assignment to v_3, i_3 and x_2 should be assigned so that they justify it, regardless of the assignments to other variables.

A node which is found to be an independent root by the static analysis of the TRG, is always an independent root, regardless of the current assignment. On the other hand, a node which was dynamically found to be an independent root, is only independent in the context of the current assignment. This is demonstrated in Figure 3(a,b).

We now describe an algorithm for statically determining if a node v_0, a root of a subgraph G, is an independent root. The algorithm performs a DFS over G, starting form v_0. For each node $v \in G$, the algorithm counts the number of edges into v which are reachable from v_0.

The score of a node v is the number of edges into v or into all of its descendants, which are not on a path from v_0. Thus, the score of v_0 is the sum of all the edges into G which are not reachable from v_0. If the score of v_0 is 0, then v_0 is an independent node. For complete static analysis, we have to perform this algorithm on each node in the TRG separately. This is an $O(|TRG|^2)$ operation, which has to be done once, prior to the solving process.

In order to dynamically find independent roots, we use the score which was calculated in the static analysis. During the solving process, edges are removed from the TRG, or added to it, in correspondence to the current assignment. When an edge into a node v is removed, or put back, we decrease/increase the score of v respectively, and notify all the predecessors of v. Each notified node updates its score, and notifies its predecessors, until all the ancestors of v are notified. A node updates its score no more than once for each change in the graph, even if it is notified of it by more than one decedent. This operation involves all the ancestors of the v.

Note that there is a tradeoff between the time spent on dynamically detecting independent roots and the time saved on solving subproblems independently.

Non-important Roots

We call a node v in the TRG is a *non-important root* if it is an independent root of a sub-graph G, where all the terminal nodes are in $\overline{I}$. As with independent roots, this property can be decided statically or dynamically. In Figure 3(a), only the terminal nodes of the input variables are non-important roots. In Figure 3(b), v_3 is a non-important root.

For a node v which is a non-important root of a subgraph G, v is, by its definition, an independent root. Therefore we can solve the All-SAT problem of G independently on the rest of the TRG. Moreover, recall that we search for partial assignments to $\overline{x}$ for which *there is some assignment to $\overline{I}$ such that they* justify the current assignment to $\overline{x'}$. Since all terminal nodes in G are in $\overline{I}$, we only have to find a single assignment to these nodes which justifies the current value of v. Consequently, it is enough to solve the SAT problem once for G, with the current value of v.

A node v that is found to be a non-important root of a sub-graph G by the static analysis of the TRG, is always a non-important root, regardless of the current assignment to the rest of the variables. Therefore, we can solve the SAT problem for G to justify the values 'true' and 'false' for v only once, on the first time that they are required, store the result, and reuse it when v is reached again.

The static detection of non-important roots is straightforward. For each v in the TRG, we perform a DFS from v, counting $v.scope$, the number of current-state variables reachable from v. If v is an independent root, and $v.scope = 0$, then v is a non-important root. This is an $O(|TRG|^2)$ operation, which has to be performed once, at the beginning of the checking process.

For a node v we use $v.scope$, calculated in the static analysis, for dynamic detection of non-important roots. During the solving process, if a current-state variable is given an assignment, we notify all its ancestors to decrease their scope, as with the independent roots. When the All-SAT algorithm backtracks and nullifies an assignment to a current-state variable, we notify its ancestors to increase their scope. At any given moment, an independent root v for which $v.scope = 0$ is a non-important root. This operation involves all the ancestors of the current-state variable, and yields an under approximation of the non-important nodes in the TRG. As with independent roots, there is a tradeoff between the time spent on dynamically detecting non-important roots and the time saved on solving SAT instead of All-SAT subproblems.

It is important to note that SAT and All-SAT subproblems that originate from independent and non-important roots can be solved within the same SAT solver without the need to duplicate the formula or parts of it. This is because our algorithm does not identify a solution when all clauses are satisfied. Instead, it identifies it when all its branches were justified and no additional branches are required.

3   Model Checking Using Hybrid Pre-image Computation

3.1   Model Checking

The algorithm presented in Figure 4 describes a backward search for model checking a formula of the form AGp. When this algorithm is implemented with BDDs, its bottleneck is the pre-image computation performed on line 7. This is because intermediate results of this operation might be one or two orders of magnitude larger than the initial and resulting BDDs. In order to avoid this problem we use our hybrid approach. $S_0, \neg P, S^*$ and new are represented as BDDs. The operations in lines 3, 4 and 6 of the algorithm are performed directly on their BDD representation. The pre-image operation, on the other hand, is performed with the hybrid All-SAT algorithm, which solves Equation 2.1.1, in

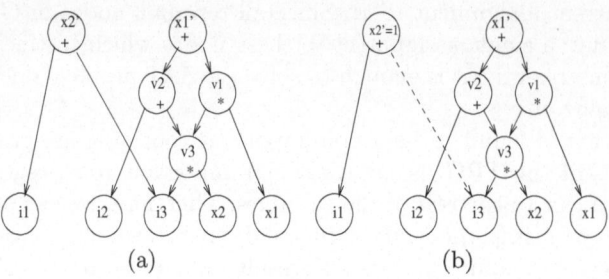

(a)   (b)

Figure 3. Independent and Non-Important Roots. (a) Only the terminal nodes are independent roots. The terminal nodes of input variables are the only non-important roots (b) After the assignment to x_2' and i_1, x_1' and v_3 are also independent roots. v_3 is also a non-important root.

<table>
<tr>
<td>
Pre-image computation (line 7) is performed only for the newly found states, in order not to find the predecessors of a state more than once.

The algorithm returns 'TRUE' if $M \models \varphi$, and 'FALSE' otherwise.
</td>
<td>

$ModelCheck(S_0, T, \neg P)$
1) $S^* \leftarrow \phi$
2) $new \leftarrow \neg p$
3) $while \ (new \neq \phi) \ \{$
4) $if \ new \cap S_0 \neq \phi$
5) $return \ FALSE$
6) $S^* \leftarrow S^* \cup new$
7) $new \leftarrow \text{pre-image}(new) \setminus S^*$
8) $\}$
9) $return \ TRUE$
</td>
</tr>
</table>

Figure 4. Model Checking Algorithm for AGp

which S' is replaced with *new*. We describe two optimizations which originate from our approach.

3.2 Dynamic Transition Relation

Given a set $S'(\overline{x'})$, in order to apply a pre-image to S', only a subset of formulae of the partitioned transition relation $T_i \subseteq T$ should be involved in the computation. This subset includes those functions that define the next value for some variables in the support of S'. The reduced transition relation is called *dynamic transition relation*, since it is defined dynamically for each S'. Dynamic transition relations are used as an optimization in BDD computation of pre-image.

We further enhance this method by using the TRG and the BDD of S' for branching in the All-SAT process. For each path π in the BDD of S', our branching procedure considers only the subset of T which is necessary for justifying the assignment induced by π. Thus, we actually use different parts of the transition relation for different subsets of states in S'. This is achieved without computational overhead.

3.3 Incremental Learning

For pre-image computation, we apply our All-SAT algorithm on T, S' (where $S' = new$) and S^*. Only T is given to the algorithm as a CNF formula. Therefore, the formulae given to the SAT solver in different iterations are identical. As a result, conflict clauses that are learned from the transition relation in one iteration can be used in subsequent ones, thus contributing to the speedup of the SAT solver.

4 Experimental Results and Conclusions

We implemented our algorithm on top of zChaff SAT solver [5], and CUDD BDD library [17]. zChaff is a state of the art SAT solver, known to be one of the fastest solvers available. CUDD is a BDD library, common in many BDD based applications. Both zChaff and CUDD are open source tools, which allowed us to interface them with the graph representation of the transition relation.

There are many optimizations common in model checking that are not incorporated in our prototype implementation. Thus, for the sake of a fair comparison, we also implemented a model checking framework that is based on BDD backward search. This framework uses partitioned transition relation, and computes the cone of influence for the properties that are checked.

For experiments we used ISCAS89 and ISCAS99 benchmarks, which are large "real-life" problems. We checked each model against a set of properties of the form AGp, using the hybrid and the BDD based tools. All experiments use dedicated computers with 1.7Ghz Intel Xeon CPU, and 1GB RAM, running Linux operating system.

The results of our experiments are presented in Table 1 and Table 2. In order to mask initialization effects, we omit the results for the smaller models which take only a few seconds to solve. For each model and property, the tables show the number of iterations of the backward search which were completed by each tool, until either the checking is completed, timeout is reached, or memory out is reached. The tables also show the computation time for each completed check, the memory usage, and the percentage of the run time that was spent on quantification.

While the hybrid tool does not consistently outperform the BDD model checker, in half of the models it does better than the BDD tool for almost all the properties checked. In some cases where checking is not completed, the hybrid tool is able to perform the same number of pre-image iterations faster, or even perform more pre-image iterations. In most of the cases where none of the algorithms outperformed the other, the hybrid algorithm required less memory than the BDD algorithm.

For many of the models, the BDD tool suffers memory exhaustion while the hybrid tool continues until timeout, and even succeeds to complete additional iterations. This demonstrates the inherent space problem of BDDs that is discussed in Section 3.1: when performing quantification, intermediate results of a single pre-image computation commonly blow up to an order of magnitude larger

Table 1. A Comparison of Model Checking Run Times. #FF is the number of state variables, #It is the number of pre-image steps completed, Quant is the percentage of time spent on quantification, and Mem is the size of the memory used by the tool in MB. Time out is set to 24hr, and Memory limit is 1GB.

| Model | # FF | Result | BDD | | | | Hybrid | | | | | | | | |
| | | | | | | | No Op | | | Op 1 | | | Op1 + Op2 | | | |
			# It	Time (s)	Mem	Quant	# It	Time (s)	Quant	# It	Time (s)	Quant	# It	Time (s)	Mem	Quant
S1269	37	fail	8	527	31	60	8	500	50	8	436	43	8	423	29	41
		fail	2	143	15	13	2	183	40	2	187	41	2	193	18	43
		fail	3	10	5	20	3	0	10	3	0	5	3	0	3	5
		fail	8	T.O.	60	99	9	6544	56	9	5791	50	9	5702	49	50
		pass	9	8064	64	99	9	6264	60	9	5580	55	9	5373	52	53
S1512	57	-	10	M.O.	> 1GB	46	10	T.O.	77	10	T.O.	71	10	T.O.	617	75
		fail	30	253	43	43	30	300	52	30	290	50	30	286	60	50
		pass	38	24	11	52	38	50	51	38	47	48	38	47	20	48
		pass	2	0	< 1MB	10	2	10	10	2	10	10	2	10	< 1MB	10
		-	30	M.O.	> 1GB	99	102	T.O.	80	102	T.O.	73	102	T.O.	243	63
		-	9	T.O.	422	95	3	T.O.	96	3	T.O.	95	3	T.O.	510	95
		-	11	T.O.	418	94	11	T.O.	93	11	T.O.	91	11	T.O.	541	90
S1423	74	fail	9	10106	266	48	9	13812	51	9	13790	51	9	13054	242	48
		fail	6	4300	312	70	6	4532	62	6	4504	62	6	4483	302	62
		fail	4	4150	247	50	4	8113	74	4	7949	73	4	8045	394	74
		-	11	T.O.	517	92	8	T.O.	99	8	T.O.	99	9	T.O.	412	99
		-	10	T.O.	498	93	7	T.O.	99	7	T.O.	99	7	T.O.	387	99
S9234	228	pass	258	8322	261	95	258	7741	87	258	5710	82	258	5337	194	81
		pass	3	1284	12	83	3	527	49	3	495	46	3	487	6	45
		fail	123	40	4	99	123	10	34	123	10	34	123	10	4	34
		-	7	T.O.	691	92	9	T.O.	75	9	T.O.	72	9	T.O.	483	69
		-	5	T.O.	721	94	6	T.O.	71	6	T.O.	70	6	T.O.	522	67
S15850	597	pass	4	12284	426	65	4	27331	72	4	24053	68	4	23125	510	67
		pass	3	7630	551	73	3	8121	76	3	7930	75	3	8043	643	77
		pass	3	3961	743	88	3	5403	92	3	5320	92	3	5394	861	90
		pass	2	0	3	17	2	0	24	2	0	0	2	0	5	0
		-	6	T.O.	640	87	6	T.O.	42	6	T.O.	41	6	T.O.	543	41
		-	4	T.O.	746	87	4	T.O.	99	4	T.O.	99	4	T.O.	674	99
S13207	669	-	12	T.O.	432	86	12	T.O.	82	12	T.O.	85	12	T.O.	462	86
		-	8	T.O.	843	76	8	T.O.	69	8	T.O.	65	8	T.O.	684	60
		-	6	T.O.	630	99	5	T.O.	84	5	T.O.	87	5	T.O.	690	86
		-	8	T.O.	455	82	8	T.O.	86	8	T.O.	85	8	T.O.	489	86
		-	5	T.O.	924	91	5	T.O.	90	5	T.O.	89	5	T.O.	751	87
S38584	1452	pass	1	2791	439	99	1	1023	99	1	914	99	1	849	210	99
		-	7	T.O.	792	94	7	T.O.	96	7	T.O.	96	7	T.O.	750	95
		-	8	T.O.	564	95	8	T.O.	96	8	T.O.	95	8	T.O.	498	95

T.O.: Time Out ⬛ Hybrid method outperforms the BDD method
M.O: Memory Out ⬜ BDD method outperforms the hybrid method

Table 2. A Comparison of Model Checking Run Times. #FF is the number of state variables, #It is the number of pre-image steps completed, Quant is the percentage of time spent on quantification, and Mem is the size of the memory used by the tool in MB. Time out is set to 24hr, and Memory limit is 1GB.

Model	# FF	Result	BDD				Hybrid									
							No Op			Op 1			Op1+Op2			
			# It	Time (s)	Mem	Quant	# It	Time (s)	Quant	# It	Time (s)	Quant	# It	Time (s)	Mem	Quant
B12	121	fail	214	16924	54	54	214	18302	65	214	17860	64	214	17808	49	64
		-	21	T.O.	98	64	17	T.O.	64	17	T.O.	65	18	T.O.	102	66
		-	32	T.O.	109	61	30	T.O.	72	30	T.O.	70	30	T.O.	133	71
		-	143	T.O.	45	27	131	T.O.	56	131	T.O.	50	131	T.O.	51	50
		pass	260	10943	150	60	260	12139	65	260	12130	65	260	11941	130	64
B14_1	245	fail	45	790	144	56	45	2166	66	45	2166	83	45	2166	312	83
		fail	32	22611	625	86	28	T.O.	84	28	T.O.	85	28	T.O.	483	82
		-	91	T.O.	124	66	84	T.O.	82	84	T.O.	80	84	T.O.	412	78
		-	6	T.O.	741	91	6	T.O.	88	6	T.O.	81	6	T.O.	720	77
B15_1	449	pass	13	4532	134	73	13	6234	75	13	5794	73	13	5730	120	73
		fail	9	6190	771	86	9	6906	79	9	6882	79	9	6649	674	78
		pass	17	8980	378	83	17	1037	79	17	868	75	17	532	400	59
		-	7	T.O.	683	90	6	T.O.	84	6	T.O.	84	6	T.O.	664	83
		-	19	M.O.	>1GB	99	19	T.O.	87	19	T.O.	84	19	T.O.	798	82
B21_1	490	fail	45	3104	250	64	45	1756	48	45	1700	46	45	1940	194	53
		-	6	T.O.	774	97	8	T.O.	78	8	T.O.	69	8	T.O.	527	67
		-	6	M.O.	>1GB	99	6	T.O.	76	6	T.O.	74	6	T.O.	618	70
		-	9	M.O.	>1GB	99	13	T.O.	82	13	T.O.	77	13	T.O.	437	74
B20_1	490	fail	37	9437	418	72	37	11030	64	37	11002	64	37	12106	491	67
		fail	19	8204	728	98	19	6310	94	19	5629	93	19	5283	534	93
		-	5	M.O.	>1GB	99	7	T.O.	95	7	T.O.	89	7	T.O.	559	87
		-	12	T.O.	681	96	15	T.O.	95	15	T.O.	90	15	T.O.	573	87
B22_1	735	pass	9	21045	826	97	9	13349	90	9	12174	89	9	10631	613	87
		-	6	T.O.	324	91	7	T.O.	90	7	T.O.	87	7	T.O.	271	85
		-	6	M.O.	>1GB	98	7	T.O.	96	7	T.O.	95	7	T.O.	685	95
		-	8	T.O.	442	99	8	T.O.	99	8	T.O.	99	8	T.O.	420	99
B17_1	1415	pass	6	39702	620	91	6	43005	84	6	42930	84	6	42882	607	84
		-	2	T.O.	714	90	1	T.O.	99	1	T.O.	99	1	T.O.	681	99
		-	3	T.O.	661	95	3	T.O.	95	3	T.O.	98	3	T.O.	494	98
B18_1	3320	fail	18	5277	97	96	18	3700	94	18	2892	92	18	2431	76	91
		-	1	T.O.	862	94	1	M.O.	99	1	M.O.	99	1	M.O.	>1GB	99
		-	-	T.O.	637	99	1	T.O.	95	1	T.O.	90	1	T.O.	729	89
B19_1	6642	fail	3	7546	340	87	3	10184	80	3	10041	80	3	11032	560	82
		-	3	M.O.	>1GB	99	4	T.O.	92	4	T.O.	90	4	T.O.	674	88
		-	4	M.O.	>1GB	99	4	T.O.	99	4	T.O.	99	4	T.O.	639	99

T.O.: Time Out   Hybrid method outperforms the BDD method
M.O: Memory Out   BDD method outperforms the hybrid method

than the eventual BDD size [18]. In contrast, the hybrid tool uses the All-SAT engine for quantification, and thus memory blowup does not become an issue, except for one model.

The table shows a strong correlation between the models for which the hybrid algorithm performed better, and the models for which it required less memory than the BDD algorithm. It is also shown that in these problems the BDD algorithm usually spent a higher percentage of the solving time on quantification than in other problems.

From the experimental results we conclude that there are models for which the hybrid method requires less memory, and achieves shorter run times. Further analysis of the models is required, and may result in characterization of a wider set of problems for which our hybrid approach is efficient.

In some of the problems where the hybrid algorithm uses more memory than the BDD algorithm, we observe that the time which was spent by the hybrid algorithm on quantification is relatively low. This means that the Boolean operations on the BDDs, other than the pre-image, required a larger part of the computation time. We believe that this is the result of the BDD order that we impose on the All-SAT solver during the quantification, which may not be optimal. Further research should be conducted on adapting our method to other data structures, possibly not canonical, in order to avoid this problem.

5   Related Work

All-SAT engines that are built on top of modern SAT solvers tend to block solutions that were already found by adding their negation to the formula during the search [1,2,6,19]. In [3] a specific order of the search prevents the solver from instantiating the same solution more than once, without adding clauses. In our work, as in [12], the negation of the solutions is kept in a BDD. When performing pre-image computation, the set of next-states is also given to our All-SAT algorithm as a BDD, which decreases the size of the problem substantially relative to clausal representation of the states. The size of the problem is also addressed in the following works: In [20], a ZBDD is used to store solutions found by an All-SAT solver. In [21], a method for managing ZBDDs is suggested. In [22], solutions are stored by using an or-inverter graph. We believe that additional research should be done in order to adapt our new hybrid algorithm and BDD based branching heuristic, described in Section 2.6, to other data structures.

When using All-SAT for image and pre-image computation in [2,7,8,9,10,11], after a solution is found, it is analyzed in order to generalize it to represent a set of solutions. In our algorithm, the branching procedure instantiates maximally justifying assignments, which represent sets of solutions, without actually instantiating assignments to all the variables, and without the overhead of generalizing them.

Hybrid SAT and non-clausal procedures were presented in [23,24,25,26,27]. In these methods, the non clausal representation of the problem is used to guide the search, learning over the non-clausal representation is performed, and some

pruning of empty subspaces is done. However, these branching heuristics are aimed at finding a single solution to the formula, and do not perform best when looking for all of the solutions. In addition, pruning empty subspaces using these procedures implies a significant computational overhead.

References

1. McMillan, K.L.: Applying SAT methods in unbounded symbolic model checking. In: Brinksma, E., Larsen, K.G. (eds.) CAV 2002. LNCS, vol. 2404, p. 250. Springer, Heidelberg (2002)
2. Chauhan, P., Clarke, E.M., Kroening, D.: Using SAT based image computation for reachability analysis. Technical Report CMU-CS-03-151, Carnegie Mellon University (2003)
3. Grumberg, O., Schuster, A., Yadgar, A.: Memory efficient all-solutions sat solver and its application for reachability analysis. In: Hu, A.J., Martin, A.K. (eds.) FM-CAD 2004. LNCS, vol. 3312, pp. 275–289. Springer, Heidelberg (2004)
4. Davis, M., Logemann, G., Loveland, D.: A machine program for theorem proving. CACM 5(7) (July 1962)
5. Moskewicz, M., Madigan, C., Zhao, Y., Zhang, L., Malik, S.: Chaff: engineering an efficient SAT solver. In: 39th Design Aotomation Conference, DAC 2001 (2001)
6. Plaisted, D.: Method for design verification of hardware and non-hardware systems. United States Patents 6(131), 078 (2000)
7. Parthasarathy, G., Iyer, M.K., Cheng, K.-T., Wang, L.: Safety Property Verification Using Sequential SAT and Bounded Model Checking. IEEE Des. Test 21(2), 132–143 (2004)
8. Lu, F., Iyer, M.K., Parthasarathy, G., Wang, L.-C., Cheng, K.-T., Chen, K.-C.: An efficient sequential sat solver with improved search strategies. In: DATE (2005)
9. Iyer, M.K., Parthasarathy, G., Cheng, K.-T.: SATORI - A Fast Sequential SAT Engine for Circuits. In: ICCAD 2003 (2003)
10. Kuehlmann, A.: Dynamic Transition Relation Simplification for Bounded Property Checking. In: ICCAD 2004 (2004)
11. Jin, H., Somenzi, F.: Prime clauses for fast enumeration of satisfying assignments to boolean circuits. In: DAC 2005 (2005)
12. Gupta, A., Yang, Z., Ashar, P., Gupta, A.: SAT-based image computation with application in reachability analysis. In: Johnson, S.D., Hunt Jr., W.A. (eds.) FMCAD 2000. LNCS, vol. 1954, pp. 354–371. Springer, Heidelberg (2000)
13. Burch, J.R., Clarke, E.M., Long, D.E.: Symbolic model checking with partitioned transition relations. In: VLSI 1991 (1991)
14. Biere, A., Cimatti, A., Clarke, E.M., Fujita, M., Zhu, Y.: Symbolic model checking using SAT procedures instead of BDDs. In: DAC 1999 (1999)
15. Fujiwara, H., Shimono, T.: On the acceleration of test generation algorithms. IEEE Trans. Computers 32(12), 1137–1144 (1983)
16. Shtrichman, O.: Tuning SAT checkers for bounded model checking. In: CAV (2000), citeseer.nj.nec.com/shtrichman00tuning.html
17. Somenzi, F.: Cudd: Cu decision diagram package release (1998), citeseer.ist.psu.edu/somenzi98cudd.html
18. Heyman, T., Geist, D., Grumberg, O., Schuster, A.: A scalable parallel algorithm for reachability analysis of very large circuits. Formal Methods in System Design 21(3) (2002)

19. Lahiri, S.K., Bryant, R.E., Cook, B.: A symbolic approach to predicate abstraction. In: Hunt Jr., W.A., Somenzi, F. (eds.) CAV 2003. LNCS, vol. 2725, pp. 141–153. Springer, Heidelberg (2003)
20. Li, B., Hsiao, M.S., Sheng, S.: A novel sat all-solutions solver for efficient preimage computation. In: DATE 2004 (2004)
21. Chandrasekar, K., Hsiao, M.S.: State set management for sat-based unbounded model checking. In: ICCD 2005 (2005)
22. Ganai, M.K., Gupta, A., Ashar, P.: Efficient sat-based unbounded symbolic model checking using circuit cofactoring. In: ICCAD 2004 (2004)
23. Barrett, C., Donham, J.: Combining SAT methods with non-clausal decision heuristics. In: PDPAR 2004 (2004)
24. Ganai, M.K., Ashar, P., Gupta, A., Zhang, L., Malik, S.: Combining Strengths of Circuit-Based and CNF-Based Algorithms for a High-Performance SAT Solver. In: DAC 2002 (2002)
25. Kuehlmann, A., Ganai, M.K., Paruthi, V.: Circuit-based Boolean Reasoning. In: DAC 2001 (2001)
26. Jin, H., Awedh, M., Somenzi, F.: CirCUs: A Satisfiability Solver Geared towards Bounded Model Checking. In: Alur, R., Peled, D.A. (eds.) CAV 2004. LNCS, vol. 3114, pp. 519–522. Springer, Heidelberg (2004)
27. Thiffault, C., Bacchus, F., Walsh, T.: Solving non-clausal formulas with dpll search. In: Wallace, M. (ed.) CP 2004. LNCS, vol. 3258, pp. 663–678. Springer, Heidelberg (2004)

Author Index